Archival Afterlives

Scientific and Learned Cultures and Their Institutions

Editor

M. Feingold (*California Institute of Technology*)

VOLUME 23

The titles published in this series are listed at *brill.com/slci*

Archival Afterlives

Life, Death, and Knowledge-Making in Early Modern British Scientific and Medical Archives

Edited by

Vera Keller, Anna Marie Roos, and Elizabeth Yale

BRILL

LEIDEN | BOSTON

Cover illustration: RS ClP/3i/35 (detail), ©The Royal Society.

The Library of Congress Cataloging-in-Publication Data is available online at http://catalog.loc.gov
LC record available at http://lccn.loc.gov/2018023150

Typeface for the Latin, Greek, and Cyrillic scripts: "Brill." See and download: brill.com/brill-typeface.

ISSN 2352-1325
ISBN 978-90-04-32429-9 (hardback)
ISBN 978-90-04-32430-5 (e-book)

This book is printed on acid-free paper and produced in a sustainable manner.

Contents

Acknowledgements

The editors would like to thank the following organizations who helped make *Archival Afterlives* possible: The University of Iowa, the University of Lincoln, the University of Oregon, the Royal Society, and the Andrew W. Mellon Fellowship of Scholars in Critical Bibliography at Rare Book School (University of Virginia). We would also like to thank all our essay contributors, Rupert Baker, Mordechai Feingold, Keith Moore, and the anonymous reviewers whose comments enriched and improved the volume. In addition, we are grateful to the audience of the original conference held at the Royal Society, *Archival Afterlives*, as well as to those conference participants who do not appear in this volume: Leigh Penman, Lauren Kassell, and Anne Goldgar. The editors contributed equally to the authorship of this introduction and to the volume as a whole.

Figures and Table

Figures

Table

Contributors

Arnold Hunt
teaches history at Cambridge, where he is currently a Lecturer in the Faculty of History and a Fellow of Girton College. He was a Curator of Manuscripts at the British Library for ten years, from 2005 to 2015. His book *The Art of Hearing: English Preachers and their Audiences 1590–1640* was published in 2010. His main interests are in the social, cultural, and religious history of Britain, 1500–1900, and he has also published widely on the history of library, archives and collecting.

Michael Hunter
is Emeritus Professor of History at Birkbeck College, University of London. He is the principal editor of the *Works, Correspondence* and work diaries of Robert Boyle and author of *Boyle: Between God and Science* (2009) as well as many other books on the intellectual history of early modern England.

Vera Keller
is Associate Professor of History at the Robert D. Clark Honors College, University of Oregon. In addition to articles published in *Ambix, British Journal for the History of Science, Configurations, Early Science and Medicine, History of Science, Isis, Journal of the History of Ideas, Nuncius, Renaissance Quarterly,* and *William and Mary Quarterly*, she is the author of *Knowledge and the Public Interest, 1575–1725* (Cambridge, 2015). She is currently completing a monograph on Cornelis Drebbel (1572–1633).

Carol Pal
teaches history at Bennington College. She is the author of *Republic of Women: Rethinking the Republic of Letters in the Seventeenth Century* (Cambridge, 2012), which was awarded the 2013 Joan Kelly Memorial Prize by the American Historical Association. She is currently editing a sourcebook in the history of medicine, and preparing a volume on Marie du Moulin for the series The Other Voice in Early Modern Europe. Her next monograph is an analysis of one of Samuel Hartlib's manuscripts in the British Library.

Anna Marie Roos
is a Reader of the history of science and medicine at the University of Lincoln. Roos has published three monographs in the history of science including *Web of Nature: Martin Lister (1639–1712), the first arachnologist* (Brill, 2011), as

well as scientific papers concerning the use of historical sources in the solving of taxonomic controversies. She is a Fellow of the Linnean Society, a Fellow of the Society of Antiquaries, and in 2017 was a Visiting Fellow at All Souls, Oxford. Currently, Roos is working on a biography of Royal Society president Martin Folkes (Oxford University Press), as well as a book for Reaktion Press entitled: *Goldfish*.

Richard Serjeantson
is a Fellow and Director of Studies in History at Trinity College, Cambridge. He serves as a deputy director of the *Oxford Francis Bacon*, for which he is editing Volume III: *Earlier Jacobean Writings, 1603–1613*, in collaboration with Angus Vine. He is the author of a variety of studies of Bacon's books and manuscripts, and also of other studies in the history of knowledge in the early-modern period.

Victoria Sloyan
is a British archivist currently based at the Wellcome Library in London. She qualified in 2011 and has previously worked at the Bodleian Library, University of Oxford and London South Bank University. Her research interests lie in the preservation and cataloguing of born-digital records and the acquisition of scientific archives.

Alison Walker
is Lead Researcher for the Sloane Printed Books Project based at the British Library. She was previously Head of the National Preservation Office, and has worked in preservation services and reference services at the British Library. She edited *From Books to Bezoars: Sir Hans Sloane and his Collections*, with Michael Hunter and Arthur MacGregor (London: British Library, 2012).

Elizabeth Yale
is Lecturer in the Department of History at the University of Iowa. In addition to articles published in *Book History*, *Isis*, and *Studies in History and Philosophy of Biological and Biomedical Sciences*, she is the author of *Sociable Knowledge: Natural History and the Nation in Early Modern Britain* (Philadelphia: University of Pennsylvania Press, 2016). At present, she is working on a book exploring archives and posthumous publication as sites for understanding science and gender in the early modern world.

Introduction

Vera Keller, Anna Marie Roos, and Elizabeth Yale

1 "Our Former Naturalists": Antiquarian Science

It was once the case that an admixture of antiquarianism appeared to be an adulteration of science, a body of knowledge based on new things, not old words.[1] The high number of antiquarians among the membership of the Royal Society in the long eighteenth century, for instance, illustrated its debilitation to some historians. Surveys of English science in the eighteenth century, relying on satirists such as Swift, Steele, and Hill, concluded that the Royal Society "from 1680 to 1780 fared ingloriously, except during Newton's presidency."[2] Today the picture is far more nuanced, thanks to decades of scholarship linking philological and historical expertise to natural study, not least in the work of Newton himself.[3] The present collection explores one crucial aspect concerning the relationship between the study of nature and the study of the past. How and why did naturalists engage with archives, and in particular, with collections of "old" science?

Thanks to current scholarship, it is now easy to see why early modern naturalists might wish to amass and winnow large collections of information concerning global natural particulars, or to secure their own reputation for the future through the establishment of a personal archive. The structure,

1 As in Walter Houghton's attempt to "winnow the 'genuine scientist' from those who encouraged only the 'dilution and distortion of the scientific mind.'" Craig Ashley Hanson, *The English Virtuoso: Art, Medicine, and Antiquarianism in the Age of Empiricism* (Chicago: University of Chicago Press, 2009), 8.

2 George Rousseau, "The eighteenth century: science," in *The Eighteenth Century*, The Context of English Literature Series, ed. Pat Rogers (London: Holmes & Meier, 1978), 158.

3 Anthony Grafton, *Defenders of the Text: The Traditions of Scholarship in an Age of Science, 1450–1800* (Cambridge, MA: Harvard University Press, 1991). Gianna Pomata and Nancy Siraisi, eds., *Historia: Empiricism and Erudition in Early Modern Europe* (Cambridge: MIT Press, 2005). Jed Buchwald and Mordechai Feingold, *Newton and the Origin of Civilization* (Princeton: Princeton University Press, 2012).

 | DOI 10.1163/9789004324305_002

organization, and intentions for papers held in archives are part of the story of the emergence of institutionalized experimental natural study, as was first demonstrated by the 1998 volume *Archives of the Scientific Revolution*, edited by Michael Hunter.[4] Yet the question remains, at a time when the battle of the ancients and the moderns raged, to what ends and using what tools did proponents of the advancement of knowledge collect, organize, edit, mine, and filch from the papers of their dead predecessors? What role did the paper remains of scientists past have to play in the creation of new knowledge?

This volume offers diverse responses to these questions, but overarching themes emerge. Most prominently, we see a large-scale transition from multiple forms of repositories to institutional ones. The collecting, organizing, and membership of those institutions thus bore outsize effects on the fates of the papers of naturalists past. The Royal Society, which kindly served as the venue for the 2015 conference upon which this volume is based, thus figures in many of these chapters. This is not because Royal Society science, already the focus of many studies, has suffered from a lack of attention, but precisely because the archival practice of the Society bore such agency for the stories told about the emergence of science and even for who counts as a naturalist, as previous reputations were absorbed, reconfigured or obliterated within new institutions. Only by tracing the assimilation of individual collections into the institution can archival afterlives be unearthed, their significance revived, and alternative stories told.

A case in point is the letter the naturalist Emanuel Mendes Da Costa (1717–1791), in his capacity as Royal Society clerk and keeper, wrote to an unknown recipient on 18 December 1757:[5]

4 Michael Hunter, ed., *Archives of the Scientific Revolution: The Formation and Exchange of Ideas in Seventeenth Century Europe* (Woodbridge: Boydell Press, 1998). More recently, see Elizabeth Yale, "With Slips and Scraps: How Early Modern Naturalists Invented the Archive," *Book History* 12 (2009): 1–36. Lorraine Daston, "The Sciences of the Archive," *Osiris* 27 (2012): 156–187; and Lorraine Daston, ed., *Science in the Archives: Pasts, Presents, Futures* (Chicago: University of Chicago Press, 2017).

5 Royal Society, Miscellaneous Manuscripts, Vol. 14, number 141, MM/14/141. As Da Costa was writing from Tooting, it is possible he was sending the letter from the home of naturalist Reverend Dr Henry Miles FRS (1698–1753) who lived there, and who was a frequent correspondent. It is also possible Da Costa was writing to his relative Joseph Salvador, a merchant who lived in Tooting.

I promis'd you to write you some Anecdotes of our former Naturalists ... in the Ashmolean Museum at Oxford, the following Collections are Absorbed, viz. Tradescants, Plotts, Lhuyd, Leigh's and some of Listers[6] in the Bodleian Library the late Dr Shaws[7]

6 Mendes Da Costa is referring to:

1. The collections of the John Tradescants, which were bought by Elias Ashmole and comprised the basis of the early museum. They are now in the Museum of the History of Science in Oxford. See Arthur MacGregor, *Tradescant's Rarities: Essays on the Foundation of the Ashmolean Museum, 1683, with a Catalogue of Surviving Early Specimens* (Oxford: Clarendon Press, 1983).
2. The collections of Robert Plot (1640–96), FRS, natural historian, antiquarian, Royal Society secretary, and the first keeper of the Ashmolean Museum and first Chemistry Professor at Oxford. None of his natural history specimens survive, although some of the archaeological collections and manuscripts are still extant. See Paul Bidwell, "The Roman Names of the Fort at South Shields and an Altar to the Di Conservatores," in *Life in the Limes: Studies of the people and objects of the Roman frontiers*, ed. Rob Collins and Frances McIntosh (Oxford: Oxbow Books, 2014), 49–58; Anna Marie Roos, *Web of Nature: Martin Lister (1639–1712), the first arachnologist* (Leiden: Brill, 2011), 263–266; Anna Marie Roos, "The Chymistry of 'The Learned Dr Plot' (1640–96)," *Osiris* 29, 1 (2014): 81–95.
3. The fossil collections of Edward Lhuyd FRS, the second keeper of the Ashmolean Museum. Sixteen specimens figured in *Lithophylacii Britannici ichnographia* have been identified in the Natural History Museum's collections at Oxford. See Edward Lhuyd, Natural History Museum, Oxford, http://www.oum.ox.ac.uk/learning/pdfs/lhwyd.pdf [Accessed 16 July 2016]. See also R.T. Gunther, *Early Science in Oxford: Life and Letters of Edward Lhwyd* (Oxford: for the subscribers, 1945).
4. Charles Leigh (1662–1701), a physician and colleague of Plot's. Leigh's *Natural History of Lancashire, Cheshire and the Peak in Derbyshire* was published by subscription in 1700. His proposals for the book may be found in Bodl. MS Ashmole 1816, ff. 15–16, and may be accessed in Early Modern Letters Online, Bodleian Library, Oxford, http://tinyurl.com/6xkmwd2
5. Martin Lister (1639–1712), who donated over 1000 books, 1100 copperplates, and whose manuscripts comprise MSS Lister 1–41 in the Bodleian Library. For the afterlives of Lister's collections, please see the chapter by Anna Marie Roos in this volume.

7 This is likely Dr Thomas Shaw FRS (1694–1751) of Queen's College, chaplain to the factory at Algiers, publishing *Travels in Barbary and the Levant* (Oxford, 1738). He also was an entomologist. The 1716 entry for the Benefactor's Book for the Ashmolean Museum notes: "Thomas Shaw BA, from Queen's College, who was much devoted to Natural History, gave this Museum a large collection of insects caught around Oxford, elegantly arranged into their various classes and families" [*THOMAS SHAW é Coll. Reginensi A:B: Historiæ Naturalis Studiosissimus, quam plurima Insecta circa Oxonium capta, et in suas Classes et Familias elegantissime disposita huic Museo D.D. Bodl.*] AMS 2; online edition: http://www.ashmolean.org/ash/objects/?mu=571 [Accessed 17 August 2016.]

> in the Sloanian or British Museum, Corteyns, Petivers, and all Plukenets Horti Sicci[8]
>
> … Francis Willoughby Esqr who travell'd with Sir Philip Skippon and Mr Ray, seems to have made no regular Collection, but I am informd there is a small Cabinet of Natural History of their Collecting at his descendants the lord Middleton's at Wollaton hall near Nottingham …[9]
>
> Lister's collection of shells, I cannot learn what became of it, he says when he offerd it to Sir Hans [Sloane] that there were three Cabinets, viz one of Sea univalves, one of Sea Bivalves, and one of River Land Shells.[10]

Da Costa assumes the existence of a group of "our former Naturalists," indicating his pride in and association with a continuing legacy of British scholars. He associates these naturalists themselves with their collections, in this case largely of objects rather than paper archives; the relationship between

8 Mendes da Costa's notes refer to the collections of William Courten (1642–1702), James Petiver (1663–1718), and Leonard Plukenet (1641–1706). For the afterlives of some of Courten's collection, see the chapter by Anna Marie Roos in this volume. For an excellent overview of his collection, see Sachiko Kusukawa, "William Courten's lists of 'Things Bought' from the late seventeenth century," *Journal of the History of Collections* 29,1 (2017): 1–17. Plukenet's and Petiver's collections were subsumed into those of Sir Hans Sloane, consisting of 265 volumes of *hortus siccus*; the volumes are presently housed in the Natural History Museum of London. Petiver's collections constitute more than 100 of these volumes.

9 A cabinet from Willughby's collection has been recently rediscovered in Lord Middleton's collection and is described in Tim Birkhead, ed. *Virtuoso by Nature: the scientific worlds of Francis Willughby* (Leiden: Brill, 2016), 275–278. The cabinet was moved to its current location in Yorkshire from Middleton Hall in 1925. It contains a bird egg collection and botanical seed specimens, as well as some dried animals such as a skink.

10 In 1707, Sloane asked Lister "what value you sett upon the cabinetts of Shells and what upon the plates, separately if you will part them, or together if you design they shall go together." (Bodleian MS Lister 37, f. 109). Lister's price was apparently too expensive, Sloane complaining "really the summe is so great, to yet under what the value may be, that I cannot any ways contrive how to think them worth so much to me." (Bodleian MS Lister 37, ff. 107r and 109r). In reality, Sloane had only offered £150 for both; though ultimately they came to an agreement about the shells, Lister bequeathed nearly 1000 copperplates to the University of Oxford. See Roos, *Web of Nature*, 413. For the afterlives of some of the shells, see the chapter in this volume by Anna Marie Roos.

those two forms of collection will be one of the themes of the present volume. The scholars are all male, although, as some of our chapters will point out, women participated in several of the collections Da Costa named. These men have been "absorbed" into the great, even national, new institutions of the day, such that it became a show of expertise for Da Costa to be able to identify and trace their individual fates. Those naturalists unlucky enough not to be included in the great passage from family collections to metropolitan repositories were left afloat in backwaters, with Da Costa, unsure, for example, of the nature of Willughby's collection, which may or may not have been still in a descendant's country estate. The fact that Willughby's descendants may have acquired a collection put together not only by Willughby but also by fellow travelers Skippon and Ray, hints at some of the tensions that may have flared when collaborative works passed to just one heir, tensions magnified many times over in the larger institutional absorption of many naturalists by an Ashmole or a Sloane.[11]

In the early days of the Royal Society, its secretary, Henry Oldenburg (1618–1677), suggested a system of manuscript registration to prevent the "usurpation" of ideas and inventions still in development.[12] The system was designed to prevent priority disputes and also thereby to act as an incentive for scholars to submit their work to the society, which promised to better protect and validate their claims than a personal collection or premature publication could. While drawing thoughtfully on techniques for the organization of knowledge, Oldenburg's system specialized in the most recent discoveries and in safeguarding future reputations. It was not designed to resuscitate or curate reputations from the past. At the high point of amalgamation of individual collections into institutions charted in Da Costa's letter, Royal Society interests turned to curating archives past as a vehicle for securing its institutional reputation. The growing predominance of fellows who were both natural philosophers and antiquarians in the Royal Society under the presidencies of Sir Hans Sloane (1660–1753) and Martin Folkes (1690–1754) led to more general efforts to recover early Society manuscripts as well as the contents of the Repository.

---

11 On Willughby's collections and collaborations, see the essays in Birkhead, ed., *Virtuoso by Nature*, especially Mark Greengrass, Daisy Hildyard, Christopher D. Preston, and Paul J. Smith, "Science on the Move: Francis Willughby's Expeditions," 142–226, and Isabelle Charmantier, Dorothy Johnston and Paul J. Smith, "The Legacies of Francis Willughby," 376–385.

12 Adrian Johns, *The Nature of the Book: Print and Knowledge in the Making* (Chicago: University of Chicago Press, 1998), 486.

Folkes actively encouraged fellows to pass along manuscript materials in their possession that elucidated the history of the Royal Society. The naturalist Reverend Dr Henry Miles (1698–1763) in a letter to microscopist Henry Baker (1698–1774)[13] of 13 March 1744/5 heeded Folkes' call by searching for the papers of early Fellows Henry Oldenburg and John Beale (1608–1683).

> … I am particularly obliged to our most worthy president [Folkes] for his complements … let him know that I have not forgot my promise of preparing some letters of Mr [Henry] Oldenburgh's & c to be lodgd in the Library of the R.S. those of Mr Oldenburg's being long since pasted in a Folio Chart book and would have been sent before the meeting of the Society in October but that I had unhappily mislaid a paper, without which I could not finish my design and a variety of interruptions have prevented my adding those of Dr [John] Beale and some others: many being write on all sides, and so near the edge, that I have found it very difficult without taking up more time than my necessary affairs would allow, to make any considerable progress: however I have taken care that they will not be lost to the Society, if I should die, and I hope soon to finish and present them myself. [14]

Indeed, the Oldenburg-Beale correspondence is, in the present Royal Society Library, not an "official" part of the Early Letters Collection, but an additional guard-book, so it seems that Miles had made good on his promise to Baker and Folkes.[15] Baker also noted to Miles that Folkes was "obliged for your Care in preserving the Letter to Mr Boyle."[16]

This was an era of close relationships between the Royal Society and the Society of Antiquaries of London; Da Costa and Miles were members of both, and Folkes even attempted to unite the two societies into one body in the 1750s. Rather than a sign of the decline of Royal Society science, the

---

13 Baker was elected FRS in 1740 and to the Society of Antiquaries and was a recipient of the Copley Medal in 1744 for his microscopic observations of salts, and a frequent contributor of papers to the *Philosophical Transactions*.

14 John Rylands Library, Henry Baker Correspondence, Eng MSS 19, volume 2, ff. 19–20.

15 Email from Keith Moore, Royal Society Librarian to Anna Marie Roos, 14 July 2016. The letters between Beale and Oldenburg are separately catalogued as Early Letters OB (Oldenburg Beale). The letters have been arranged chronologically and indexed as stated in a note by Henry Miles dated 20 August 1745.

16 Henry Miles to Henry Baker, 26 March 1745, John Rylands Library, Henry Baker Correspondence, Eng MSS 19, volume 2, f. 33.

proposed amalgamation indicated an era of Fellows who could draw on antiquarian concerns and skills to curate the history of the institution by recovering older scientific papers. Miles served as editor and transcriber (with Thomas Birch) of Robert Boyle's manuscripts, purchased from Boyle's apothecary.[17]

This concern with institutional reputation was a double-edged sword, however, as Miles and Birch are also known for throwing away many of Boyle's private papers after his death, having dismissed them as inconsequential, embarrassing, or too personal, according to a common eighteenth-century editorial practice. Folkes himself in his role of editor of the biblical chronology of Newton also largely sanitized it of any heretical content, and "the material that Newton had been preparing to publish was deliberately shorn of most of its context in the history of ancient religion."[18]

These eighteenth-century naturalists' engagements with the papers and collections of their seventeenth-century forebears highlight the two parallel strands the present volume follows: we consider both: (1) how and why naturalists, antiquaries, and archivists have attended to the material record of the scientific past (the papers of the dead, as it were); and (2) their efforts to preserve, transmit, and make use of that record, three activities that could come into conflict with each other. These efforts have taken a number of forms over the years, as this volume explores. Naturalists and archivists have placed papers in institutional archives. They have remediated manuscript records through print editions (in the eighteenth century) and digital records through other digital forms (in the twenty-first century). Via these transmission processes, they have attended to the qualities and affordances of the material substrata on which their records are stored both in order to preserve those records and to demonstrate the authenticity of texts published from them. At every turn along these pathways of transmission, alternate possible afterlives beckon, dramatizing the contingencies implicit in the physical dispersal, categorization, and use of papers after the death of the author.

17 Marie Boas Hall, "Henry Miles FRS (1698–1763) and Thomas Birch (1705–1766)," *Notes and Records of the Royal Society* 18 (1963): 39–44.

18 Scott Mandelbrote, *Duty and Dominion, Footprints of the Lion: Isaac Newton at Work.* On-line Exhibition, Cambridge University Library, http://www.lib.cam.ac.uk/exhibitions/Footprints_of_the_Lion/duty_dominion.html [Accesssed 16 July 2016]. See also Jed Z. Buchwald and Mordechai Feingold, *Newton and the Origin of Civilization* (Princeton: Princeton University Press, 2012).

1.1 *Thinking Archivally about Early Modern Science and Medicine*

If naturalists claimed all knowledge as their province, natural philosophy was a paper empire. Early modern naturalists generated, collected, and shared massive amounts of paper. Inspired by calls for the wholesale reform of natural philosophy and schooled in humanist note-taking practices, they generated correspondence; reading notes (in margins, on scraps, in notebooks); experimental and observational reports; and drafts (rough, partial, fair) of treatises intended for circulation in manuscript or further replication in print. In our own day, naturalists' materials, ensconced in archives, libraries, and (occasionally) private hands, are the foundation of a history of science that has taken a material turn towards paper, ink, pen, and filing systems as technologies of communication, information management, and knowledge production.[19] Reading information history with the gaze of the information age,

19 Stephen Clucas, "Samuel Hartlib's 'Ephemerides,' 1635–1659, and the Pursuit of Scientific and Philosophical Manuscripts: The Religious Ethos of an Intelligencer," *Seventeenth Century* 6 (1991): 33–55; Mark Greengrass, "Samuel Hartlib and Scribal Communication," *Acta Comeniana* 12 (1997): 47–62; Andrew Warwick, *Masters of Theory: Cambridge and the Rise of Mathematical Physics* (Chicago: University of Chicago Press, 2003); Ann Blair, "Note Taking as an Art of Transmission," *Critical Inquiry* 31 (2004): 85–107; Matthew Eddy, "Tools for Reordering: Commonplacing and the Space of Words in Linnaeus's *Philosophia Botanica*," *Intellectual History Review* 20 (2010): 227–252; Andrew J. Mendelsohn, "The World on a Page: Making a General Observation in the Eighteenth Century," in *Histories of Scientific Observation*, ed. Lorraine Daston and Elizabeth Lunbeck (Chicago: University of Chicago Press, 2011), 396–420; Elizabeth Yale, "Marginalia, Commonplaces, and Correspondence: Scribal Exchange in Early Modern Science," *Studies in History and Philosophy of Biological and Biomedical Sciences* 42 (2011): 193–202; Staffan Müller-Wille and Isabelle Charmantier, "Natural History and Information Overload: The Case of Linnaeus," *Studies in History and Philosophy of Biological and Biomedical Sciences* 43 (2012): 4–15; Richard Yeo, *Notebooks, English Virtuosi, and Early Modern Science* (Chicago: University of Chicago Press, 2014); Staffan Müller-Wille and Isabelle Charmantier, ed. *Worlds of Paper*, a special issue of *Early Science and Medicine* 19 (2014): 379–503; Yale, *Sociable Knowledge: Natural History and the Nation in Early Modern Britain* (Philadelphia: University of Pennsylvania Press, 2016); Matthew Eddy, "The Interactive Notebook: How Students Learned to Keep Notes During the Scottish Enlightenment," *Book History* 19 (2016), 86–131; Renée Raphael, *Reading Galileo: Scribal Technologies and the* Two New Sciences (Baltimore: Johns Hopkins University Press, 2017). On early modern learned culture as paperwork more generally, see Ann Blair, *Too Much to Know: Managing Scholarly Information Before the Modern Age* (New Haven: Yale University Press, 2010).

current literature looks to notebooks and card files as "paper tools" and "paper machines" through which users framed their worlds.[20]

The history of scholarly (and within that, scientific) note-taking and archival practices has developed alongside a growing field investigating the growth of archives as instruments of social, religious, and political control, tools for administering churches, religious orders, city-states, commercial empires, and colonies.[21] In this body of work, scholars index the rise of institutional power, especially as embodied in the early modern state, through archival techniques. These include the physical management of papers through bagging, labeling, pinning, and boxing; the organization of papers using indices,

---

20 Bernhard J. Dotzler, *Papiermaschinen: Versuch über Communication & Control in Literatur und Technik* (Berlin: Akademie, 1996); Ursula Klein, "Paper Tools in Experimental Cultures," *Studies in the History and Philosophy of Science* 32 (2001): 265–302; Anke te Hessen, "The Notebook: A Paper-Technology," in *Making Things Public: Atmospheres of Democracy*, ed. Bruno Latour and Peter Weibel (Cambridge: MIT Press, 2005), 582–589; Marcus Krajewski, *Paper Machines: About Cards and Catalogs, 1548–1929*, trans. Peter Krapp (Cambridge: MIT Press, 2011).

21 Antoinette Burton, ed., *Archive Stories: Facts, Fictions, and the Writing of History* (Durham: Duke University Press, 2005); Miles Ogborn, *Indian Ink: Script and Print in the Making of the English East India Company* (Chicago: University of Chicago Press, 2007); Ann Laura Stoler, *Along the Archival Grain: Epistemic Anxieties and Colonial Common Sense* (Princeton: Princeton University Press, 2009); Kathryn Burns, *Into the Archive: Writing and Power in Colonial Peru* (Durham: Duke University Press, 2010); Arndt Brendecke, "'*Arca, archivillo, archivo*': The Keeping, Use and Status of Historical Documents about the Spanish *Conquista*," *Archival Science* 10 (2010): 267–83; Filippo de Vivo, "Ordering the Archive in Early Modern Venice (1400–1650)," *Archival Science* 10 (2010): 231–248; Markus Friedrich, "Archives as Networks: The Geography of Record-keeping in the Society of Jesus (1540–1773)," *Archival Science* 10 (2010): 285–298; Nicholas Popper, "From Abbey to Archive: Managing Texts and Records in Early Modern England," *Archival Science* (2010): 249–266; Francis X. Blouin, Jr. and William G. Rosenberg, *Processing the Past: Contesting Authority in History and the Archives* (Oxford: Oxford University Press, 2011); Jacob Soll, *The Information Master: Jean-Baptiste Colbert's Secret State Intelligence System* (Ann Arbor: University of Michigan Press, 2011); Cheryl Beredo, *Import of the Archive: U.S. Colonial Rule of the Philippines and the Making of American Archival History* (Sacramento: Litwin Books, 2013); Kirsten Weld, *Paper Cadavers: The Archives of Dictatorship in Guatemala* (Durham: Duke University Press, 2014); Liesbeth Corens, Kate Peters, and Alexandra Walsham, eds., *The Social History of the Archive: Record-keeping in Early Modern Europe, Past and Present*, Supplement 11 (2016); Anthony Grafton, "Matthew Parker: The Book as Archive," *History of Humanities* 2,1 (2017): 15–50. For a recent overview, see Elizabeth Yale, "The History of Archives: The State of the Discipline," *Book History* 18 (2015): 332–359.

catalogues, and registers; and the regulation of access via material and social means, locks and keys as well as the movement of records between private homes and public spaces.

Historians of archives also look to the increasing roles that documents played in everyday political, civic, and economic situations. As Andy Wood and Adam Fox have traced, whereas medieval English villagers might have settled disputes over land boundaries by recourse to the memory of the oldest man in the village, early moderns relied on paper records, often maintaining their own copies of these in their homes.[22] In the Iberian Atlantic world and on the Italian peninsula, as Kathryn Burns and Laurie Nussdorfer have shown, the notary played an all-important role, becoming an obligatory *point de passage* for legal and economic transactions.[23] The modern expansion of written record keeping in Europe can be traced back into the medieval period; in some ways, the introduction of paper to Europe in the late middle ages only accelerated already existing trends.[24]

In the words of the archivists and archival scholars Joan M. Schwartz and Terry Cook, "Archives, then, are not passive storehouses of old stuff, but active sites where social power is negotiated, contested, and confirmed."[25] To "social power," we would add scientific and medical knowledge. The essays in this volume bring together, extend, and complicate current work on paper technologies and archives. In contrast to the literature on state archives, we focus on scholarly notes and papers: what we might call private papers. With historians of archives and archivists, we see these as sites at which knowledge and power are contested. In tracing their archival afterlives, we see early modern science, and its modern narratives, afresh. Collectively, the essays in this volume suggest that whatever else they were, the naturalists, experimentalists, chymists, and physicians of the early modern British world were also archivists and

---

22 Adam Fox, "Custom, Memory and the Authority of Writing," in *The Experience of Authority in Early Modern England,* ed. Paul Griffiths, Adam Fox, and Steve Hindle (London: MacMillan Press, 1996), 89–116. Andy Wood, *The Memory of the People: Custom and Popular Senses of the Past in Early Modern England* (Cambridge: Cambridge University Press, 2013).

23 Laurie Nussdorfer, *Brokers of Public Trust: Notaries in Early Modern Rome* (Baltimore: Johns Hopkins University Press, 2009); Kathryn Burns, *Into the Archive: Writing and Power in Colonial Peru* (Durham: Duke University Press, 2010).

24 M.T. Clanchy, *From Memory to Written Record: England, 1066–1307* (3rd ed., Malden, UK: Wiley-Blackwell, 2012).

25 Joan M. Schwartz and Terry Cook, Introduction, "Archives, Records, and Power," *Archival Science* 2 (2002): 1.

antiquaries.[26] The new sciences came into being through the use and manipulation of archival collections.

In the essays collected here, we see how the history of archives can reorient our approach to the history of science. Early modern naturalists understood building, using, disseminating, and preserving collections of personal and institutional papers as integral to the development of the new sciences. Scholarship in the history of archives alerts us further to how fragmentary and contingent archival pasts were. It also pushes us to look for the ways in which naturalists strategically manipulated and transformed their collections of papers, amassing knowledge and power through the deployment of their archives. We read with and against "the archival grain" described by anthropologist Ann Laura Stoler in order to recover the stories that archives tell, intentionally and unintentionally.[27]

In excavating archival stories from the history of early modern science, we inevitably disturb old ground. As noted above, many of these essays explore the archives of figures affiliated, in one way or another—as posthumous philosophical inspiration, secretary, fellow, or benefactor—with the Royal Society of London. On the one hand, this suggests the entanglement of the archive and the renovation of knowledge. As the East India Company (which Robert Boyle served as a member of the Court of Committees and in which the Royal Society bought stock) promoted trade through the use and maintenance of paper recordkeeping systems, the Royal Society promoted experimental knowledge.[28] Paperkeeping habits, and personnel overlapped, too, between the Royal Society and Charles II's State Paper Office, with Joseph Williamson serving as president of the former and Keeper of the latter.[29] But the fellows and early secretaries of the Royal Society not only built up institutional

26 Though not the focus of this volume, cases of archival and scientific entanglement can also be found across the history of early modern mathematics and astronomy—the latter a "science of the archive," in the words of Lorraine Daston, since the time of the ancient Babylonians. Daston, "The Sciences of the Archive," *Osiris* 27 (2012): 164; Adam Perkins, "The Flamsteed Papers in the Archives of the Royal Greenwich Observatory," in Frances Willmoth, ed., *Flamsteed's Stars: New Perspectives on the Life and Work of the First Astronomer Royal (1646–1719)* (Woodbridge, Suffolk: The Boydell Press, 1997), 217–224; Sarah Dry, *The Newton Papers: The Strange and True Odyssey of Isaac Newton's Manuscripts* (Oxford: Oxford University Press, 2014).

27 Stoler, *Along the Archival Grain*.

28 Miles Ogborn, *Indian Ink: Script and Print in the Making of the English East India Company* (Chicago: University of Chicago Press, 2007), XV–XXIII.

29 Nicholas Popper, "Archives and the Boundaries of Early Modern Science," *Isis* 107 (2016): 86–94.

archives (or personal collections that aided them in their immediate duties); they created an institution that would absorb the collections of individuals, past, present, and future. Fellows also collected and deployed papers as individuals, working with and against each other, as Elizabeth Yale and Vera Keller's chapters highlight.

The Royal Society's archival legacy, then, is one of overlapping paper regimes, with tendrils that reach out into other collections, to be occasionally "rediscovered," and if funds are sufficient to the acquisition, brought home.[30] Given this, the collections of the Royal Society, and of individuals we now associate with it, whether as intellectual precursors or fellows, can be made to tell surprising stories. Carol Pal, in her essay, uncovers the "anti-archives" of learned seventeenth-century women who never expected that their papers would be preserved. Similarly, as the chapter by Anna Marie Roos shows, the antiquarian William Huddesford's (1732–72) interest in the naturalist Martin Lister ensured the survival of the illustrations and ephemera Susanna and Anna Lister produced for their father's scientific works. Their afterlife permits us to understand the sisters' central role in the production of Lister's conchological work, the *Historiae Conchyliorum* (1685–92).

The story of "British" scientific and medical archives was an international one, a point obscured by the eighteenth- and nineteenth-century development of the national libraries and archives that now house some of these materials.[31] Papers traveled with people: as Richard Serjeantson recounts, for example, one of Francis Bacon's "composition books" followed William Boswell to the Hague, to be published by the Dutch Scholar Isaac Gruter. In her tale of afterlives, Vera Keller reveals the Huguenot physician Theodore de Mayerne's posthumous influence on discussions of color, dyeing works, and experimental methodology in the early Royal Society. For collectors like James Petiver and Hans Sloane, letters, printed texts, and objects traveled through the channels of international correspondence, settling in their museums archival deposits from around the vast globe.

30 This process is most recently signified by the Royal Society's acquisition of the Hooke Folio, but Hans Sloane, in purchasing stray papers and books from Robert Hooke's library as they came on the market in the early 18th century, also participated in it. See Lisa Jardine, *Temptations in the Archives: Essays in Golden Age Dutch Culture* (London: UCL Press, 2015), 42–44; Giles Mandelbrote, "Sloane's Purchases at the Sale of Robert Hooke's Library," in Giles Mandelbrote and Barry Taylor, eds., *Libraries within the Library: The Origins of the British Library's Printed Collections* (London: The British Library, 2009), 98–145.

31 On the tightening ties binding national archive and nation state in the nineteenth century, see Blouin and Rosenberg, 32.

The reverse operation, applying the history of science to the history of archives, is equally productive. For historians of archives and knowledge management, as well as (the editors hope) archives professionals, these essays open up new questions and fields of inquiry. Perhaps the most important of these is the relationship between individuals and institutions in the early modern world. Historically, archives have reflected official processes and transactions—a notion reflected in the word's Greek root, ἀρχεῖον and codified by Hilary Jenkinson in *A Manual of Archive Administration* in 1922, if not before.[32] And, as noted above, the history of archives has tended to concentrate on the papers produced within developing state and religious bureaucracies. This is a valuable approach: the acceleration of paper recordkeeping fed and mirrored the growth of early modern institutions.

Yet, in the history of record keeping and record use within the history of early modern learning, we see lives lived in connection to early modern institutions—the Elizabethan and early Stuart states, the Royal Society, the Ashmolean Museum—but not subsumed by them. Scientific and antiquarian records pooled and flowed through intimate genealogies, from husband to wife, master to secretary, colleague to colleague and friend to friend, and naturalists' archives were not intimately tied to institutions, much less to states. The natural philosopher emerged gradually as a public individual in his own right, through his instantiation in posthumously printed editions and in archives. The sense that some private papers were worth preservation in public institutions (such as the Ashmolean Museum and the Royal Society Archive) and even preservation by the state, as in the British Museum, emerged in the history of knowledge in the eighteenth century—before it arose in the nineteenth, and even the twentieth centuries, in the history of contemporary literature.[33] Collectively, these essays document a crucial chapter in the history of what

32 Hilary Jenkinson, *A Manual of Archive Administration Including the Problems of War Archives and Archive Making* (Oxford: Clarendon Press, 1922), 2–11. See also Jacques Derrida, *Archive Fever: A Freudian Impression*, trans. Eric Prenowitz (Chicago: University of Chicago Press, 1996).

33 We distinguish here between eighteenth-century attitudes towards medieval literary manuscripts, such as those collected by Robert Cotton and the Harleys, and any papers produced by, say, Francis Bacon's contemporary William Shakespeare. See Anthony Rota, "The Collecting of Twentieth-Century Literary Manuscripts," *Rare Books and Manuscripts Librarianship* 1 (1986), 39–53; Thomas F. Staley, "Literary Canons, Literary Studies, and Library Collections: A Retrospective on Collecting Twentieth-Century Writers," *Rare Books and Manuscripts Librarianship* 5, 1 (1990): 9–21; Amy Hildreth Chen, *Archival Bodies: Twentieth-Century British, Irish, and American Literary Collections* (Emory University, Ph.D. Diss., 2013), esp. 209–217. Many thanks to Dr. Chen for sharing her dissertation with us.

Francis Blouin and William Rosenberg have called "collecting institutions" and "identity archives," archives dedicated to a particular event or process or to a group defined by a racial, ethnic, gender, or sexual identity.[34] Such archives cannot be neutral or "objective" records of the past, and scientific collections are no exception to the rule. The scientific world erected both "great men" and large institutional archives and libraries at the same time—developments not wholly unconnected, as Anna Marie Roos's essay, in particular, shows.

But our investigations of archives do not have to reinforce the "great man" narrative. Carol Pal shows us how to read against the archival grain of Samuel Hartlib's papers, recovering the "anti-archives" of learned women such as Bathsua Makin (c.1600–1680), Dorothy Moore (c.1612–1664), and Katherine Jones, Lady Ranelagh (1615–1691). Though they were all too aware that their achievements were destined to be left unrecorded in more intentional archives, at least these figures have been captured, fragmentarily yet fortunately, in the amber of Hartlib's scientific outpouring. In the present, as Victoria Sloyan discusses in her essay, with new documentation strategies being developed by archivists, archives document scientific collectivities.

As a framework for historical analysis, then, "archival afterlives" considers how naturalists engaged in archival processes and attitudes, as well as the descent and use of documents within specific institutions. With a focus on papers and what becomes of them over time, however, we do limit ourselves to tracing a particular kind of material.[35] This allows us to see posthumous publication as a kind of archiving (albeit one inattentive to the future auction value of holograph manuscripts). Perhaps the challenges facing Francis Bacon's seventeenth-century editors were not unlike those facing the archivists in the Wellcome Library's Collecting Genomics project. Preservation necessarily entails some loss of an original using or reading experience, as digital records need to be constantly migrated to new formats, and will necessarily be encountered by future readers through physical devices that do not exactly mimic the computers on which they were created. We also see archival processes in

34 Blouin and Rosenberg, 133–34, 156–158.

35 Following the advent of modern critical theory, particularly as articulated by Michel Foucault, humanities scholars have tended to use "the archive" more broadly, to refer to any body of sources—print or manuscript, novel, newspaper, document, or letter—mined to make an argument. See Michel Foucault, *The Archaeology of Knowledge and the Discourse of Language,* trans. A.M. Sheridan Smith (New York: Pantheon Books, 1972), 126–131; and, as an example, Anjali Arondeka, *For the Record: On Sexuality and the Colonial Archive in India* (Durham: Duke University Press, 2009). We do not intend to follow this usage.

mixed media. Alison Walker takes us into, for example, seventeenth-century medicine's "archives of practice," the interleaved and annotated pharmacopeia that Hans Sloane collected from other physicians' libraries. In these hybrid objects, the book touches the archive.

The history of archives has concentrated largely on paper records, whether folded in boxes, pinned in bags, or bundled and rolled. In considering the history of archives in relation to the history of early modern science, we can recover the close relationships, now obscured, that once existed between archives, studies, libraries, and museums.[36] Essays in this volume, particularly Arnold Hunt's essay on James Petiver's papers and collections and Anna Marie Roos's essay on Martin Lister's conchological work bring to the fore the relation between papers and objects, archives and collections. In the natural sciences, an archive could be an instrument for navigating a collection of natural objects and vice versa: neither was intelligible without the other. These materials are frequently divided in modern archival and museum settings—which can, as Hunt shows, lead to confusion about how an originator ordered and used his collections.

In classical archival scholarship, dating to the nineteenth century, an archive's *fonds*, its organizational structure, is an image of the structures and power relationships that inhered in the institution that kept the archive.[37] In these essays, we see that in scientific collections of papers, whether institutional or personal in origin, the *fonds* can reveal the logic of experimental knowledge production. As Vera Keller shows in her essay, naturalists associated with the Royal Society, canny archivists *avant la lettre*, understood this, using the material traces of an archive's organization (folds in paper, the placement of monograms) to reconstruct experimental method. Sometimes, on the other hand, later generations ignored or misinterpreted these traces of organization. In some cases, systems of classification had changed, making an old order

---

36 William Clark, "On the Bureaucratic Plots of the Research Library," *Books and the Sciences in History*, ed. Marina Frasca-Spada and Nick Jardine (Cambridge: Cambridge University Press, 2000), 190–206; Helen Watanabe-O'Kelly, "Kunstkammer, Library and Chamber of Anatomy: The Management of Knowledge at the Dresden Court in the Early Modern Period," *Ways of Knowing*, ed. Mary Lindemann (Boston: Brill, 2004), 53–65; Barbara Segelken, *Bilder des Staates: Kammer, Kasten und Tafel als Visualisierungen staatlicher Zusammenhänge* (Berlin: Akademie, 2010); Glenn Adamson, "The Labor of Division: Cabinetmaking and the Production of Knowledge," in *Ways of Making and Knowing: The Material Culture of Empirical Knowledge*, ed. P.H. Smith, A.R.W. Meyers, and H.J. Cook (Ann Arbor: University of Michigan Press, 2014), 243–279.

37 Blouin and Rosenberg, 29–31.

seem confused to the point of insensibility: as Hunt recounts, this is one reason James Petiver's papers have bewildered generations since Linnaeus. Similarly, Anna Marie Roos suggests that changes in rock and fossil classification systems may have been partly behind Emanuel Mendes de Costa's low estimation of Edward Lhuyd's skill as a lithologist.

1.2 *Thinking Archivally in Early Modern Science and Medicine*

Archival Afterlives invites the historian of archives and the historian of science to view the history of knowledge in terms of scholars' social and material tactics for authenticating, using, and preserving records. As we think archivally about early modern science and medicine, we catch glimpses of early modern naturalists themselves thinking archivally within early science and medicine. As this volume illustrates, many naturalists were self-reflexive about their tactics, with alertness to the material nature of their documents that would grow over the course of the century. Such material bibliographic concerns grew out of interlacing studies of antiquity and of nature. Antiquarianism drew attention to issues of loss in the historic record as well as to techniques of authenticating, falsifying, and preserving documents and other materials.

As they collected, transferred, discussed, and published documents, virtuoso-archivists found ways of authenticating their documents and thus rendering them, if not authoritative sources, at least starting points for further discussion and experimentation. As Richard Serjeantson argues in his account of Francis Bacon's afterlives, at the beginning of the seventeenth century, manuscript collectors sought signs of the reliability of a text rather than a document. The destruction of the original manuscript in the publication of Bacon's works illustrates what little regard the holograph manuscript attracted, other than as an indication of the text's authenticity.[38] Over time, however, naturalists increasingly valorized the materiality of original texts. In her chapter, Vera Keller describes the views of John Beale (1608–1683) concerning the physical and psychological effects of the color and chemical composition of particular inks, regardless of the text; Beale proved nostalgic even for the lost scribal arts of Catholic illuminated books of devotion. According to Serjeantson, around the turn of the eighteenth century, "a new respect for the manuscript as a document of interest in its own right had demonstrably come into being."

38 This was par for the course in printing more generally. Anthony Grafton, *The Culture of Correction in Renaissance Europe* (London: British Library, 2011).

Larger international trends were at play in the developing valuation and historicization of original sources and their material qualities. English naturalists drew on a long tradition of investigating ancient techniques of book production and preservation, dating back over a long history of forgers and their critics.[39] Sixteenth-century histories of invention and of loss brought the study of material bibliography to the fore, as scholars sought to chart how varying literary technologies mapped on to the rise and fall of learning.[40] In England, tracing the varying course of knowledge motivated individuals such as Sir Francis Kynaston (1587–1642), in his commentary on Chaucer, to explore ancient techniques of inscription, such as the figures inscribed with a stylus on bark in an ancient example he saw in the Bodleian or a thirteen-hundred-year-old Saxon manuscript belonging to Patrick Junius, as well as modern techniques of printing, engraving, and etching.[41] In his epic 1656 poem on the career of King David, Abraham Cowley, future member of the Royal Society, puts on view "all the various Forms" through which "Letters sav'd themselves from Death of old":

> Some painfully engra'vd in thin wrought Plates,
> Some cut in Wood, some lightlier trac'd on Slates;
> Some drawn on fair Palm Leaves, with short-liv'd Toil,
> Had not their Friend the Cedar lent his Oil.

---

39 Anthony Grafton, *Forgers and Critics: Creativity and Duplicity in Western Scholarship* (Princeton: Princeton University Press, 1990) and, in an English context, Grafton, "Matthew Parker."

40 Guido Pancirolli, *Rerum Memorabilium iam olim deperditarum*, trans. and ed. Heinrich Salmuth (Amberg: Forster, 1599), "de librariis, sive bibliothecis," 111–119; "de Papyro," 240–241; "De characteribus literarum," 269–290. Many English naturalists discussed here, such as Kynaston, Cowley and Evelyn, relied on Pancirolli, even though the commentary on papyrus added by Pancirolli's editor Heinrich Salmuth was taken largely from Melchior Guiland's work on Pliny's accounts of papyrus. J.J. Scaliger had subjected Guiland to severe criticism. Anthony Grafton, "Rhetoric, Philology and Egyptomania in the 1570s: J.J. Scaliger's Invective against M.Guilandinus's Papyrus," *Journal of the Warburg and Courtauld Institutes* 42 (1979): 167–194. Salmuth responded to Scaliger in his preface to a new edition, Melchior Guiland, *Papyrus*, ed. Heinrich Salmuth (Amberg: Schönfeld, 1613). Although Polydore Vergil's history of invention is today better known than Pancirolli's is, the way that Pancirolli paired invention with loss made him a central touchstone for bibliographic debates over the material supports that either preserve knowledge or led to its destruction. On Vergil and the early modern genre of heuromatography, see Catherine Atkinson, *Inventing Inventors in Renaissance Europe: Polydore Vergils's De Inventoribus Rerum* (Tübingen: Mohr Siebeck, 2007).

41 Bodleian Library, MS. Add. C. 287, 169, 205, and 523; 204.

Some wrought in Silks, some writ in tender Barks;
Some the sharp Stile in waxen Tables markes;
Some in Beast Skins, and some in Biblos Reed;"[42]

The poem appeared with Cowley's own extensive scholarly notes, drawing on Guido Pancirolli's discussions of ancient forms of the book and including citations of Ovid and other classical authorities on cedar oil.[43] In 1672, John Aubrey (1626–1697) FRS began collecting historical specimens for compiling a "Chronologia Scriptoria" for dating scripts; he and other fellows of the Royal Society, such as Saville Professor and cataloger of the Bodleian manuscripts, Edward Bernard, would admire Mabillon's *De re diplomatica.*[44]

Naturalists mined older technologies of the book not only for techniques of identification and dating of ancient works, but as recoverable tools that could safeguard their own works for the future. In a work on insects edited by Theodore de Mayerne, Thomas Muffet (1553–1604) noted that the ancient books of Numa, legendary for their longevity, had been preserved in antiquity through cedar, which might act as an effective defense against book worms.[45] In later editions of his study of trees, John Evelyn (1620–1706) FRS also noted that cedar oil "best preserves the monuments of the learned, books and writings." He debated interpretations of classical sources on the use of cedar on paper, previously discussed by his friend Cowley.[46] The upheaval of the Interregnum, and even more so of the Great Fire of London, encouraged investigation into techniques of long-term preservation in the Royal Society

---

42 Abraham Cowley, "Davideis," in *Poems* (London: Humphrey Moseley, 1656), 18.

43 Ibid, 41–42. Cowley's literary executor, Thomas Sprat, edited the 1668 posthumous edition of Cowley's works, adding "all that I could find in his Closet, which he had brought to any manner of perfection," while careful to omit anything that might harm his reputation, as he claimed Cowley had instructed in his will. Thomas Sprat, "An Account of the Life and Writings of Mr. Abraham Cowley," in Abraham Cowley, *Works*, ed. Thomas Sprat (London: Martyn, 1668), [Av].

44 Kelsey Jackson Williams, *The Antiquary: John Aubrey's Historical Scholarship* (Oxford: Oxford University Press, 2016), 85. Jean Mabillon, *De re diplomatica* (Paris, 1681). Bernard published his own systematic chart of the historical development of Near Eastern alphabets. Edward Bernard, *Orbis eruditi literaturam à charactere Samaritico hunc in modum savente Deo deduxit Eduardus Bernardus* (Oxford, 1689).

45 Thomas Muffet, *Insectorum, sive minimorum animalium theatrum*, ed. Theodore de Mayerne (London: Cotes, 1634), 274.

46 John Evelyn, *Silva,* 5th Edition (London: Walthoe, 1729), 137. In the first edition of *Sylva* (London: Martyn and Allestry, 1664), 62, Evelyn noted that "*cedro digna*" meant "worthy of eternity," but he did not discuss cedar's efficacy as a preservative for paper.

and sister institutions. Edward Lhuyd not only sent a sample of paper made from the recovered ancient substance, asbestos, to the Royal Society in 1684, but experimented with printing on it.[47]

Alongside their antiquarian interests, Fellows also engaged in collecting the scientific works of their own recent past. Naturalists applied bibliographic tools developed for the study of antiquity and more distant history to papers of the relatively recently deceased, increasingly allocating attention to handwriting (looking for holograph materials), to provenance, and to copying and recopying marks of authenticity, such as a particular writer's monogram.[48] In his translation of Mayerne's papers for the Royal Society, Robert Hooke not only reproduced Mayerne's monogram, manicules, and marginalia and even his colored ink, but also noted how the paper had been folded, as Keller discusses in her chapter. From the Society's earliest days, figures such as Sir Robert Moray, Sir Christopher Wren, and John Collins envisioned the Society as a repository for the private papers of earlier English naturalists, dating back to the sixteenth and early seventeenth centuries.[49] John Aubrey, who memorialized past "hints" or incomplete ideas and connected them to current research, praised the Society's role in "having putt discourse in paper and brought it to use."[50]

A persistent counter-narrative of loss and destruction accompanies our stories of archival preservation and use. Destruction could be deliberate—a matter of controlling the flow of sensitive information, when a particular communication was meant for the moment—rather than for posterity, or for others' eyes. As Arnold Hunt has written elsewhere, when communicating on intimate or politically risky topics, correspondents might command recipients to burn or return letters to protect them.[51] With print the privileged medium of wide dissemination, donating material to an archive could be a way of both

47 Edward Lhuyd, "An Account of a Sort of Paper Made of Linum Asbestinum Found in Wales in a Letter to the Publisher, from Edward Lloyd of Jesus Coll. Oxon," *Philosophical Transactions,* 14 (1684): 823–824. Rachel Maines, *Asbestos and Fire: Technological Tradeoffs and the Body at Risk* (New Brunswick: Rutgers, 2013), 34.

48 See, for example, Yale, *Sociable Knowledge*, 139–140.

49 Mordechai Feingold, "Of Records and Grandeur: The Archive of the Royal Society," in *Archives of the Scientific Revolution*, 171–184 (on pp. 176–177).

50 Kate Bennett, "John Aubrey, Hint-Keeper: Life-Writing and the Encouragement of Natural Philosophy in the pre-Newtonian Seventeenth Century," *The Seventeenth Century*, 22,2 (2007): 358–380 (on p. 364).

51 Arnold Hunt, "'Burn This Letter': Preservation and Destruction in the Early Modern Archive," in James Daybell and Andrew Gordon, eds., *Cultures of Correspondence in Early Modern Britain* (Philadelphia: University of Pennsylvania Press, 2016), 189–209.

preserving and hiding it—as John Aubrey knew when he placed the volumes of his too-candid-for-print *Brief Lives* in the Ashmolean.[52] Documents were lost in the day-to-day business of office moves and household moves, both in seventeenth-century London and amongst the genomicists discussed by Victoria Sloyan in her essay on the Wellcome Library's Collecting Genomics Project. In Serjeantson's story of Bacon's manuscripts, holograph copies were lost not only to the printing shop recycling bins; the collection shed pieces as it passed through the uncertain hands of a chain of executors. Such losses were commonplace, but political dislocation and religious upheaval made them more likely.[53] With a deep historical memory grounded in stories told by his parents and grandparents, John Aubrey saw the dissolution of the monasteries, the Interregnum, and the Revolution of 1688 as links in a chain of cultural destruction. Fears of loss drove his efforts to preserve his and his contemporaries' papers.[54] Even as papers were increasingly institutionalized in the eighteenth century, they could still be lost via misfiling, slipping onto the wrong shelf or into the wrong collection, as the case of Francis Bacon's private memorandum book, unknown for almost two centuries, shows. Institutional boundaries continued to be porous: open to negotiation by interested parties, as Elizabeth Yale shows in her exploration of Hans Sloane's use of deceased colleagues' papers. Into the present, the archival appraisal process necessarily entails losses that can be quite steep, as documents deemed redundant, marginal, or outside a collecting institution's mission have no place on the shelves or on the computer servers.[55]

2 The Plan of the Volume

Our chapters proceed chronologically across the seventeenth and eighteenth centuries, with a present-day coda by archivist Victoria Sloyan and a commentary on the whole by Michael Hunter. As we move forward through time, from the archival afterlives of Francis Bacon to the institutionalization of Hans Sloane's collection, we find our subjects increasingly valorizing the past. Richard

52 Kate Bennett, "John Aubrey and the Printed Book," *Huntington Library Quarterly* 76 (2013): 393–411 (on p. 405); Bennett, "John Aubrey, Hint-Keeper," 369.

53 A private chain of transmission plus political pressures contributed to posthumous losses in Samuel Hartlib's archive. See Leigh T.I. Penman, "Omnium Exposita Rapinae: The Afterlives of the Papers of Samuel Hartlib," *Book History* 19 (2016): 1–65.

54 Yale, "With Slips and Scraps," 1–2, 19–20.

55 Blouin and Rosenberg, 144–145.

Serjeantson's chapter provides a seminal overview of the entire archival afterlife of the collection of Sir Francis Bacon's own manuscripts that were in Bacon's possession at his death. Going beyond the usual paleographic tools to identify Baconiana, Serjeantson painstakingly analyzes the lists that editors, biographers, and intelligencers made of the Bacon manuscripts in their possession or known to them. These lists were records, but also promise of future knowledge. In doing so, he gives us a complete and surprising genealogy of their travels, concluding that the great majority of surviving manuscripts do not derive from Bacon's faithful editors William Boswell or William Rawley, but rather from the negligent Sir Thomas Meautys. As a creditor, Meautys had acquired legal responsibility for disposing of Bacon's estate and heedlessly kept a box of Bacon's manuscripts until his death in 1649. By escaping the press and editorial intervention, the documents survived more by benign neglect then any sense of scholarly responsibility. Until the late seventeenth century, Serjeantson further illustrates, printing usually meant destruction for holograph manuscripts, but thereafter the emergence of antiquarian collectors ensured that autograph documents "could be published without perishing." Serjeantson's chapter well analyzes the relationship between print and manuscript collections, and their changing character and value over the early modern era.

Carol Pal highlights a collection during the period of its initial formation, circulation, and use. In the early twentieth century, Dorothy Stimson dubbed the London-based Prussian Samuel Hartlib (c.1600–1662) an "intelligencer" in an analogy with political intelligencers, shadowy figures criss-crossing Europe, uncovering secrets, collecting intelligence, negotiating treaties and otherwise performing services for their political masters.[56] While Hartlib was far more sedentary, he likewise labored strenuously uncovering, collecting, and communicating information, negotiating learned and patronage relationships, performing miscellaneous learned tasks, and mediating publications. By shepherding works to the press, he often contributed to the fame of others, while becoming relatively obscure himself; although well-known during his life, Hartlib is one of the few central figures in this volume for whom no portrait survives. Hartlib's career has drawn attention to so has drawn attention to the importance the importance of intelligencing in scientific communication and archival build-up more generally.[57] What Carol Pal shows, however, is that Hartlib, through his intelligencing, not only collected and shepherded into print the products of more famous figures, but

56 Dorothy Stimson, "Hartlib, Haak and Oldenburg: Intelligencing," *Isis* 31,2 (1940): 309–326.

57 Vera Keller and Leigh T.I. Penman, "From the Archives of Scientific Diplomacy: Science and the Shared Interests of Samuel Hartlib's London and Frederick Clodius's Gottorf," *Isis* 106,1 (2015): 17–42.

also swept up and preserved the efforts of figures whose careers as scholars were not otherwise destined to be memorialized, namely women. While these figures were highly active, collaborative, and renowned during their lifetimes, they themselves feared that, unremembered by history, they would have no afterlife outside of the domestic sphere. Had Hartlib's papers been edited for the press or collected for the purposes of institutional memory, these learned women might have been erased. The accidental survival of Hartlib's working papers happily preserved some memory of their achievements.

Pal opens with the counterexample of Princess Elisabeth of Bohemia (1618–1680) who actively attempted to control her manuscript legacy, although as Pal notes, she is only now beginning to be reconsidered as a philosopher. She continues with a range of figures. The educator Bathsua Makin (c.1600–1680) had few direct connections to Hartlib, but appears fragmentarily in Hartlib's omnivorous collection of learned intelligence and few places elsewhere. Dorothy Moore (c. 1612–1664), by contrast, was closely connected to Hartlib and like him an active political negotiator on behalf of learning, both of her own and of other women such as Princess Elizabeth and Katherine Jones, Lady Ranelagh (1615–1691), her aunt by marriage. Lady Ranelagh, sister to Robert Boyle, had a public career that spanned fifty years. She collaborated so closely with Hartlib across the full range of his endeavors that Pal characterizes her as a "co-intelligencer."

While few in the seventeenth century advocated for the purposeful collection and preservation of women's learning, many did so for another group whose achievements were believed to be at risk of loss—artisans. Long before the Royal Society's History of Trades program, many including Hartlib sought out artisanal writings for the purposes both of immediate use and for composing formal and authoritative histories of trades. Sir Theodore de Mayerne, in competition with Hartlib, was one such individual. His own collections, both of artisanal writings and of courtly figures, went on to inform the Royal Society's History of Trades program at its height. Vera Keller's chapter traces how such afterlives might be quickened by social animus and deployed as social currency by those who enjoyed access to desirable papers, as was the case for Sir Theodore de Vaux, who introduced Sir Theodore de Mayerne's papers to the Royal Society in dribs and drabs for over three decades. Rivalry with Sir Thomas Povey's advocacy of artistic research in the Society may have animated de Vaux. Mayerne's papers attracted particular attention in 1666 when de Vaux brought in a large group of Mayerne papers dealing with inks, pigments, and dyes, including holograph works of international artisans, Mayerne's notes upon them, and Mayerne's records of his own extensive experimental practice improving dyes. A special committee collectively examined and translated Mayerne's papers on color, which proved particularly important for Robert Hooke who returned to Mayerne's papers during his own dyeing experiments in 1669. Mayerne's papers served not only as sources of artisanal secrets, but as exemplars of experimental paper technologies at a time when the Society was debating its methods.

Elizabeth Yale's contribution begins where Vera Keller's ends, exploring the Royal Society's continuing interest in, and use of, the papers of the dead into the eighteenth century. Yale examines Hans Sloane's publication of Ashmolean curator Edward Lhuyd's account of his Celtic travels in *Philosophical Transactions* in the years immediately following Lhuyd's death. In doing so, she offers another lens through which to understand Hans Sloane's collecting activity, complementary to Alison Walker's: he collected the papers and correspondence of the recently deceased, amalgamating them into his own repository, with an eye towards how those papers could be deployed in the Royal Society setting. Sloane published Lhuyd's letters, which describe the philologist's natural historical, linguistic, and antiquarian progress as he traveled Ireland, Scotland, and Wales, in the context of his ongoing dispute with John Woodward for control of the Royal Society, which turned specifically on the character of *Philosophical Transactions* under Sloane's editorship. Woodward felt that Sloane allowed the journal to descend into triviality, opening it up to mockery and satire -- in the form, for example, of William King's 1700 *Transactioneer.* Their dispute came to a head in 1710, when Woodward mocked Sloane's account of a bezoar stone, read at a meeting of the Society. Sloane grimaced at Woodward, Woodward publicly insulted Sloane, and Woodward refused to apologize. As a result, he was ejected from the Royal Society Council, the institution's governing body. The records of this meeting, both in the rough minutes and in the fair copy, give further evidence of Sloane's archival strategies for maintaining his own prestige: the fair minutes are preserved in the Royal Society archives, but Sloane transferred the rough minutes, in which the fellows worked out precisely what Sloane had done, how Woodward had responded, and what was to be done about it, into his own archive. The public, institutional record preserves only the decision made, representing it as a straightforward, impartial action against Woodward. If the boundary between public and private archive was a troubled one, it was troubled in Sloane's favor.

James Petiver (ca. 1665–1718), studied by Arnold Hunt, offers another example of a figure whose archival remains and posthumous reputation were refashioned by Sloane, who acquired what Hunt describes as Petiver's "collection of collections." Hunt unearths James Petiver's tactics for acquiring, organizing, managing, and networking his collections of correspondence, manuscript notebooks, printed works, and natural specimens. An apothecary, Petiver did not enjoy Sloane's resources. He did possess the ability to spin a web of contacts, both of personal relationships and *quid pro quo* exchanges that elucidate the connections between the objects in his collection and manuscript and printed sources. As Hunt argues, Petiver's subsequent reputation as an unmethodical collector was largely due to subsequently forgotten note-taking techniques developed to meet the challenges of pre-Linnean collecting. Sixteenth-century collectors such as Gesner and Aldrovandi regularly employed a literal "cut and paste" technique, cutting up previous works, and re-arranging paper slips, glued or tied into a work that could be easily disassembled and re-assembled.[58] Petiver thus followed a long-lived tradition. Attempting to negotiate diverse taxonomic systems, Petiver kept his textual works—both manuscript and print—in an imperfect state, well suited for later adjustment, cross-reference, and re-arrangement. This lack of finality, which struck later scholars as haphazard, was a widespread strategy for dealing with a rapidly changing field and proliferating set of categorical variables. Petiver's methods and the world of pre-Linnean record-keeping techniques they represent were subsequently obliterated by Sloane's carefully crafted reputation as a public figure, Hunt notes, as well as by "the disaggregation of Petiver's objects," manuscripts and printed works across the three "'Sloane institutions' (the British Museum, the British Library and the Natural History Museum)," and the rise of Linnean method.

The overweening influence of Hans Sloane reached even more unexpected corners. Although, as Anna Marie Roos's essays shows, the preservation of Martin Lister and Edward Lhuyd's ephemera was through the more traditional means of acquisition by the Bodleian Library, Sloane's collection was personal. Personal ties thus governed its dissemination and use before it was formally incorporated into the British Museum in 1753. As Roos demonstrates, Sloane's control of a vast collection of *naturalia* meant that leading virtuosi such as Martin Lister borrowed specimens from his museum, and Lister's daughters Susanna and Anna illustrated several of Sloane's shells for their father's *Historiae Conchyliorum* (1685–92). These conchological specimens continue, themselves, to have afterlives. Several of the Sloane shells are still extant in the Natural History Museum in London, which as Roos's essay shows, allows for an analysis of the migration of knowledge of nature from one medium to another, from object to drawing to printed image, as well as the circulation

58 Blair, *Too Much to Know*, 213–29. Fabian Krämer and Helmut Zedelmaier, "Instruments of Invention in Renaissance Europe: The Cases of Conrad Gesner and Ulisse Aldrovandi," *Intellectual History Review* 24,3 (2014): 321–341.

of knowledge and the establishing of credibility and taxonomic type characteristics in scientific (visual and textual) discourse and illustration. We see that establishment of these characteristics not only involve visual power of apprehension, but tactile ones. With the specimens still extant, we are able to appreciate that Susanna and Anna Lister conveyed the experience in their illustrations of what it was like not only to look at the shell, but also to hold it and to identify its type characteristics. The Sloane shells themselves were archives with posthumous fates in both printed and physical form, and even now continue to be consulted to answer questions about taxonomy and biological diversity.

In Sloane's library, the printed book itself was an archive, a repository for knowledge and experience built up over a lifetime. In her essay, Alison Walker explores the marginalia in Hans Sloane's collection of printed books. Sloane did not just buy books that happened to be annotated: he bought specific books, and collections of books, for their annotations. Sloane purchased, for example, copies of the London *Pharmacopaeia* previously owned and annotated by prominent physicians, including Theodore de Mayerne and Christopher Merrett, with records of their own medical practices as well as recipes for specific remedies they had used in their careers. De Mayerne, and others, had their printed *Pharmacopaeia* bound interleaved with blank sheets for their own notes. For the physician, the printed book was a space of practice: deliberately crafted through binding as an interactive technology for developing and documenting their work, it perdured in use through generations.

Archivist Victoria Sloyan takes us on a foray into a medical and biological archive of the present, documenting the work of the Wellcome Library's Collecting Genomics project. One signal theme of the literature on the history of archives, both scientific and state, is attention to the absences in the archives: historical actors or systems whose lives or functioning were never recorded to begin with, or were lost over the course of a long afterlife. For Collecting Genomics, a proactive collecting policy minimizes such lacunae. The Collecting Genomics team has adopted a "documentation strategy based approach": rather than collect records only from scientific superstars, they collect records from individuals who have been involved in human genome research at different stages in their careers, often initiating contact with them at mid-career, rather than waiting until a scientist's retirement. They also collect material relating to each scientist's teaching and researching responsibilities, which will allow future historians to reconstruct the broader environment in which key breakthroughs in genomics occurred.

Sloyan's essay allows us to see afresh the making of an archive as a creative act. An individual archive may form through a lifelong geological process of layering—individual documents drift down into the file cabinets, fossilizing there, over the decades—but it is also an active process, a collaboration between the individual scientist and the Collecting Genomics team. Sloyan's essay further suggests to the historian that closer collaborations with archivists would be fruitful in two ways: (1) to better understand what has been preserved and why, in any particular collection; and (2) that we might work together to minimize gaps in the archive at the point of curation.

Reading Sloyan's contribution against others in the volume, we see important differences between early modern and contemporary science, in terms of the relationship between scientific practice and the creation and use of scientific archives. Both activities are now more rigidly institutionalized, and sharper divisions are drawn between professional archivists and professional scientists. We see these divisions in embryo, perhaps, in the eighteenth century. Though the available structures for publication are in flux between print and digital media, journal publication dominates scientists' vision of how they communicate their results. Before being contacted by the archivists, the molecular biologists often had given little thought to the survival of their archives, believing that anything worth knowing about their work was already contained in their published papers. Early modern naturalists made no such assumptions.

Yet even today, as in the seventeenth century, significant domains of scientific practice go unarchived—or may be difficult for future scholars to access, if they are. In an era when "citizen science" is performed from the home computer in university-driven crowd sourced projects, it is important to understand the history of "domestic science" in all its complexity.[59] Similarly, research carried out in a corporate setting, or under a public-private sponsorship may or may not be recorded in an accessible archive. Vast drafts of scientific material may be lost entirely, maintained in corporate archives that are made public selectively, if at all.[60] And regardless of the institutional setting of scientific

59 See, for example, the projects hosted on the Zooniverse platform (www.zooniverse.org; accessed 28 August 2017). Volunteers perform transcription and data analysis on projects spanning astronomy, biology, medicine, physics, paleontology, meteorology, history, and literature.

60 See Soraya de Chadarevian, "The future historian: reflections on the archives of contemporary sciences," *Studies in History and Philosophy of Biology and Biomedical Sciences* 55 (2016): 54–60. This is part of a special issue dedicated to "Navigating 'big biology': challenges in historicizing and documenting contemporary life sciences," edited by Christine Aicardi and Miguel García-Sancho. For the neglect of the work of industrial scientists, see Steven Shapin, "Who is the industrial scientist? Commentary from academic sociology and from the shop-floor in the United States, ca. 1900-ca. 1970," in *The Science–Industry*

research, the collection of papers for posterity is influenced by individual and institutional prominence, class, gender, and duration in the scientific career. Absent a "documentation strategy based approach" like the one followed by Collecting Genomics, historians of the future may be driven towards focusing on today's stars, missing aspects of the diverse, and dispersed, cultures of science.

Today's researchers and archivists face many of the same questions as the virtuoso-archivists of the seventeenth and eighteenth centuries. In attending to what they collected, what has been lost, and what remains, whether by intention or accident, *Archival Afterlives* draws a new map of the territory of early modern learning, one in which absences as well as presences are delineated. Through this imaginative excavation of the past, "we will grieve not, rather find strength in what remains behind."[61] Reconceptualizing the history of early modern science and medicine through the lens of "archival afterlives," loss becomes something more than absence: it becomes historical evidence in and of itself. Many of our early modern subjects were keenly attuned to tracing what knowledge had been lost and why.[62] In the shifting patterns of loss, destruction, survival, and institutionalization, we too follow the path of scientific endeavor through history.

Nexus: History, Policy, Implications: Nobel Symposium 123, ed. K. Grandin, N. Wormbs, and S. Widmalm (Sagamore Beach, MA: Science History Publications, 2004), 337–363. This volume also has essays that address challenges of preserving scientific data in electronic formats.

61 William Wordsworth, "Ode: Intimations of Immortality from Recollections of Early Childhood," in *The Poetical Works of William Wordsworth*, ed. E. de Selincourt and H. Darbishire (5 vols. Oxford: Oxford University Press, 1940–9), vol. 4.

62 Vera Keller, "Accounting for Invention: Guido Pancirolli's Lost and Found Things and the Development of *Desiderata*," *Journal of the History of Ideas* 73,2 (April, 2012): 223–245.

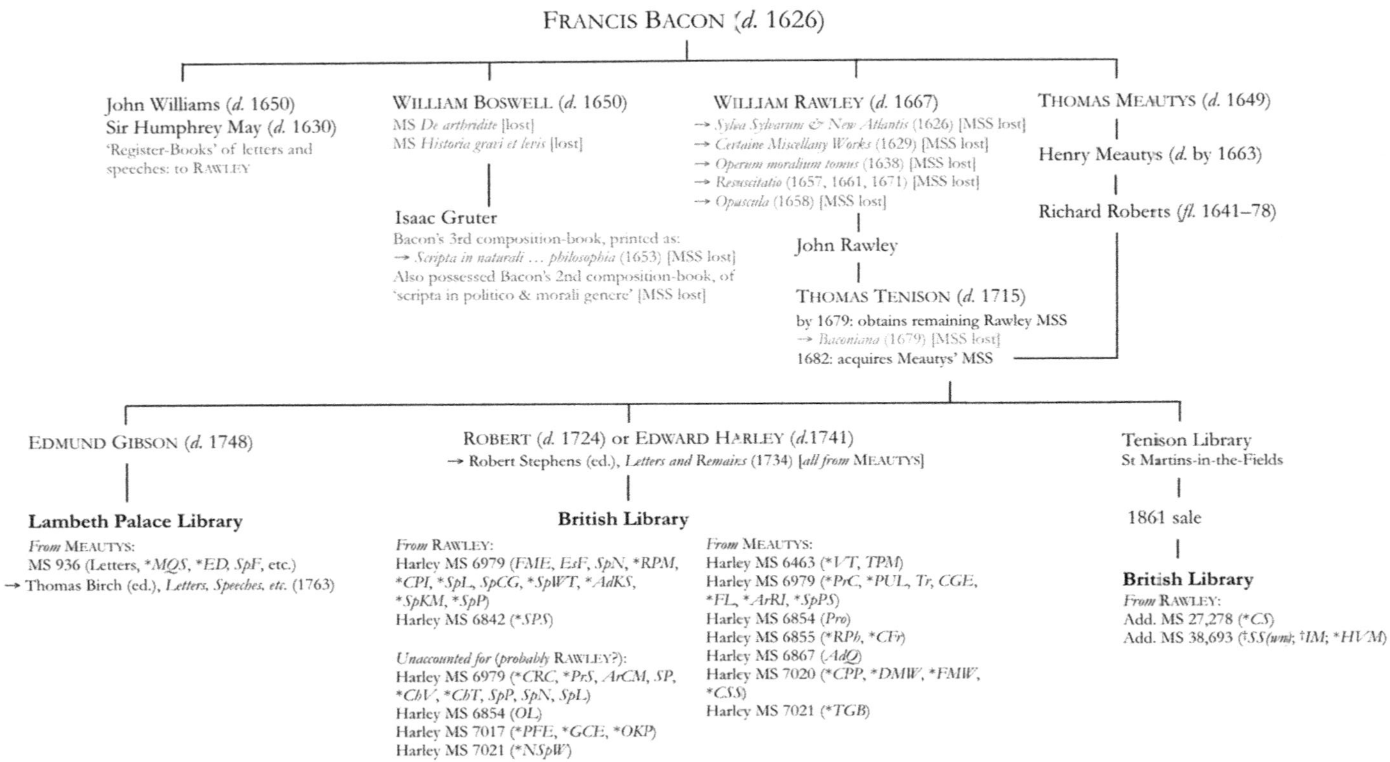

FIGURE 1.1 *Summary account of the fortunes of Bacon's posthumous manuscripts.*
ABBREVIATIONS OF BACON'S WRITINGS ARE THOSE EMPLOYED BY THE *OXFORD FRANCIS BACON*.

CHAPTER 1

The Division of a Paper Kingdom: The Tragic Afterlives of Francis Bacon's Manuscripts

Richard Serjeantson

1 Introduction

Time, Francis Bacon liked to say, is like a river, which carries things down to us that are light and frothy, but into which weighty and solid things sink.[1] It is an appropriate image with which to begin an account of the posthumous fortunes of his manuscripts. We have, as it happens, an unusually large number of these. Yet as we shall see, much more has been lost than has been preserved. This is true more generally for the great majority of figures from his century. Destruction, not preservation, is the default condition for this period. Yet the story of that destruction, and of efforts to restrain the devouring ravages of time, can tell us a number of things about the nature of early modern manuscript: about mechanisms of preservation, both material and legal; about the relationship between manuscript and print; and, perhaps most importantly, about changing patterns of manuscript collecting across the seventeenth and early-eighteenth centuries.

The survival of an author's manuscript archive, in whole or in part, permits quite different questions to be posed of them than their printed books can ever allow. Yet although Francis Bacon's printed books have consistently attracted readers, it is fair to say that the scholarly opportunities presented by his surviving manuscript archive have only recently begun to be taken up. Many avenues of inquiry are therefore open. Bacon's manuscripts will allow us to understand in detail the painstaking ontogeny of his *Great Instauration*. They enable us to reconstruct his own highly-structured archive.[2] They will permit us to look over his shoulder at his working methods and habits of composition.[3] They will allow us to understand the ways in which Bacon encouraged the circulation of

---

1 Francis Bacon, *The Advancement of Learning* [1605], ed. Michael Kiernan, *The Oxford Francis Bacon* [hereafter OFB], IV (Oxford: Clarendon Press, 2000), 29 (see also 231).

2 Angus Vine, "Francis Bacon's Composition Books," *Transactions of the Cambridge Bibliographical Society* 14 (2008): 1–31.

3 Graham Rees, "Introduction," to Francis Bacon, *Philosophical Studies, c.1611–c.1619*, OFB, VI (Oxford: Clarendon Press, 1996), XXXI–XXXV, LIX–LXV.

 | DOI 10.1163/9789004324305_003

his correspondence from an early stage in his public life.[4] They will enable us to reconstruct his close involvement with the professional scribal worlds around Parliament, the Inns of Court, and the book-trade.[5] Finally, Bacon's manuscripts will in due course allow us to map the changing function and personnel of his secretariat (the term is not too strong) throughout his life. But in order to undertake these studies, it will also be helpful to understand the fortunes of his posthumous archive.

We have been placed in a position to embark upon this inquiry by the heroic labours of Peter Beal. In his *Index of English Literary Manuscripts* (1980), and latterly in the even fuller *Catalogue of English Literary Manuscripts* (2014), we now possess something approaching a full list of surviving manuscripts of Bacon's writings.[6] Only copies of Bacon's speeches remain unsystematically catalogued. And when Beal's work is combined with that of Alan Stewart and his collaborators on Francis Bacon's surviving letters, we have an even more comprehensive conspectus of the documentary situation.[7] Taken together, these reference-works permit us to undertake here an inquiry that has never previously been attempted: an overview of the entire archival afterlife of Bacon's collection of his own manuscripts in his possession at his death. (A consequence, however, is that I will have very little to say here about the enormous quantity of professional and private copies of Bacon's writings that were made without any connection to him, both in his own lifetime and extensively afterwards.)[8]

This is not to suggest, of course, that several of Bacon's previous editors have not already confronted the question of how a particular manuscript, or set of manuscripts, has come to survive. From Thomas Birch in the 1760s, through James Spedding in the mid-nineteenth century, down to the contemporary editors of the on-going *Oxford Francis Bacon*—Graham Rees, Michael Kiernan,

4 For a case study, see Andrew Gordon, "'A Fortune of Paper Walls': The Letters of Francis Bacon and the Earl of Essex," *English Literary Renaissance* 37 (2007): 319–336.

5 Richard Serjeantson and Thomas Woolford, "The Scribal Publication of a Printed Book: Francis Bacon's *Certaine Considerations Touching … the Church of England* (1604)," *The Library* 10 (2009): 119–56.

6 Peter Beal, *Index of English Literary Manuscripts*, vol. I: *1450–1625*, part 1: *Andrews–Donne* (London: Mansell, 1980), 17–52; Peter Beal, *Catalogue of English Literary Manuscripts, s.n.* "Francis Bacon, Baron Verulam, Viscount St Albans (1561–1626)" (www.celm-ms.org.uk; hereafter *CELM*).

7 Alan Stewart, ed., *Francis Bacon Correspondence Project* (www.livesandletters.ac.uk/projects/correspondence-francis-bacon) (hereafter *FBCP*).

8 These are catalogued in *CELM* alongside manuscripts containing Bacon's own handwriting.

Alan Stewart, and Angus Vine—scholars have sought to explain how the documents they are editing have survived, and this synthesis draws extensively on their individual findings.

How can items from Bacon's archive be identified? The primary way, of course, is to find his own handwriting on a document. I use this expression advisedly. Bacon's writings are not, on the whole, written out entirely in his own hand. From an early period of his life he developed the habit of using scribes—initially professionals, latterly also those whom we might call gentleman amateurs—to make fair copies of his work. But Bacon would then correct, revise, and extend these copies, covering their elegant transcripts with his own vigorous improvements. These revisions are usually sufficiently extensive, and also characteristic, for there to be no room for doubt about the attribution. This method of identification has been used since Robert Stephens' searches for Baconian items among the Harleian manuscripts in the earlier eighteenth century (on which more anon); it was developed, with conspicuous success, by James Spedding in his work among the State Papers during the 1850s.[9]

But to this venerable and essential procedure it is possible to add a supplementary one, which is the key to the whole affair. As we shall see, seventeenth-century interest in Bacon's writings was sufficiently strong that a number of different people made lists of extant Bacon manuscripts. These lists, once they have been systematically studied, are the means by which it is possible to reconstruct the flow of Bacon's manuscripts down the river of time in the century following his death.

Most of Bacon's own surviving manuscripts are now to be found in a single place: the Harleian collection in the British Library (formerly the British Museum). They flowed into the safe harbor of Robert and Edward Harley's collection in the early years of the eighteenth century. But as we shall see, they did not all arrive in the same vessel. Once berthed there, moreover, they found themselves somewhat indiscriminately mingled together. It is this confluence of previously separate journeys that has hitherto hindered our understanding of the genealogy of their travels. For the reconstruction offered here of Francis Bacon's archival afterlife shows that his posthumous manuscripts had a convoluted, even a tragic, history. Upon the death of their author, his paper kingdom was divided between three different heirs; it would turn out to be the least of his inheritors who best proved his love for his former master.

---

9 Robert Stephens, "Preface," to *Letters and Remains of the Lord Chancellor Bacon* (London, 1734), III. James Spedding, *The Letters and the Life of Francis Bacon*, 7 vols (London: Longman, 1861–74). By their nature, Baconian manuscripts in the State Papers passed out of his hands during Bacon's lifetime, and hence fall outside the scope of this investigation.

2 On Bacon's Death: 1626–27

Francis Bacon's compositional process is a distinctive one. He seems to have begun by drafting a work either in his own hand or to dictation. He would then have an interim version copied out fairly by a scribe. This fair copy would then be defaced by additions, corrections, and deletions in Bacon's own hand. The process would then be repeated: scribal fair-copy; authorial tinkering. But Bacon seems to have been fairly rigorous about destroying earlier drafts of his writings as they became superseded by the interim fair-copies made for him by scribes. This means that in cases where one of his writings survives in manuscript, it tends to survive in only a single authorial manuscript.

During his lifetime, Bacon maintained an extensive and orderly manuscript archive of his own writings. His papers were organised according to carefully-defined principles, which he had established by at least as early as 1608, when his personal archive already contained no fewer than twenty-eight different notebooks. In particular, he divided his manuscript "libri" into five main categories, of which the first (and no doubt most important) were six loose-leaf folio "composition books," in theology, politics, natural philosophy, logic and rhetoric, orations, and letters.[10] By virtue of the compositional process described immediately above, these volumes therefore contained what might be called "working fair-copies" of Bacon's original writings.

It is clear that the outlines, at least, of Bacon's careful archival organization remained in place at his death on 9 April 1626. His will further informs us that his papers were kept in a variety of "Cabinetts," "Boxes," and "Presses."[11] These repositories may, however, have been physically divided between his principal residences: the neighbouring houses of Gorhambury and of Verulam, at St Albans, where he retired when he was banished from the verge of the court in 1621; and his chambers at Gray's Inn, to which he was able to return after his banishment was lifted.[12] In the course of his life, and especially in its final five years, Bacon published many of his writings. But many more remained to see the light of print across the seventeenth and eighteenth centuries.

---

10 On the principles by which Bacon ordered his manuscripts in 1608, see BL, MS Add. 27,278, fols. 10v–14r (the "Commentarius Solutus," printed in Spedding, *Letters and Life*, 4:59–62); see further Vine, "Composition Books," 6–8, and Angus Vine, "Commercial Commonplacing: Francis Bacon, the Waste-book, and the Ledger," *English Manuscript Studies* 16 (2011): 197–218.

11 National Archives, Kew (hereafter TNA), PROB 1/33/1–2 (also printed in Spedding, *Letters and Life*, 7:539–45).

12 Markku Peltonen, "Bacon, Francis, Viscount St Alban (1561–1626)," in *Oxford Dictionary of National Biography* (oxforddnb.com; hereafter *ODNB*).

Bacon's unpublished manuscripts were at the forefront of his mind as death approached. He first elaborated his wishes in the curt will he drew up on 10 April 1621:

> My compositions unpublished, or the fragments of them, I require my servant Harris to deliver to my Brother Constable,[13] to the end, that if any of them be fit in his judgment to be published, he may accordingly dispose of them. And in particular I wish the Elogium I wrote in felicem memoriam [Reginæ *inserted*] Elizabethæ may be published.[14]

This will was superseded by a later "last" will, which survives in two slightly different versions. In both of these, Bacon disposes of his manuscripts before his other goods—after, that is, his soul, which he left to God, and also his reputation, which he theatrically bequeathed "to foreign nations, and the next ages." The earlier version of the "last" will requires:

> my Servant, *Henry Percy*,[15] to deliver to my Brother *Constable*, all my Manuscript-Compositions, and the Fragments also of such as are not Finished; to the end that, if any of them be fit to be Published, he may accordingly dispose of them. And herein I desire him, to take the advice of Mr. *Selden*,[16] and Mr. *Herbert*, of the *Inner Temple*,[17] and to publish or suppress what shall be thought fit. In particular, I wish the Elegie, which I writ in *felicem Memoriam Elizabethæ*, may be Published.[18]

---

13 Sir John Constable, Bacon's brother-in-law.

14 British Library (hereafter BL), MS Add. 4259, fol. 111r, a copy in the handwriting of John Locker (on whom see further below, at n. 141 onwards). The document Locker was copying has not been located. See further Spedding, *Letters and Life*, 7:228–29.

15 Henry Percy is left £100 in the final version of Bacon's will (TNA, PROB 1/33/1–2). In an autograph letter of 26 Jan. 1626, sent from Gray's Inn to the Secretary of State, Edward Conway, Bacon made a petition on behalf of Percy, there described as "my good frend and late servant" (TNA, SP 16/19/49; Spedding, *Letters and Life*, 7:549). See further Lisa Jardine and Alan Stewart, *Hostage to Fortune: The troubled life of Francis Bacon* (London: Gollancz, 1998), 501.

16 John Selden (1584–1654), scholar and legal historian.

17 Edward Herbert (*c.*1591–1657; knighted 1641), the uncle of George Herbert and of Edward Herbert, first Baron Cherbury. He defended Selden against the Crown in his trial of 1629.

18 Thomas Tenison, *Baconiana* (London, 1679), 203–4.

A later and more polished last will, dated 19 December 1625, removes the references to Henry Percy, John Selden, and Edward Herbert. But it now adds a new element:

> Alsoe whereas I haue made vpp two Register-Bookes the one of my Orations or speeches the other of my Epistles or Letters whereof there may be vse, And yet because they touch vpon busines of state they are not fitt to bee putt into the handes but of some Councellor, I doe deuise and bequeath them to the right ho*noura*ble my verie, good Lord the Lord Bushopp of Lincolne and the Chancellor of his Ma*iestie*s Dutchy of Lancaster.[19]
>
> Alsoe, I desire my Executors espetiallie my brother Constable and also Mr Bosvile p*re*sently after my decease, to take into theire handes all my papers whatsoeuer, w*hi*ch are either in Cabinetts Boxes or Presses, and them to seale vpp vntill they may att theire leasure pervse them.[20]

Thus the inheritors of Bacon's manuscripts should have been his brother-in-law, Sir John Constable, about whose "communication in studies" with him Bacon had earlier written;[21] and William (later Sir William) Boswell (*c.*1583–1650) ("Mr Bosvile"), who was evidently a trusted friend at the end of Bacon's life. More generally, this will should, in principle, have served to order the dispersal of Bacon's estate following his death.

It did not. Bacon died owing enormous debts, which his estate could not begin to meet.[22] Moreover, every single one of the six men whom Bacon named as his executors declined to serve in that capacity. This is made clear in the letters of administration that were eventually issued by the prerogative court

---

19 That is, John Williams and Sir Humphrey May. See further Bacon's undated letter to Williams on this matter (*FBCP*, bcn0796), printed from Bacon's own copy in Tenison, *Baconiana*, 195.

20 TNA, PROB 1/33/1–2. *CELM*, *BcF 654, indicates that this document contains the presence of Bacon's autograph. But the names of the witnesses are in the hand of its scribe, and I suggest that the signature "Fr st alban" is too. The handwriting of the codicil disinheriting Bacon's wife is also continuous with the main document, further indicating that it is a later copy.

21 Francis Bacon, *Essaies* (London, 1612), sig. A4r (dedication to Constable).

22 The auditor of Bacon's estate, Francis Phelips, stated the total sum owing to be £22,371 1*s* 3*d.* His bill is printed in Francis Bacon, *Opera omnia*, ed. John Blackbourne (London, 1730), 2:565–68. (But note that Jardine and Stewart, *Hostage to Fortune*, 513, offer a different sum, of £19,658 4*s* 4*d.*) Bacon's wealth at death, by contrast, has been calculated at a mere *c.* £7,000 (Peltonen, "Bacon, Francis").

of the Canterbury province on 30 July 1627, which dispense each of those appointed by Bacon, and instead confer the office of winding up Bacon's affairs upon two of his creditors instead. The first of these creditors was Sir Robert Rich, a Master in Chancery; the second was Bacon's former gentleman retainer, Thomas Meautys, who had rashly lent him money during his final sickness, and whom we shall meet again. Thus Bacon's estate was administered as if he had died intestate, and Rich and Meautys became ultimately responsible in law for the distribution of his goods.[23]

In principle these goods would have included Bacon's manuscripts. Yet although they are named first in his will, it seems most doubtful that they were regarded as having any financial value—unlike the leases, "rich hangings," or table-carpets that Bacon also itemises in his will. Perhaps this fact explains William Boswell's later testimony to Samuel Sorbière that the bequest of Bacon's papers was the only one of his numerous legacies that was ever executed. Boswell also claimed that he himself inherited "all" of Bacon's papers. If we assume that Constable gave over his rights in the manuscripts to Boswell then this statement may be strictly true—even if Sorbière's subsequent claim that Bacon left four hundred thousand pounds for an "imaginary College" after the model of his *New Atlantis*, is not.[24] Yet although Boswell may have possessed all the manuscripts in law, he did not do so in fact. In the event, Bacon's posthumous archive was divided (as we shall now see) between three different people, all of whom appear in the will in some capacity. Boswell, named as literary executor, was one of them. William Rawley (*c.*1588–1667), one of Bacon's chaplains, a beneficiary of £100, and the will's first witness, was another. And

23 So I reconstruct a complex situation that has attracted a several different and inconsistent interpretations. The executors named by the will were Sir Humphrey May, Sir Richard Hutton, Sir Thomas Crewe, Sir Francis Barnham, Sir John Constable, and Sir Euball Thelwall. The letters of administration are printed from the register of the Canterbury prerogative court in Bacon, *Opera omnia*, 2:563–64. See also James Spedding, "Preface," to Francis Bacon, *Works*, ed. James Spedding, Robert Leslie Ellis, and Douglas Denon Heath, 7 vols (London, 1858–61), 3:3, and Spedding, *Letters and Life*, 7:551; and compare with the accounts offered by Rees, "Introduction" to *OFB* VI, LXX; Jardine and Stewart, *Hostage to Fortune*, 513; and Doina-Cristina Rusu and Christoph Lüthy, "Extracts from a Paper Laboratory: The Status of Francis Bacon's *Sylva Sylvarum*," *Intellectual History Review* 27 (2017): 171–202.

24 *Sorberiana*, ed. François Graverol (Toulouse, 1691), 51 (art. "Bacon"): "Mr. Boswel me dit qu'il avoit eu particuliere connoissance avec ce rare homme, qui lui laissa par testament tous ses papiers, qui fut la seule chose éxécutée de plus d'un milion de legats qu'il avoit fait par galanterie. Il leguoit 400. mille livres à un Collège imaginaire, dont il dresse le plan en son *Nova Atlantis*."

the last was the loyal creditor Thomas Meautys (*c.*1590–1649), whose debt is given priority in a codicil to the will.[25] These were the three men who would play the role of variously faithful daughters to Bacon's over-hopeful Lear.

3 The Division of Bacon's Manuscripts, 1627–1667

3.1 *The Boswell Portion*

The later of Bacon's "last" Wills charges his executors to take possession of his manuscripts and seal them up pending further investigation of their contents by Sir John Constable and William Boswell. Constable seems to have been happy to allow Boswell to take up his share of the burden, for no more is heard of him, whereas Boswell evidently took possession of an important portion of Bacon's manuscripts. Alan Stewart and Graham Rees have reconstructed the fortunes of these documents. A successful university scholar who latter became a diplomat, Boswell was known to be a sound judge of the political significance of unpublished manuscripts: in the aftermath of the assassination of the Duke of Buckingham he assessed the potentially seditious library of Sir Robert Cotton for the Privy Council.[26] Across his lifetime, Boswell himself became an increasingly significant collector of manuscripts, including ones by John Dee, whom he greatly admired, as well those he had inherited from Bacon.[27] Though his Bacon manuscripts no longer survive, some of the works they contained, do, for after Boswell's death in 1650 at The Hague, where he had served as Charles I's ambassador, they were published. Their editor was the Dutch scholar Isaac Gruter (1610–1680).

Gruter had certainly seen the original Bacon manuscripts that had been in Boswell's possession, a point he was keen to emphasize both in private and in public. In May 1652 he wrote to William Rawley about manuscripts "from boxes from Boswell's collection" that were either in Bacon's own handwriting, or in Rawley's with revisions by Bacon.[28] And he began his preface to the edition itself by noting

25 TNA, PROB 1/33/2.

26 Rees, "Introduction," to *OFB* VI, LXX–LXXXV, drawing upon unpublished research by Alan Stewart.

27 In his Ephemerides for 1639 Samuel Hartlib noted that Boswell possessed "a nest" of manuscripts by John Dee and Nathaniel Bacon. Sheffield University Library (hereafter SUL), HP 30/4/26A, consulted here and subsequently in the publication by Mark Greengrass, Michael Leslie, and Michael Hannon, eds., *The Hartlib Papers* (Sheffield: Humanities Research Institute, 2013) (hrionline.ac.uk/hartlib).

28 See Thomas Tenison, *Baconiana: Or Certain Genuine Remains of S*r *Francis Bacon* (London, 1679), 222 (Isaac Gruter to William Rawley, The Hague, 29 May 1652), on those

that it derived "from manuscript volumes which their author had carefully corrected and in various places altered."[29] His descriptions here square precisely with what we know of Bacon's habitual composition practices. But the edition itself was printed, not from Bacon's originals, but from copies of Boswell's manuscripts that Gruter himself had made, as both Gruter's preface and his letter to Rawley make clear.[30]

Gruter published the works in a popular little octavo volume entitled *Scripta in universali et naturali philosophi* that was put out by Lodewijk Elsevier at Amsterdam in 1653. The only piece in it that did not derive from Boswell was the Latin translation of Thomas Bodley's letter to Bacon from 1608 critiquing his philosophy, which Gruter tells us he made himself from a volume of Bacon's letters.[31] The majority of the original works in Gruter's volume are known from it alone. Gruter's volume includes a number of writings dated by Rees to the period c.1611–c.1619: the *Phænomena universi*, the *De fluxu et refluxu maris*, the *Descriptio globi intellectualis*, the *Thema cœli*, and the *De principiis atque originibus*.[32] It also includes a number of more miscellaneous writings of which the majority probably date from a slightly earlier period: the *Indicia vera de interpretatione naturæ*; the *Cogitationes de natura rerum*; the *Aphorismi et consilia, de auxiliis mentis*; the *De interpretatione naturæ sententiæ XII*; and the *De interpretatione naturæ procemium*.[33] Finally, the Boswell/ Gruter *Scripta* also contains a few writings that are also known either in whole or in part from other manuscript sources: these include the *Temporis Partus Masculus*, and, most importantly, the *Cogitata et visa*.[34]

As James Spedding realised, the cache of natural philosophical manuscripts that Gruter published in 1653 corresponds precisely to the third of the six

---

manuscripts: "quæ ex *Boswelliani* Musei scriniis chartaceis penes me exstant vel propria manu descripta, vel alterius apud vos, sed *Baconi* manum & limam experta."

29 Isaac Gruter, "Lectori Salutem," in Francis Bacon, *Scripta in naturali et universali philosophica* (Amsterdam, 1653), sig. *5r: "… ex Manuscriptis Codicibus, quos accurate recensuerat & varie emendarat author."

30 See Gruter, "Lectori," sig. *5r; Tenison, *Baconiana*, 223 (Gruter to Rawley, 29 May 1652): "apographa me redactis ex legato Boswelliano."

31 Gruter, "Lectori," sig. *5r. This letter had been printed, from a publicly circulating manuscript volume of items of Bacon's correspondence, in *The Remaines of the Right Honorable Francis Lord Verulam* (London, 1648), sigs. L4v–M3v.

32 All of these texts are published in Bacon, *Philosophical Studies* (OFB VI).

33 These writings are scheduled for publication in volume 5 of the *Oxford Francis Bacon*.

34 Queen's College, Oxford, MS 280, pp. 205–233 (CELM, *BcF 289), a handsome presentation manuscript of the *Cogitata et visa* now in the archive of Thomas Barlow (1609–1691), may be the copy that Bacon sent to Sir Thomas Bodley in 1607.

different "composition books" that Bacon maintained across his life. Bacon's own title for this collection was "Scripta in Naturalj et vniuersalj philosophiâ" (writings on natural and universal philosophy), and this is precisely the wording that Gruter uses his volume of unpublished Baconiana. Indeed, he explicitly states in his preface that "the title which stands at the front of the book, and which includes the entire subject of the book, divided into its various treatises, derives from Bacon himself."[35]

This being so, it is likely that the order of the texts in the *Scripta* reflects in some way the order in which they existed in Bacon's archive. Our best evidence here is Gruter's explanation for why he gave the name of "Philosophical Attempts" (*Impetus Philosophici*) to the collection of items in the latter part of his volume:

> Everything that follows from the piece entitled *True Indications of the Interpretation of Nature* to the end I have called "Philosophical Attempts," a name I noted in my conversations with that great man [*i.e.* Boswell] when he mentioned these documents to me. For he never used to call the things that were associated with the earlier items (each of which has its own title) anything else.[36]

James Spedding suggested that this implied that "the five pieces which stand first under separate titles ... were found copied out in a book; and that the rest ... were in separate papers, tied up with it," drawing attention to the method of archival organisation Bacon had laid out in 1608.[37] It seems very likely that was correct.[38] If so, then Bacon's archive of unpublished *Writings on Natural and Universal Philosophy* stood at his death along the lines indicated in Table 1.1. It is important to bear in mind, however, that many, and perhaps all of the items in it had either been abandoned or had been superseded by the printed publications of Bacon's lifetime.

35 BL, MS Additional 27,278, fol. 10v (Spedding, *Letters and Life*, 4:59). Gruter, "Lectori," sig. *5r: "Titulus, quem frons libri præfert, & totum complectitur opusculi, in varias dissertationes secti, argumentum, ab ipso *Verulamio* est."

36 Gruter, "Lectori," sig. *5v: "Quicquid sequitur ab eo loco, cujus inscriptio est in ipso contextu *Iudicia* [recte *Indicia*] *vera de interpretatione Naturæ* usque ad finem, donavi eo nomine *Impetus Philosophici*, quod ex familiaribus Viri magni colloquiis notassem, cum de istis chartis mecum ageret. Non aliter enim appellare solebat quicquid prioribus per titulos suos separatis connecteretur." Cf. Rees, "Introduction," to *OFB* VI, LXXXIV.

37 See above, n. 10.

38 Spedding, "Preface," 7–8. See also Rees, "Introduction," to *OFB* VI, LXXXIV; Vine, "Composition Books," 7.

Yet the texts that Gruter printed in the *Scripta* were not the only ones whose manuscripts he had seen. In his 1652 letter to Rawley, Gruter also mentions that he had once had "in my hands" a "large and complete" manuscript of Bacon's *Historia Gravis et Levis*, which nonetheless consisted only of headings which Bacon had not worked up.[39] This document, and indeed the work itself, is now lost.[40]

Even more strikingly, Gruter also appears to have possessed Bacon's second composition-book—the one containing "Scripta in Politicis et Moralibus" (writings on politics and morals). In 1651 Gruter wrote to Thomas Browne (1605–1682) that, "for the last few weeks I have had in my possession manuscript *opuscula* of Bacon of Verulam—some of them moral and political, others natural-philosophical—corrected by the author himself."[41] In the preface to the 1653 *Scripta*, Gruter elaborated slightly on this point:

> Friendly reader: you owe all these unpublished works ... to that most worthy man William Boswell, who was left them by Bacon himself, together with other ones written on political and moral subjects, which, thanks to that late man's gift, are now preserved in my hands and will not be hidden very much longer.[42]

Gruter's classification of these manuscripts (*in politico & morali genere*) again corresponds exactly to a title that we know Bacon himself had used in 1608 when describing the structure of his own archive.[43] Indeed, it seems very likely that the Boswell manuscripts Gruter had access to were prefaced by two Latin title-pages, perhaps written in Bacon's own hand—like that of the equivalent

39 Tenison, *Baconiana*, 223 (Gruter to Rawley, 29 May 1652): "De *Gravi & Levi* in manibus habui integrum & grande volumen."

40 On this lost work, see further Rees, "Introduction," to *OFB* VI, LXXXIV–LXXXV, and Graham Rees, "Introduction," to Francis Bacon, *The Instauratio Magna: Last Writings*, The Oxford Francis Bacon XIII (Oxford: Clarendon Press, 2000), XXV.

41 Rees, "Introduction," to *OFB* VI, LXXVII, translating MS Rawlinson D. 391, fol. 101r (Isaac Gruter to Thomas Browne, 24 June 1651): "Nunc a pauculis septimanis in manibus mihi sunt manuscripta opuscula Baconi Verulamii, ab ipso authore emendata, partim politica et moralia, partim physiologica." See also Vine, "Composition Books," 20.

42 Gruter, "Lectori," sig. *5v: "Omnia autem hæc inedita ... debes, Amice Lector, Nobilissimo *Guil. Boswello*, ad quem ex ipsius Baconi legato pervenerant, cum aliis in politico & morali genere elaboratis, quæ nunc ex dono τοῦ μακαρίτου penes me servantur non diu premenda." See further Spedding, "Preface," 7; Spedding, *Letters and Life*, 4:59; Vine, "Composition Books," 6–7.

43 BL, MS Add. 27,278, fol. 10v (Spedding, *Letters and Life*, 4:59): "2 Scripta in Politicis et Moralibus."

TABLE 1.1 *Bacon's third composition-book, reconstructed*

Title-page:	
Scripta in Naturalj et vniuersalj philosophiâ	
In the composition-book proper:	
Cogitata et visa de interpretatione naturæ	[scheduled for *OFB* V]
Descriptio globi intellectualis	[*OFB* VI, pp. 95–169]
Thema Cœli	[*OFB* VI, pp. 171–93]
De Fluxu et refluxu maris	[*OFB* VI, pp. 63–93]
De principiis atque originibus	[*OFB* VI, pp. 195–267]
Associated papers of "impetus philosophici:"	
Indicia vera de interpretatione naturæ	[for *OFB* V]
Phænomena universi	[*OFB* VI, pp. 1–61]
Scala intellectus	[for *OFB* V]
Prodromi, sive anticipationes philosophia secunda	[for *OFB* V]
Cogitationes de natura rerum	[for *OFB* V]
Filum labyrinthi, sive inquisitio legitima de motu	[for *OFB* V]
Aphorismi et consilia de auxiliis mentis	[for *OFB* V]
De interpretatione naturæ sententiæ XII	[for *OFB* V]
De interpretatione naturæ procœmium	[for *OFB* V]
Topica inquisitionis de luce et lumine	[for *OFB* V]

fifth composition-book, entitled "Orationes, Instrumenta, Acta," which travelled by a different journey.[44] But Gruter's promised second volume never appeared.[45] And although he was clear that he printed the *Scripta* from copies, his two references to possessing Boswell's manuscripts after his death suggest that it is at his door that their loss should be laid.

44 It is preserved at BL, MS Harley 6797, fol. 11r. The title is attested in BL, MS Additional 27,278, fol. 10v (Spedding, *Letters and Life*, 4:59). See further Vine, "Composition Books," 10, 28.

45 On this point, see also John Locker, "Postscript," to Bacon, *Letters and Remains*, 519.

Although Gruter says nothing about them, others thought that William Boswell owned further Bacon manuscripts still. In the latter part of 1655 the intelligencer Samuel Hartlib noted that John Aubrey (1626–1697) was "writing the life of Verulam." At the same time he recorded that "Mr Schle*zer* mentioned to have heard of a MS. of his [*sc.* Bacon's] De Arthritide."[46] Hartlib expanded on this note in a letter that he wrote to John Worthington, the Master of Jesus College, Cambridge, in December of that year. Noting his pleasure that Jesus "had the honour to have so famous a gentleman" as Sir William Boswell as one of their fellows, he went on to ask whether, amongst a bequest of books to his old college, Boswell might have given "a MS of my Lord Verulam's de Arthritide a most elaborate tract." He has been assured (Hartlib writes), "that S*ir* W*illiam* Boswell had it in his keeping, but hitherto it cannot be found." Hartlib's searches for Bacon's "elaborate" treatise on gout, like those of Bacon's modern editors, were fruitless.[47]

According to one contemporary testimony, probably reporting the judgment of William Rawley, Sir William Boswell "got the best" of Bacon's *Nachlass*.[48] This being so, it is fortunate that Gruter printed part of it in the *Scripta*, thereby giving Bacon's growing readership in the Low Countries, England, the Empire, and elsewhere, access to some of his earliest surviving studies in natural philosophy. But all the evidence suggests that the texts Gruter printed in the *Scripta* were only a moiety of what Boswell possessed.

Hence the story of Boswell's Bacon manuscripts, as indeed of his manuscript collections as a whole, is ultimately a tragic one. Boswell was a remarkably well-connected figure, and he had put together an unusually choice selection of manuscript treatises. Samuel Hartlib noted precisely this in 1639: Boswell, he wrote, "hase a World of little Tr*eatises* very Excellent ones in several kindes w*hi*ch hee hase beene a gathering these many *years*. These hee causes now to bee bound vp together that they bee not lost."[49] Vanity!

---

46 SUL, HP 29/5/54B (Ephemerides, 1655). "Mr Schlezer" is Friedrich Schlezer, "a pupil of [Joachim] Jungius" and a "godly and learned young man," who visited Hartlib in London in the 1650s. (Stephen Clucas, "Samuel Hartlib's Ephemerides, 1635–59, and the Pursuit of Scientific and Philosophical Manuscripts: The Religious Ethos of an Intelligencer," *Seventeenth Century*, 6 (1991): 47–8).

47 *The Diary and Correspondence of John Worthington*, ed. James Crossley (Manchester: Chetham Society, 1847–55), 1:66–8 (Hartlib to Worthington, 12 Dec. 1655).

48 *Hartlib Papers*, Ephemerides, 1639 (SUL, HP 30/4/4B). Hartlib's intermediary source was John Sparrow, on whom see below, at n. 65.

49 *Hartlib Papers*, Ephemerides, 1639 (SUL, HP 30/4/20B), headed "Libri select*i* S*ir* W*illiam* B*oswell*."

No manuscript from Boswell's collection, whether by Bacon or anyone else, is now known to survive.

3.2 *The Rawley Portion*

If Sir William Boswell and Isaac Gruter were less than wholly faithful stewards of Bacon's textual legacy, the next recipient of a portion of his archive was not. William Rawley entered Bacon's household in 1618 as the first of the chaplains that Bacon's newly ennobled status entitled him to maintain, and remained with him when all the rest had left.[50] As Rawley had served Bacon in life, so he continued to serve him in death—as he himself put it.[51] Presumably with Boswell's consent, Rawley came into the possession of a significant quantity of Bacon's posthumous manuscripts.[52] Certain indications suggest, however, that the portion he received may have been regarded as the least significant part of the archive. Nonetheless, over the next thirty years, Rawley would make himself the most important editor of Bacon's posthumous works, publishing at least five, and (as we shall see) possibly six different volumes of Baconian *Nachlass*. As Graham Rees and Pete Langman have emphasised, our knowledge of Bacon and his writings owes a great deal to Rawley's intermediation.[53] He both enlarged Bacon's posthumous canon and carefully defined it against the perceived threat of apocrypha and unauthorised editions.

Rawley already had experience of editing Bacon's writings, for it was in that capacity that he had written the brief preface to the folio edition of the *De augmentis scientiarum* in 1623.[54] The first posthumous work that Rawley

---

50 Bacon's will mentions "Mr Peterson my late Chapline," in connection with his funeral sermon (TNA, PROB 1/33/1). On Rawley, see further Rees, "Introduction," to *OFB* XIII, LXXIII–LXXIX; Angus Vine, "'His Lordships First, and Last, Chapleine:' William Rawley and Francis Bacon," in *Chaplains in Early Modern England: Patronage, literature and religion*, ed. Hugh Adlington, Tom Lockwood, and Gillian Wright (Manchester: Manchester University Press, 2013), 123–140.

51 Cf. Tenison, *Baconiana*, 214–15 (Rawley to Elie Diodati, 9 Mar. 1633): "... Cui [*sc.* Bacon] olim inservisse, atq*ue* etiamnum inservire ...."

52 On this point, see also Michael Kiernan, "Introduction," to Francis Bacon, *The Historie of the Raigne of King Henry the Seventh and Other Works of the 1620s*, The Oxford Francis Bacon VIII (Oxford: Clarendon Press, 2011), CXIX–CXX.

53 Rees, "Introduction," to *OFB* XIII, LXXVIII–LXXXIII, assesses the respective editorial merits of Gruter and Rawley; see also A. P. Langman, "'Beyond the Old World and the New': Authority and Knowledge in the Works of Francis Bacon, with Special Reference to the *New Atlantis*" (Ph.D. diss., University of London, 2006).

54 William Rawley, "Lectori S*alutem*," in Francis Bacon, *De Dignitate & Augmentis Scientiarum* (London, 1623), sig. ¶2r–v. (This preface appears never to have been translated into English.)

published from Baconian manuscripts was the folio volume containing the *Sylva Sylvarum* and *New Atlantis* in 1626/7.[55] Rawley's dedication of this book to Charles I presented himself as "one that was trusted with his *Lordships Writings*, euen to the last."[56] Yet in discharging this trust by printing the *Sylva*, its manuscript was destroyed. Only a few fragments related it now survive.[57] Not for the last time, the appearance in print of a work by Bacon led to the destruction of its manuscript.

The next work of Bacon's that Rawley was responsible for printing was the collection of items that appeared in the *Certaine Miscellany Works* of 1629. Again, none of the manuscripts that contributed to make up that volume now appear to survive.[58] And in 1638 Rawley's next Baconian publication appeared, in the form of a volume of *Moral and Civil Works*. This imposing folio constitutes an important document in the canon of Bacon's writings, for according to the claim on the title-page the principal works in it were "given their Latinity (for the most part) by their honorable author himself."[59] If so, then we must suppose that Rawley also possessed the manuscripts containing Bacon's deathless Latin, before they, too, perished by virtue of having been printed.

Certain copies of this capacious folio of 1638 also include sheets of the *Novum organum* which had been printed eighteen years earlier, in 1620, lacking however the famous engraved frontispiece. These sheets had presumably

---

55 See further David Colclough, " 'The Materialls for the Building': Reuniting Francis Bacon's *Sylva Sylvarum* and *New Atlantis*," *Intellectual History Review* 20 (2010): 181–200.

56 Francis Bacon, *Sylva Sylvarum* (London, 1626), sig. ¶1v.

57 See n. 108, below.

58 Francis Bacon, *Certaine Miscellany Works*, ed. William Rawley (London, 1629). This volume contains copies of "Considerations Touching a War with Spain" (written 1624); "An Advertisement Touching an Holy Warre" (written 1622); the "Offer ... of a Digest ... of the Laws of England" (made to James I in 1622); and the fragmentary introduction of the "History of the Reigne of King Henry the Eighth" (written 1623). The relations of surviving manuscripts to some of the printed texts are analysed in Kiernan, "Introduction," to *OFB* VIII, CVIII and CIX–CXV.

59 Francis Bacon, *Operum moralium et civilium tomus*, ed. William Rawley (London, 1638), title-page: "Ab ipso Honoratissimo Auctore, præterquam in paucis, Latinitate donatus." (Referring to the Latin versions of the *History of Henry VII*; *Essays*; *De sapientia veterum*; *Holy War*; and *New Atlantis*.) The 1648 Amsterdam reprint of the *Novus Atlantis* similarly claims that it was "written in Latin by that distinguished man, Francis Bacon" (*Latine conscriptum, ab illustri viro Francisco Bacone*). But we also have Aubrey's testimony that Thomas Hobbes translated "severall of his Essayes into Latin" (John Aubrey, *Brief Lives*, ed. Andrew Clark (Oxford: Clarendon Press, 1898), 1:331).

remained unsold or otherwise undistributed after 1620 and had travelled with Bacon's papers, probably in Rawley's possession, before being put to use in 1638.[60] Hence although this book did not draw upon any unpublished manuscripts, it too—or at least, those copies of it which contain the *Novum organum*—should be regarded as an integral part of Bacon's archival afterlife. An archive may, after all, include copies of an author's unsold books, as well as his manuscripts.

Rawley's possession of unpublished material from Bacon's archive aroused the interest of others. The French pastor Elie Diodati wrote to him from London in 1633, noting that he "eagerly awaited" the edition of Bacon's "works" that Rawley promised.[61] William Rawley even lent his Bacon manuscripts to certain inquirers. One of these was the Oxford scholar Gilbert Watts (*d.* 1657).[62] Watts played an important role in the posthumous dissemination of Bacon's writings by virtue of the lavish edition he published in 1640 of his fine translation back into English of Bacon's *De augmentis scientiarum*. But Watts did not use any unpublished materials for this volume, which was made from a copy of the 1623 folio. Nor did he ever publish the apology for Bacon's philosophy or the English translation of the *Novum organum* that he advertised in that translation.[63] But Watts evidently maintained his interest in Bacon's writings, because in 1655 Samuel Hartlib recorded that "Dr Rawlins [*i.e.* William Rawley] hath a very thick Book ... (w*hi*ch will make a fol*io*) of L*o*rd Verul*a*m containing Miscellany Speeches w*hi*ch hee keepes vnpublished." Hartlib's note continues: "Dr Wats had it for some times in his keeping but hee was called vpon by the s*ai*d Dr and restored the same."[64]

60 Rees, "Introduction," to *OFB* XIII, LXXVII; Graham Rees, "Introduction," to *The Instauratio magna Part II: Novum organum and Associated Texts*, The Oxford Francis Bacon XI (Oxford: Clarendon Press, 2004), XCIX; Graham Rees and Maria Wakely, *Publishing, Politics, and Culture: The King's Printers in the Reign of James I and VI* (Oxford: Oxford University Press, 2009), 116 n. 99.

61 Tenison, *Baconiana*, 217–18 (Diodati to Rawley, London, 4 Apr. 1633): "Promptam ejus Operum Editionem, quam polliceris, tam avidè expecto."

62 See further Hugh de Quehen, "Watts, Gilbert (*d.* 1657)," *ODNB*.

63 Francis Bacon, *Of the Advancement and Proficience of Learning; or the Partitions of the Sciences. Nine books*, trans. Gilbert Watts (Oxford, 1640). Watts mentions his "*Justification*" of Bacon at sig. ¶4^{r}, and the translation of the *Novum organum* at sig. 2¶1^{v}.

64 *Hartlib Papers*, Ephemerides, 1655 (SUL, HP 29/5/57A). Hartlib also mentions here that "Dr Wats hath translated the 1. Book of [his *deleted*] Novum Organon Verul*amii*."

Furthermore, prior to England's troubles of the 1640s, a young barrister and graduate of Trinity College, Cambridge, called John Sparrow made "A Catalogue of the extant Manuscripts of Bacon, L*or*d Verulam viscount St Alban" from documents then in the possession of William Rawley.[65] Rawley himself also made a further and fuller list of Baconian manuscripts. A copy of this list, together with Sparrow's shorter one, survives among the papers of a physician, also from Trinity College, called Daniel Foote.[66] Samuel Hartlib sent copies of the first list to interested inquirers, including Justinian Isham and Lord Robartes.[67] Rawley seems to have been putting his Baconian manuscripts in order in preparation for the next set of editions he would publish.

It is generally supposed that there was then a twenty-year hiatus between Rawley's volume of Bacon's *Opera* in 1638 and his next Baconian publication in 1657.[68] But in 1651 there appeared an otherwise anonymous manual of prayers and meditations entitled *Gleanings of Refreshment in Gods Vineyard*, which contains four items ascribed to Bacon. Bacon's former servant Thomas Bushell later asserted that the "worthy and faithfull" Dr Rawley had published this little book.[69] It is notable for containing the only known text of a tiny treatise attributed to Bacon entitled "The Summe of the Bible." In 1639 Justinian Isham saw a manuscript of it that was owned by Sir Christopher Hatton, and was not impressed: "The Abridgement of *th*e Bible, w*hi*ch I made such Account to see was nothing to that I expected, not aboue 30 lines &

65 Sparrow's list survives in two copies: one, made by Samuel Hartlib in his Ephemerides in 1639 (*Hartlib Papers*, SUL, HP 30/4/5A); the other, undated and in an unidentified hand, is among the papers of Daniel Foote (BL, MS Sloane 429, fol. 243r). Hartlib's copy is edited in Clucas, "Hartlib's Ephemerides," 50–51. On Sparrow, see K. Grudzien Baston, 'Sparrow, John (1615–1670)', *ODNB*.

66 BL, MS Sloane 629, fols. 244r–245v. This list is identifiable as Rawley's by its reference to Bacon's Latin discourse *In felicem Memoriam Elizabethæ* "turned into English by my selfe, since my Lords death" (245v): Rawley's translation was published in his *Resuscitatio* (n. 72, below). On Foote, see David Thorley, "Daniel Foote, M.D., of Cambridge: The Evidence in Print and from the Sloane Collection," *eBLJ* art. 15 (2013): 1–30. It seems that Hartlib also saw a copy of this longer list: see *Hartlib Papers*, Ephemerides, 1639 (SUL, HP 30/4/27B): "Sparrow imparted a full Catal*ogue* of Verul*ams* MS. w*hi*ch Dr Raleigh hase."

67 See the Appendix, below.

68 For instance by Rees, "Introduction," to *OFB* XIII, LXXVIII.

69 Thomas Bushell, *An Extract by Mr. Bushell of his Late Abridgement of the Lord Chancellor Bacons Philosophical Theory in Mineral Prosecutions* (London, 1660), sig. H1r. On Bushell, see George C. Boon, "Bushell, Thomas (*b.* before 1600, *d.* 1674)," *ODNB*.

whither writt by Verula*m* tis doubtfull; there is nothing neere that, in it, as in *the* Preface to most Bibles."[70] "The Summe" does not appear on either of the surviving lists of Rawley's Baconian manuscripts, but it is not completely out of the question that it constitutes a new addition to the canon of Bacon's writings.[71]

In 1657 Rawley finally disburdened himself of an extensive edition of many of the unpublished English writings by Bacon that he had in his possession in the volume he entitled *Resuscitatio*.[72] As copy for most of the letters in this volume he used the "Register Book" of his correspondence that Bacon had prepared towards the end of his life.[73] Lastly, in 1658, Rawley completed his debt to Bacon's memory by editing a smaller volume of Latin *Opuscula*, mostly containing unpublished natural philosophical material.[74] It has been suggested that the implication of Rawley waiting until Boswell was dead before he published these works implies that he was conscious that he possessed their manuscripts illicitly. But this does not quite square with the testimony of Samuel Hartlib in 1639, that Boswell had promised to give Rawley his own manuscripts in order to publish them.[75] It seems more likely that these books finally appeared, as Rawley indeed says, because of his growing consciousness of his own mortality. Rawley knew, after all, his master's own thoughts on the matter, which Bacon had expressed to John Williams at the end of his life, and which Rawley articulated in the preface to the *Resuscitatio*. Almost certainly alluding to the two "Register-Bookes," Rawley noted that Bacon had aimed not "at the *Publication* of them, but at the Preservation onely; And *Prohibiting* them from *Perishing*." But the events to which the letters and speeches related were now

70 *The Hartlib Papers*, Isham to Hartlib, 2 Sept. 1639 (YALE/16, *i.e.* Beinecke Library, Yale University, Osborn Collection, MS File 16792). Isham had previously heard it was "a MS. of Verulam containing an Epitome of the Histories of the Bibel." *Hartlib Papers*, Ephemerides, 1639 (SUL, HP 30/4/12B).

71 *Gleanings of refreshment in Gods vineyard. … With a confession of faith and a sum of the Bible, by the right honorable Francis L. Verulam, Viscount S. Alban* (London, 1651), sigs. B4v–B5r. See further Richard Serjeantson, "Francis Bacon and 'The Summe of the Bible'," *Notes and Queries* 64 (June 2017), 318–21.

72 Francis Bacon, *Resuscitatio, or, bringing into publick light severall pieces of the works, civil, historical, philosophical, & theological, hitherto sleeping*, ed. William Rawley (London, 1657).

73 See above, at n. 19. Rawley refers to his use of "his Lordships Register-Book, of Letters" in *Resuscitatio*, sig. 3M1r.

74 Francis Bacon, *Opuscula*, ed. William Rawley (London, 1658).

75 See n. 98, below.

sufficiently distant that Rawley felt he would no longer be treading "too near, upon the *Heels* of *Truth*" by publishing them.[76]

Between them, Rawley's two volumes of *Resuscitatio* and *Opuscula* contain many of the works that appear on the lists he and Sparrow had previously made. But they do not contain everything. One such item is entitled "Historia*m* Generatio*n*is H*omi*nis" (the history of human reproduction). Rawley describes this as "Some of the Exper*i*me*n*ts ... reserved out of the Natural History"—that is, left out of the *Sylva Sylvarum*—"bec*ause* they were not fit to bee published in English;"[77] Hartlib confirmed that "p*ro*pter obscenitatem" (on account of their indecency) they "will hardly bee published except it bee in Latine."[78] But in the event these papers were never published, even in the decent obscurity of a learned language, and are now, alas, entirely lost. Hence we may never know the ins and outs of Bacon's experiments into that touchy subject.

Less morally demanding was a different natural history, the *Historia densi et rari*. In a letter of 1652, Isaac Gruter had mentioned to William Rawley that among the copies he had made from Boswell's manuscripts was an incomplete version of Bacon's *Historia densi et rari*.[79] Rawley had evidently previously told Gruter that he possessed a fuller and later version of that work.[80] Gruter had hoped to print the *Historia densi* among the other treatises in the 1653 *Scripta*, but in the event he did not do so. Mostly likely Rawley had replied to tell him that he planned to print it himself, which he duly did in the *Opuscula*. A consequence, however, is that not only Rawley's manuscript, but also Gruter's copy, are now lost.[81]

By the time William Rawley died in 1667, he had therefore published many—though not all—of the Bacon manuscripts that had been in his possession. As with the Boswell manuscripts, the fact of being printing was fatal for the original documents themselves. But not everything was printed. There

76 Bacon to John Williams [early 1626?] (*FBCP*, bcn0796), in Tenison, *Baconiana*, 195. William Rawley, "To the Reader," in Bacon, *Resuscitatio*, sig. (a)4r–v.

77 *Hartlib Papers*, Ephemerides, 1639 (SUL, HP 30/4/5A). This description is not found in the equivalent list in the Foote papers, BL, MS Sloane 429, fol. 243r.

78 *Hartlib Papers*, Ephemerides, 1639 (SUL, HP 30/4/4B).

79 Tenison, *Baconiana*, 223 (Gruter to Rawley, 29 May 1652): "In apographis meis ... reperies historiam *Densi & Rari*, sed imperfectam, opusculo licet longiusculè producto."

80 Tenison, *Baconiana*, 223 (Gruter to Rawley, 29 May 1652).

81 There is a further manuscript of an early draft of the *Historia densi* in the Bibliothèque nationale de France, coll. Dupuy, nº 5, fols. 7r–23v in the hand of Nicolas Rigault (*CELM*, BcF 295.5; Rees, "Introduction," to *OFB* XIII, LIX–LXVIII); it does not correspond to Gruter's description of the copy he saw.

is one particularly important collection of manuscripts that appears to derive from Rawley's portion. These are those documents contained within the larger and more miscellaneous volume that is now Harleian MS 6797, which Angus Vine, in a miraculous feat of codicological reconstruction, has identified as containing the contents of Bacon's fifth Composition Book, consisting of his speeches, proclamations, and reports, as it stood in about 1609.[82] All of the items in this Composition Book appear in the *Resuscitatio*. And their texts are close to, though not always identical with, those in the *Resuscitatio*.[83] How then did the manuscripts survive? They bear no marks of an encounter with the print-shop.

The answer may lie with the "Register Book" of his speeches that Bacon prepared at the end of his life, alongside that of his letters. This Register Book must have been made directly from the Composition Book whose contents now survive in the Harleian collection, but which Bacon, and also Rawley, then took to be superseded. If, as seems likely, Rawley possessed this second Register Book and used it to print the *Resuscitatio*, this would account for the survival of the fifth Composition Book: its contents would merely have constituted superseded duplicates. The irony is that, by virtue of being contemporary documents containing Bacon's own revisions, the surviving manuscripts are of much greater scholarly interest than any fair copy would have been. But it is not often that the cards fall in one's favour in this way.

3.3 *The Meautys Portion*

If Rawley, by his loving editorial labours, plays the role of Cordelia to Bacon's Lear, then that of heartless Regan is filled by the third of the three men who divided Bacon's archive between them: Thomas Meautys. Here we have a new story to tell. It has long been evident that both Boswell and Rawley possessed portions of Bacon's *Nachlass*, and indeed they have commonly been regarded as its only two inheritors. But as will now become clear, not only Boswell and Rawley, but also Thomas Meautys, took possession of a most important portion of his former master's archive.

Since Stephen Clucas first brought it to light, scholars have known of Samuel Hartlib's testimony about the fortunes of Bacon's posthumous manuscripts.

---

82 Vine, "Composition Books," esp. 8–19.

83 Bacon's "Certaine Considerations Touching the Plantation in Ireland," BL, MS Harley 9796, fol. 122r, is misdated, in his own hand, to 1606. This erroneous date is repeated in *Resuscitatio*, 2K4r, whose text must therefore ultimately derive from that manuscript, but probably (as is suggested here) via an intermediate "Register Book" copy.

Reporting what he had been told by John Sparrow, Hartlib wrote in his Ephemerides for 1639 that Bacon's manuscripts "were divided amongst his Chapl*ain* Raleigh, Sir W. Boswell who got the best and Mutis one of his Servants who remained longest with him, whom hee also made Executor."[84] Hartlib repeated this statement later on in the same year: "Mr Mutis, Sir W. B*oswell* and Dr Raleigh haue them all."[85] Graham Rees circumspectly suspended judgment on the accuracy of Hartlib's testimony: "Meautys may in fact have possessed some Bacon manuscripts but we only have Samuel Hartlib's hearsay evidence for this."[86] Yet other evidence proves that Hartlib's information was indeed quite correct.[87]

This evidence confirms that, until his death in 1649, Meautys possessed a box of manuscripts containing several very important pieces of writing by Bacon of which we would otherwise be entirely ignorant. They include some of his earliest surviving philosophical writings. Among them is the slim quarto volume containing the revealing early English version of Bacon's philosophical system, the *Valerius Terminus*, together with part of the Latin *Temporis Partus Masculus*.[88] Meautys' box also contained the English *Filum Labyrinthi*—a version of the Latin *Cogitata et visa*, of which Boswell had Bacon's copy. Confusingly, this document, like several others deriving from Meautys' portion, is now bound up in Harley 6797, alongside documents that came from Rawley.[89] And Meautys also had another English work with a Latin title, the *Sequela cartarum sive Inquisitio Legitima de calore et frigore*, written, unusually, entirely in Bacon's own hand.[90] The political writings preserved in the Meautys box included the important political treatise *Of the True Greatnes of the Kingdome of Brittaine*, which Bacon would later set aside in order to transform it into the less nationally-specific essay of 1612 entitled "Of the True Greatness of Kingdoms."[91] Meautys also possessed the sole known draft of the speculative Proclamation that Bacon drew up upon the accession of James I, but which was never published at

84 *Hartlib Papers*, Ephemerides, 1639 (SUL, HP 30/4/4B). Clucas, "Hartlib's Ephemerides," 41.

85 *Hartlib Papers*, Ephemerides, 1639 (SUL, HP 30/4/27B).

86 Rees, "Introduction," to *OFB* VI, LXXIII n. 20.

87 This evidence takes the form of an inventory made by the discoverer of Meautys' collection in 1682, and of letters reporting its discovery; see §4(II), below, and esp. n. 113.

88 BL, Harley MS 6463, pp. 1–70, 70–73 (*CELM*, BcF *285 and BcF 307).

89 BL, MS Harley 6797, fols. 139r–146v (*CELM*, BcF *214).

90 BL, MS Harley 6855, fols. 52r–60r (*CELM*, *BcF 111).

91 BL, MS Harley 7021, fols. 25–42 (*CELM*, *BcF 232)

the time;[92] and various fugitive documents relating to the Mint and to the Court of Wards.[93]

No less importantly, Meautys' box also contained numerous copies of letters sent by Bacon (some in Bacon's own hand, others in the hands of various of his scribes, including Meautys himself), which now constitute the single most important surviving volume of Bacon's letters, MS 936 in Lambeth Palace Library.[94] It is from this collection, for instance, that we know of Bacon's letter to Isaac Casaubon; of his correspondence with Gondomar; and of Toby Matthew's letter to Bacon enclosing a copy of a letter by Galileo.[95]

Taken as a whole, Thomas Meautys' chest of manuscripts constitutes a collection of extraordinary richness. Without it, our understanding both of Bacon's intellectual development and of his political life would be greatly diminished. One may well imagine the excitement of someone interested in such things upon coming across it. Moreover, unlike both the Boswell and Rawley collections, the contents of this box of manuscripts were never printed in the seventeenth century. By escaping the press then, the documents themselves survived. Yet by the time they were first printed, in the eighteenth century, their association with Meautys had been entirely lost. From then until now, the provenance of the documents in his box has been obscure. Yet the implications are clear: the great majority of surviving Bacon manuscripts do not derive from the diligent William Rawley, but rather from the cautious Sir Thomas Meautys.[96]

*

92 BL, MS Harley 6797, fols. 13r–14v (not in *CELM*).

93 BL, MS Harley 7020, fols. 156–166 (not in *CELM*).

94 Lambeth Palace Library (hereafter LPL), MS 936. A catalogue of this volume may be found at lambethpalacelibrary.org.uk, and see also the listing in *FBCP*.

95 LPL, MS 936, items 272 (Casaubon), 168, 184C, 248, 282 (Gondomar), 31 (Matthew). The enclosed copy-letter of Galileo, which is listed in Henry J. Todd, *A Catalogue of the Archiepiscopal Manuscripts in the Library at Lambeth Palace* (London, 1812), 215, is now missing (a point I owe to Mordechai Feingold).

96 A loose end remains. In early 1653, Samuel Hartlib recorded in his Ephemerides that "My Lady Mitton is said to have the remainder of all Verul*ams* MS. *whi*ch Aubrey will labor to get from her" (*Hartlib Papers*, SUL, HP 28/2/49B); see also Kelsey Jackson Williams, "Training the Virtuoso: John Aubrey's Education and Early Life," *Seventeenth Century* 27 (2012): 167. I have not been able to identify this person. Is it possible that "Mitton" is a corruption of "Meautys," and hence that the entry refers to Sir Thomas Meautys' widow, Anne? Yet in 1651 Lady Anne Meautys had married Sir Harbottle Grimston, and should thereafter have been known as Lady Anne Grimston.

We are therefore now in a position to put uncertainty about the posthumous fortunes of Bacon's manuscripts to rest. Hartlib's information was good—unsurprisingly, since it derived from William Rawley himself. Bacon's *Nachlass* was indeed divided between three different men.

William Boswell, whom Bacon had specifically named in his will as the person best suited to determine the fate of his manuscripts, "got the best." Thanks to Isaac Gruter, Bacon's composition-book of "natural and universal philosophy" was eventually printed; but the other composition-book, containing "writings in politics and morals," perished.

Thomas Meautys, who sometimes copied Bacon's letters for him during the 1620s, and who as a creditor had acquired legal responsibility for disposing of Bacon's estate, obtained the majority of Bacon's draft letters, as well as various other manuscripts in both natural philosophy and politics. It is possible, indeed, that his portion consisted of the manuscripts which Bacon had left in his chambers at Gray's Inn, which Meautys took possession of following Bacon's death.[97] But although he, and his heirs, preserved them in a dedicated box, they did nothing with them.

William Rawley, finally, acquired the residue of Bacon's archive. Though the documents he possessed may have been ones that had been rejected by the other two men, there is some evidence that they recognised Rawley's willingness, and his ability, to have Bacon's *opera posthuma* printed; in 1639 Samuel Hartlib recorded that "Sir W. Bosw*ell* and Mutis have p*ro*mised to Raleigh to give him all w*ha*t they have to publish them."[98] In the event, however, neither seems to have done so. But it is likely that Rawley did obtain possession of the two Register Books of Bacon's letters and speeches that he had left to Bishop Williams and Sir Humphrey May, and that he used them in printing the *Resuscitatio*, alongside the many other works by Bacon that he conferred to the world with the authority of his close personal connection to the deceased Lord Chancellor.

Indeed, by virtue of Rawley's energy and care in enlarging Bacon's canon of printed works, it became possible by 1670 for the stationer Charles Molloy to suppose that it was Rawley who had obtained custody of "all" of Bacon's manuscripts.[99] Rawley's diligent publications might indeed give this impression, but

---

97 This suggestion is owing to A. Chambers Bunten, *Sir Thomas Meautys, Secretary to Lord Bacon, and his Friends* (London: Page & Thomas, 1918), 38, whose reasoning on this point is not obviously contaminated by his unsubstantiated views regarding the authorship of the plays of Shakespeare.

98 *Hartlib Papers*, Ephemerides, 1639 (SUL, HP 30/4/4B).

99 Charles Molloy, "To the Reader," in *The Second Part of the Resuscitatio or a Collection of several pieces of the works of ... Francis Bacon* (London, 1670), sig. A4r.

it was not correct. Yet the misapprehension is understandable. Boswell's portion was lost in the Low Countries. And the box containing Thomas Meautys' portion of the Bacon manuscripts was still waiting to be found.

4 Thomas Tenison, Consolidator

Following the death of William Rawley in 1667, an important new figure enters the convoluted story of Francis Bacon's archival afterlife; a collector and editor who, though he destroyed as much as he preserved, nonetheless helped ensure the survival of those portions of Bacon's archive that still remain. This person was the scholar, cleric, and future Archbishop of Canterbury, Thomas Tenison (1636–1715). In the river of time down which Bacon's manuscripts flowed, Tenison forms an important, though a temporary, confluence.

4.1 *The Rawley Manuscripts and Baconiana (1679)*

In the course of his time as a fellow of Corpus Christi College, Cambridge in the mid-1660s, Thomas Tenison had become a friend of William Rawley's younger son (also William), who was also a fellow there until his early death, a year before his father, in the plague-year of 1666. By the later 1670s, this connection had given Tenison access to, and probably ultimately ownership of, William Rawley's portion of Bacon's manuscripts.[100] In 1679 Tenison published from these documents the collection he entitled *Baconiana.* His stated criteria for publication turned upon the status of a document's handwriting. "Nothing here is offered," wrote Tenison in his lengthy introduction,

> which was not written either by his [i.e. Bacon's] own Hand, or in Copies transcrib'd by the most faithful Pen of his Domestic *Chaplain*, Dr. *William Rawley*: A Person whom his Lordship chiefly us'd in his Life-time, in Writing down, Transcribing, Digesting, and Publishing his Composures.[101]

On this basis Tenison filled his *Baconiana* with editions of a number of manuscript items that Rawley had declined to print (and a few other items that did not derive directly from the Rawley collection), divided into the

100 Tenison, "An Account," 79, 89–90; Thomas Birch, "Preface," to Francis Bacon, *Letters, Speeches, Charges, Advices, &c.* (London, 1763), sig. A6r–v; Spedding, "Preface to the Physiological and Medical Remains," in Bacon, *Works*, 3:797.

101 Tenison, "An Account," 79.

categories of "Politico-moralia," "Physiologica" (i.e. natural philosophy), "Medica," "Theologica," and "Bibliographica." It is to this volume that we owe unique copies of some rather miscellaneous writings on natural philosophical and medical subjects, and also an early draft of the *Abecedarium novum naturæ*.[102]

Some of the items that Tenison published appear to have existed in more than one manuscript. In particular, Tenison wrote of Bacon's *Articles of Questions, touching Minerals* that "having by me three Copies, I publish them by that one on which his Lordship had endorsed with his own Hand, *This is the clean Copy*."[103] Again, however, none of the manuscripts that he drew upon to print the volume survive. (Nor do either of the two other copies of the *Articles … touching minerals*.) Tenison also printed in the *Baconiana* a number of interesting letters, some of them addressed to William Rawley rather than to Bacon himself. The manuscripts of these letters, too, have perished. Even in the hands of so keen a collector as Tenison, then, the fact of a work's being printed still constituted a death-sentence to original documents. It is not just that the act of printing led to the destruction of the manuscript that was sent to the print-shop: it is that the existence of a printed version meant that not just the "clean Copy" of a work, but also any others, were no longer fit to be kept.

Tenison was, by definition, printing manuscripts that Rawley had declined to publish, a ticklish point that he faced squarely in his preface. But even Tenison did not print everything he had found. In particular, he speaks of "Some few imperfect Papers, about his Lordship's private Affairs, or of very little moment in Philosophy" being "still kept where they ought to be, in private Hands." It is clear from this, as well as from certain surviving documents, that already in 1679 Tenison knew of certain papers that he had chosen to hold back from the press.[104]

The most remarkable of these is Bacon's private memorandum book of July 1608, the "Commentarius Solutus." It likely (though not absolutely certain) that Rawley had inherited this as part of his portion, but what is clear is that at some point in his life it came into Tenison's possession, and this allusion suggests that it was indeed from Rawley that Tenison acquired it. Tenison's discretion regarding its contents, however, had a curious consequence. In the course of Tenison's lifetime the "Commentarius Solutus" became separated

102 See Rees, "Introduction," to *OFB* VI, LXXIII.

103 Tenison, "An Account," 96.

104 Tenison, "An Account," 80–82 (quotation at 82).

from his other Bacon manuscripts, and made its way into the public library that Tenison established for the use of his parish of St Martin-in-the-Fields in 1684.[105] There it remained, probably unread and certainly unknown to Bacon's several editors, until the resourceful James Spedding "chanced upon it" (as he modestly put it—no manuscript is identified by chance) in March 1848.[106] After the Tenison Library was wound up by Act of Parliament, and its contents auctioned off in 1861, the "Commentarius Solutus" was sold, and shortly afterwards made its way to the British Museum.[107] On such slender threads hang the survival of the single most revealing document in Bacon's entire biography.

A similar path was taken by a three fragmentary manuscripts of writings by Bacon that are now to be found in a composite volume of Thomas Tenison's miscellaneous papers. These documents are presumably the last remnants of Rawley's miscellaneous collection of Baconian natural philosophy. They survived by escaping print in Tenison's *Baconiana* and then perhaps became misplaced in Tenison's public library. From there they made their way in the 1861 sale into the collection of Sir Thomas Phillipps (1792–1872). The volume was sold again in a Phillipps family sale at Sotheby's in 1913, after which it passed into the Additional Manuscripts of the British Museum—the latest Bacon material (so far) to do so.[108]

105 W. Lee, "Archbishop Tenison's Library," *Notes and Queries*, 3rd ser., 8 (1865): 322–4; William O'Sullivan, "A Finding List of Sir James Ware's Manuscripts," *Proceedings of the Royal Irish Academy* 97C (1997): 73. A catalogue of manuscripts in this library was published by Edmund Gibson, *Librorum manuscriptorum in duabus insignibus Bibliothecis, altera Tenisoniana, Londini, altera Dugdaliana, Oxonii, Catalogus* (Oxford, 1692); it makes no reference to any manuscripts of Bacon. See also P. Hoare, "Archbishop Tenison's library at St Martin-in-the-Fields: the building and its history," *London Topographical Record*, 29 (2006), 127–50.

106 Spedding, *Letters and Life*, 4:8.

107 British Library, MS Add. 27278 (*CELM*, *BcF 153); Vine, "Composition Books," 5. This manuscript also contains (fols. 17^r–22^v) a version in Bacon's hand of his *Inquisitio legitima de motu* (*CELM*, *BcF 303). The catalogue of the library prior to sale was issued as the "Catalogue of the Valuable Library formed by Archbishop Tenison (Sotheby and Wilkinson)" (1861). It was auctioned 3–8 June 1861. A copy of this catalogue, anotated with prices obtained, at a sum total of £1410 7s, is at London Metropolitan Archives, ACC/2692/209.

108 British Library, MS Add. 38693, fols. 30r–48v, 50r–52v (notes relating to the *Sylva Sylvarum*), 29r–v (a portion of the *Inquisitio de magnete*), 49r (a fragment of the *Historia vitae* in Bacon's own hand): *CELM*, BcF 283, BcF 302, and *BcF 297, respectively. See further Graham Rees, "An Unpublished Manuscript by Francis Bacon: *Sylva Sylvarum* Drafts and other Working Notes," *Annals of Science* 38 (1981): 377–412.

These instances demonstrate that, at his death, Tenison definitely possessed certain Bacon manuscripts that had previously belonged to Rawley. Did he then also own the manuscripts that went into the *Baconiana*, or did they remain the property of William Rawley's heirs? Here the evidence is unclear. Tenison speaks of his "worthy Friend" John Rawley, William Rawley's elder son, as "the Executor of the said Reverend Doctor" and the person "by whose care most of these Papers have been preserved for the public Good." He goes on to note that he had received certain "Bundles" of manuscripts together with three catalogues that William Rawley had made of them.[109] And, as we have seen, he speaks of manuscripts about Bacon's "private Affairs" remaining in "private Hands." Yet he does not state unequivocally that those hands are his.

On balance, however, it seems most likely that Tenison did come to possess the remains of the Rawley portion—including not only the items that went into *Baconiana*, but also the items from Bacon's fifth Composition Book that are now to be found in Harleian MS 6797, and also the "Commentarius Solutus." Of these, those in the first category were printed, and then perished, probably in the print-shop; the second survived to make their way into a further collection; and the last were hidden away in his own library, only to emerge again in the revolutionary year of 1848.

4.2 *The Rediscovery of Meautys' Box*

Remarkably, however, Rawley's cache of Bacon manuscripts was only the first such collection that came into Tenison's possession. For a few years after he had published the *Baconiana*, a further and in certain ways even richer collection of Baconian manuscripts found its way into Tenison's hands. He mentioned the discovery in a letter to the then Archbishop of Canterbury, William Sancroft, in December 1682. It is this letter that proves that, just as Hartlib had reported, Thomas Meautys did indeed possess a significant portion of Bacon's posthumous archive:

> I did forgett, on Tuesday, to acquaint your Grace that I had (by a strange Providence) latelie found out, in this Town, a great many Originall Papers of the L*or*d Bacon. When I have look'd them over & sorted them I will be bold to p*re*sent y*ou*r Grace w*i*th a Catalogue of them. They came to

109 Tenison, "An Account," 79–80, 89.

> me from *th*e Executor[110] of *th*e Executor[111] of S*i*r T. Meautys, who was his L*ordshi*ps Executor.[112]

Fortunately, Tenison duly produced the catalogue he promised Sancroft, and it now survives alongside his original letter in Thomas Tanner's collections of ecclesiastical correspondence from the period. It is this catalogue that enables us to know what writings were contained in Thomas Meautys' box (Fig. 1.2).[113]

Tenison was clearly conscious that he had made an exciting discovery, for he immediately began to make plans with Richard Chiswell, the stationer who had published the *Baconiana*, to print a further volume of material drawn from collection that had belonged to Thomas Meautys:

> I have now lookd over all *th*e books & Papers in *th*e box. In *th*e books there are Copies of Essaies, Maxims of Law &c. all printed alreadie. But they contain somethings fitt to be printed: & they & *th*e Letters will make a*n* handsome folio w*hi*ch I doubt not but will turn to Account. For *th*e L*ett*ers here are divers of S*i*r T. Meautys &c worth nothing. but there are more then 40 L*ett*ers to the d*uke* of Bucks. & some of *th*e d*uke* of Bucks to him. There are 8 or 10 to K*ing* James. they are 3 or 4 to Gondamor, & Gondamours Answer to one of them. There are 2 or 3 L*ett*ers to Bis*ho*p Williams, & 2 fro*m* him. There is L*or*d Bacon's L*ett*er to Casaubon, in Latine. There is 1 Essay nev*er* printed. All w*hi*ch will be well acco*m*pted.

110 Richard Roberts, "a Surrey clergyman," was the executor of Henry Meautys' estate. See John P. Ferris and Rosemary Sgroi, "Meautys, Henry (*c.*1585–by 1663), of St. Julian's Hosp., St. Albans and Westmill, Herts.," *History of Parliament* (www.historyofparliamentonline.org).

111 Probably Henry Meautys, who (after some delay) inherited his brother's estate. See S. A. Baron, "Meautys, Sir Thomas (*c.*1590–1649)," *ODNB*.

112 OB, MS Tanner 35, fol. 147r (Thomas Tenison to William Sancroft, 18 Dec. 1682). After he had published his volume of Bacon's *Letters … & c.* in 1763, Thomas Birch obtained a copy of this letter from Richard Rawlinson, which survives among his papers at BL, MS Add. 4259, fol. 134r. He printed it in the (posthumous) second edition of his edition, which forms part of *The Works of Francis Bacon*, vol. 6 (London, 1803), sig. B8r–v. Yet its implications have never been fully considered, neither by Spedding, who knew of it (see *Letters and Life*, 2:2–3), nor by Rees, who did not. See also Stewart, "Introduction," to *OFB* I, 3.

113 OB, MS Tanner 35, fols. 145r, 146r. There is a copy of this list among Thomas Birch's papers at BL, MS Add. 4259, fols. 134v–135r.

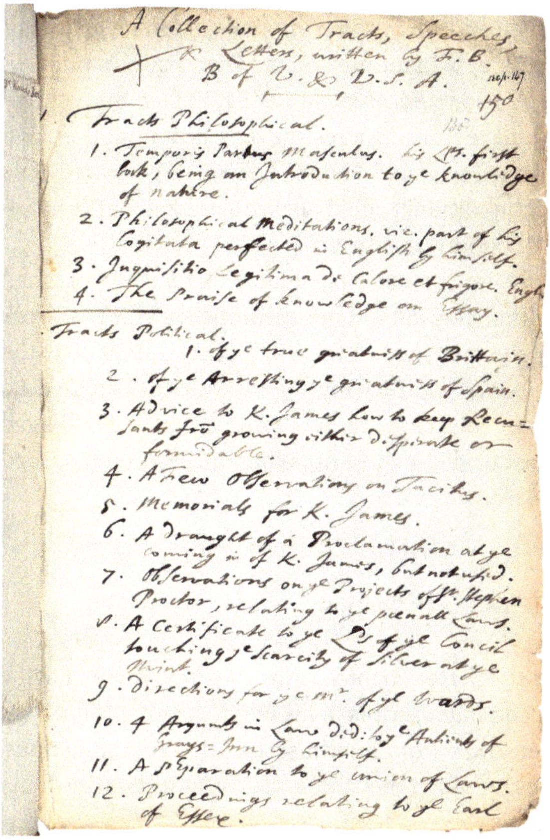
A Collection of Tracts, Speeches, & Letters, written by F. B. B of V. & V. S. A.
150
1. Tracts Philosophical.
1. Temporis Partus masculus. his Lps. first book, being an Introduction to ye knowledge of nature.
2. Philosophical Meditations, viz. part of his Cogitata perfected in English by himself.
3. Inquisitio Legitima de Calore et frigore. Engl.
4. The Praise of knowledge an Essay.
Tracts Political.
1. of ye true greatness of Brittain.
2. of ye Arresting ye greatness of Spain.
3. Advice to K. James how to keep Recusants from growing either desperate or formidable.
4. A few Observations on Tacitus.
5. Memorials for K. James.
6. A draught of a Proclamation at ye coming in of K. James, but not used.
7. Observations on ye Projects of Sr. Stephen Proctor, relating to ye penall Laws.
8. A Certificate to ye Lds of ye Concil touching ye Scarcity of Silver at ye Mint.
9. directions for ye Mr. of ye Wards.
10. 4 Arguments in Law did: by ye Antients of Grays-Inn by himself.
11. A Preparation to ye union of Laws.
12. Proceedings relating to ye Earl of Essex.

FIGURE 1.2 *Thomas Tenison's list of Thomas Meautys' Bacon manuscripts.*
BODLEIAN LIBRARY, MS TANNER 35, FOL. 145R.

> After *the* Holydays I will methodize all, & putt all L*ett*ers of *th*e same date together (for as yet they are in co*n*fusion) & then we will take further resolutions about them.[114]

This letter is revealing about the state of Meautys' Bacon archive as it then stood. It contained both "books"—that is, bound manuscript volumes—and loose "Papers." It also contained numerous letters, in a state of "confusion."

Tenison's plans to print a further volume of Baconiana from the Meautys portion came to nothing, however. Perhaps he lacked the energy to carry the work through. Yet this may be an uncharitable explanation, given that shortly

114 BL, MS Add. 4259, fol. 136 (Tenison to Chiswell, Dec. 1682).

afterwards Tenison produced an edition of some unpublished manuscripts of Sir Thomas Browne. A different explanation is offered by Chiswell's publication of a "Second Edition" of Tenison's *Baconiana* in 1684. In fact, this volume merely consists of a new title-page fronting a reissue of sheets from the 1679 imprint. Perhaps the unpublished works of Francis Bacon were less vendible than Tenison had hoped, and the stationer could not see a market for the handsome folio that the scholar confidently envisaged.

Whatever the reason may be for the non-appearance of Tenison's planned book, it is perhaps fortunate that Tenison did not continue his editorial work on Bacon's manuscripts. For the evidence of the 1679 *Baconiana* is that the print-shop would again have been fatal to any manuscript that crossed its threshold. Instead, and in sharp contrast with the portions that had belonged to Boswell and to Rawley, most of the manuscripts that once sat undisturbed in Meautys' box still now survive.

5 The Dissolution of Tenison's Collection

Archbishop Tenison lived just long enough to crown George I. But having brought together so many Bacon manuscripts in his lifetime, with his death in 1715 they again began to flow away from him in several different directions. The details of the transactions by which they did so are obscure; but the directions in which different portions went, at least, are clear.

By the time of his death Tenison had acquired a substantial collection of manuscripts, together with "a vast number" of books.[115] (He had a particular liking for documents relating to Cambridge University.[116]) His dispositions for this collection in his will are elaborate, though unfortunately not very specific. In this document he requests his executors to ensure that some of his papers (unspecified) be "burnt and destroyed." He requests that other books and manuscript already deposited in the "Public Library" at Lambeth remain there. And other (also unspecified) documents again are to be deposited there by Edmund Gibson and his library-keeper Benjamin Ibbot, saving that "as many of them as are useless or otherwise unfit" were to be "immediately destroyed." Finally, Tenison gave the "rest and residue of my said Books and Manuscripts which shall be then left and remaining" to the beneficiaries of his will. These were

115 *Letters of Humfrey Wanley: Palaeographer, Anglo-Saxonist, Librarian, 1672–1726*, ed. P. L. Heyworth (Oxford: Clarendon Press, 1989), 332.

116 See, e.g., Sarah Knight, "Milton's Student Verses of 1629," *Notes and Queries* 57 (2010): 37–39.

Tenison's cousin, Edward Tenison (1673–1735); his kinsman, George Fage (*fl.* 1691–1728); and the religious controversialist, Thomas Herne (*d.* 1722).[117] Tenison's printed and manuscript books were to be "equally divided" among them by three further people: "Dr Edmund Gibson, Mr Benjamin Ibbot and Mr Robert Clavering my domestick Chaplaynes or any two of them."[118] Somewhere amidst this complex set of bequests there sat the many Bacon manuscripts that Tenison had acquired from the heirs of both William Rawley and Thomas Meautys.

5.1 *The Lambeth Manuscripts*

Tenison's stipulations go some way towards explaining the current resting-places of Bacon's manuscripts, for a number are indeed now to be found in Lambeth Palace Library. They are found, above all, in Lambeth MS 936, a remarkably rich collection principally of Bacon's draft letters together with a few other papers, dating from between 1603 and his death, in the handwriting of Bacon himself and of various of his regular scribes, including Thomas Meautys. This volume contains many (though not all) of the letters that Tenison mentions in his letter to Chiswell, including the letters of Thomas Meautys, the letters to James I, the letters to Gondomar, the letters to and from John Williams, and the letter to Casaubon.[119] The contents of this volume, then, which has been central to the study of Bacon since some of its contents were first printed in 1763, derive from Thomas Meautys. But how did they get to Lambeth?[120]

MS 936 is part of collection that entered the archiepiscopal library in 1749 from the estate of Edmund Gibson, then the rector of Lambeth, who was charged by Tenison in his will with depositing certain of his manuscripts in the Lambeth library, and with distributing the remainder to Tenison's beneficiaries. These fourteen volumes came to be known as the *Codices Gibsoniani* or Gibson Papers.[121] The first scholar to print any of them was Thomas Birch, in his 1763 edition of Bacon's *Letters and Papers*. His preface to that volume

---

117 All of these men were alumni of Tenison's college, Corpus Christi, Cambridge. See further Robert Masters, *The History of the College of Corpus Christi and the B. Virgin Mary* (Cambridge, 1753), 19, 399, 400–3. Fage was prebendary of Tervin, 1709–1728, in succession to Edward Tenison (Joyce M. Horn, ed., *Fasti Ecclesiae Anglicanae 1541–1857*, vol. 10, *Coventry and Lichfield Diocese* (London: Institute of Historical Research, 2003), 64).

118 TNA, PROB 11/550/123, fols. 22v–23r (will of Thomas Tenison, proved 23 Jan. 1716).

119 See above, n. 114, and further the catalogue at archives.lambethpalacelibrary.org.uk.

120 On this question, see also Stewart, in *OFB* I, 500–1.

121 LPL, MSS 929–942.

offers a concise history of the collection, but does not mention its derivation from Meautys, of which Birch was at that time unaware. In it, he suggests that Tenison's Bacon papers, together with "the rest of his manuscripts, not already deposited in the library at Lambeth, were left by him ... to his chaplain, Dr. Edmund Gibson, ... and to Mr. (afterwards Dr.) Benjamin Ibbot."[122] But as we have seen, this is not exactly what Tenison's will had stipulated: Gibson and Ibbot were charged to distribute Tenison's manuscripts, not to take possession of them.

In the case of the "Gibson Papers," however, that is precisely what happened. Gibson's motives for detaining the manuscripts are unknown, but he did at least remedy the irregularity. Following his death in 1748, Gibson's executors delivered them, according to his instructions, to the then Archbishop of Canterbury, Thomas Herring (1693–1757). Birch goes on to describe their condition:

> ... as they lay undigested in bundles, and in that condition were neither convenient for use, nor secure from damage, his Grace the present Archbishop directed them to be methodized and bound up in volumes with proper indexes, which was done by his learned librarian, Andrew Coltee Ducarel.[123]

Henry John Todd, writing in 1812, fleshes this account out with a little more detail, noting that following their deposit in 1747 the Gibson papers "continued in a confused state till June 1758," when the newly appointed archbishop, Thomas Secker, "directed them to be methodized by Dr. Ducarel ... and they were bound in fourteen volumes, as they now appear."[124] It was perhaps this process that led to a few miscellaneous Francis Bacon letters finding their way into volumes of the Gibson papers in Lambeth Palace other than MS 936.[125]

122 Birch, "Preface," sig. A7r.

123 Birch, "Preface," sig. A7r–v.

124 Henry J. Todd, *A Catalogue of the Archiepiscopal Manuscripts in the Library at Lambeth Palace* (London, 1812), III. See further *OFB* I, 1012.

125 For instance, LPL, MS 941, item 58, consists of an incomplete autograph letter by Bacon to James I; it is completed by LPL, MS 936, item 145, from which it has evidently at some point become separated. The sixteen volumes of Anthony Bacon papers also held in Lambeth Palace Library (MSS 647–662) appear to have arrived independently from those of his brother Francis, having been "purchas'd, at the expence of ... archbishop Tenison, by Dr. White Kennett" (Thomas Birch, *Memoirs of the Reign of Queen Elizabeth*, vol. 1 (London: A. Millar, 1754), 2).

It is therefore the case that fewer of the manuscripts now reposing in safety in the bowels of the Church of England derive from William Rawley than has previously been thought. One at least, may indeed owe its presence there to Tenison's friendship with Rawley's family: William Rawley's notebook, containing anecdotes, recipes, proverbs—and a list of Bacon's published writings.[126] But the key Bacon letters in the Lambeth collection in fact derive not from Dr William Rawley, but from Sir Thomas Meautys.

5.2 *The Harleian Acquisition*

The story of the transmission of Bacon's archive is drawing to a close, but one last crucial (though still somewhat obscure) chapter remains to be told. The Lambeth manuscripts constitute an important portion of Francis Bacon's archive. But it is a portion that is essentially limited to copies that Bacon himself kept of his outgoing letters. It does not include the many different speeches, legal writings, and philosophical works that also flowed into Thomas Tenison's hands from both William Rawley and Thomas Meautys.

Fortunately, these documents do also survive largely entire. They do so in a single collection: that of Robert Harley, first earl of Oxford (1661–1724), and his son, Edward (1689–1741), the second earl. But because these manuscripts are dispersed throughout numerous different volumes of the Harleys' collections, their association first with Meautys, and then with Tenison, has hitherto been lost. The various manuscripts containing Bacon's holograph are conspicuous by their absence from Cyril Ernest Wright's authoritative account of the sources of the Harleian collection, *Fontes Harleiani* (1972).[127] Yet it is now clear that the great majority of Baconian holograph material now in the Harleian collection ultimately derives from the collections of archbishop Tenison.

What is not clear is precisely how Tenison's manuscripts made their way into Harley's collection. A characteristically parsimonious hint appears in a letter from the Harleys' librarian, Humfrey Wanley, to Edward Harley in January 1716: "as to the late A.Bishop's books I will still endeavor to gett them

126 LPL, MS 2086. On this document, see Angus Vine, *In Defiance of Time: Antiquarian Writing in Early Modern England* (Oxford: Oxford University Press, 2010), 122–24. A reproduction is available via *Scriptorium: Medieval and Early Modern Manuscripts Online*, gen. ed. Andrew Zurcher (scriptorium.english.cam.ac.uk).

127 Cyril Ernest Wright, *Fontes Harleiani: A Study of the Sources of the Harleian Collection of Manuscripts Preserved in the Department of Manuscripts in the British Museum* (London: British Museum, 1972).

cheaper for you; & yet you shall have the choise of the whole."[128] Was this when Tenison's manuscripts also entered the Harleian collection? Or was it from Gibson, the detainer of Bacon's letters? Or did they come from one of Tenison's three legatees, who were supposed to receive the residue of his manuscripts: Edward Tenison, George Fage, or Thomas Herne?

The question remains open.[129] What is clear is that none of the Tenison manuscripts in the Harleian collection bear Humfrey Wanley's punctilious inscriptions. For instance, as we have seen, it seems clear that a fair number of the documents now bound into codex Harley 9797 were formerly in the possession of William Rawley; that others again formerly belonged to Thomas Meautys; and that Tenison consolidated them. These items are silent about their provenance. But a few *further* items in the volume evidently arrived in it by different routes again. One of these is the copy of a composite work (containing Bacon's *Preparation for the Union of the Laws*), in a scribal hand unconnected with Bacon's secretariat, which was given by the Irish antiquary Thomas O'Sullivan and inscribed by Wanley "2 March 1722/3."[130] Another is the workmanlike professional copy of Bacon's "Humble Submission" to the Lords (a very widely-copied piece), which opens the volume, and which Wanley has inscribed: "Bought of Mr. G. Pauls Landlady."[131]

The first datable testimony to Harley's possession of the Bacon manuscripts that were formerly in the possession of Thomas Tenison occurs in the preface to a book that appeared in 1734 entitled *Letters and Remains of the Lord Chancellor Bacon*.[132] This work was begun by Robert Stephens (1665–1732), historiographer royal, and according to Humfrey Wanley "a Relation" of Lord Harley.[133] Stephens' preface, written at some point between the publication of John Blackbourne's edition of Bacon's *Opera omnia* in 1730 and Stephens' own

128 Wanley, *Letters*, 333 (Wanley to Edward Harley, 26 Jan. 1716). The context suggests this promise relates specifically to printed books from Tenison's collection.

129 I do not find it answered in *The Diary of Humfrey Wanley*, ed. C.E. Wright and Ruth C. Wright, 2 vols (London: Bibliographical Society, 1966).

130 BL, MS Harley 6797, fols. 54–78 (at 54r); mentioned in *Diary of Wanley*, 1:192, but erroneously identified by the editors as MS Harley 6688 (1:192 n. 8 and 2:514).

131 BL, MS Harley 6797, fols. 1–10 (at 10v).

132 This volume was shortly afterwards reissued with a new title-page as *Letters, Memoirs, Parliamentary Affairs, State Papers, &c. With some Curious Pieces in Law and Philosophy. Publish'd from the originals of the Lord Chancellor Bacon* (London, 1736).

133 *Diary of Wanley*, 1:224. Stephens presented the first earl with what is now MS Harley 1251, a collection of later-medieval devotions and images of saints. Much earlier, Stephens had published a collection of *Letters of Sir Francis Bacon* (London, 1702), which were taken from a collection "preserved by the care of a very worthy *Gentleman*" (sig. A4r),

death in November 1732, offers some sort of explanation for how he came to undertake the volume:

> the Right Honourable the Earl of *Oxford*[134] was pleased to put into my hands, some neglected manuscripts and loose papers, to see whether any of the Lord *Bacon*'s compositions lay concealed there, that were fit to be published. Upon the perusal, I found some of them written, and others amended with his Lordship's own hand, and believed that all of them had been in the possession of Dr. *Rawley*, his Lordship's Chaplain, and faithful Editor of many of his Works. I found, that several of the Treatises had been published by him, and that others, certainly genuine, which had not, were fit to be transcribed, and so preserved, if not divulged.[135]

In fact, as we have seen, none of the physical documents that Rawley had sent to the press do seem to survive in the Harleian collection. But there are a sufficient number of duplicated items in Harley 6797 that one can see how Stephens might have come to such a conclusion. However, in one crucial respect, Stephens' conclusion regarding the provenance of the manuscripts he had identified was wrong. They had not "all" been in the possession of William Rawley; in fact, the majority had come from Thomas Meautys. Moreover, we observe that by this point Tenison's role as the source of the documents has either been forgotten, or is being suppressed.

Stephens died in 1732, by which time he had seen through the press about half of the volume that eventually appeared. What he had edited were mostly letters, apparently copied from exemplars in the same collection from which he had printed an earlier 1702 volume of Bacon *Letters*, and which are described as being "transcribed from a fragment of his Lordship's Register."[136] He also printed just a few non-epistolary items from the Harleian collection: the unfinished treatise *Of the True Greatness of*

who remains unidentified. This volume opened the door to the much later one. Of the letters in this volume that Stephens scrupulously indicated had been printed "from the original" (all but one being from Bacon to to George Villiers) I have traced the surviving manuscript of only one: Huntington Library, MS FBL 4 (Bacon to Villiers, 22 Sept. 1617; holograph, signed, sealed). On Stephens, see further Patricia Brewerton, "Stephens, Robert (1665–1732)," *ODNB*.

134 That is, Edward Harley, the second earl.

135 Stephens, "Preface," III.

136 See Stephens, "Preface," III, and Bacon, *Letters and Remains*, sig. c3r (contents-page). This fragment has not been identified, if indeed it still exists.

Britain;[137] Bacon's *Speech at the arraignment of Lord Sanquhar*;[138] and the text docketed by Bacon himself "Notes of a Speach touching the Warr" (with Spain).[139] Only one manuscript he printed, the *Charge against the Countess of Shrewsbury*, cannot now be located.[140]

As a consequence of Stephens' untimely death, the bulk of the many important non-epistolary items in the volume were therefore edited by a different person. This was John Locker (1693–1760), who appears to have been commissioned to complete the work following Stephens' death by his widow, Mary Stephens, the author of its dedication to George II.[141] Locker did his work well. He may have benefited from transcriptions made by Stephens, but it is clear from his comments and also from his editions themselves that he edited the remaining items in the volume directly from "Originals" in Harley's library.[142] The brief postscript that Locker wrote for a volume published under Stephens' name, but for which he did at least half the work, offers an interesting and plausible account of the fortunes of Bacon's manuscripts, drawing attention to the loss of the register-books, and to the destruction of the papers committed to Sir William Boswell. Nonetheless, Locker goes on, "is not impossible, that many of the genuine copies are yet in being," and adds that he himself had recently "met with a fair copy of the Discourse on *Ireland*, which corrects all others either manuscript or printed."[143] He concludes by offering his services as an editor of any further manuscripts that might come to light, and indeed his subsequent work on Bacon's manuscripts laid the foundations for Thomas Birch's collected edition of Bacon's *Works* in 1765.[144]

137 Bacon, *Letters and Remains*, sigs. 2C1r–2E4v.

138 Bacon, *Letters and Remains*, sigs. 2F1r–2F3r, printing BL, MS Harley 6854, fols. 99r–100v (not in *CELM*).

139 Bacon, *Letters and Remains*, sigs. 2G1r–2G4r, printing BL, MS Harley 7021, fols. 181r–188v (not in *CELM*).

140 Bacon, *Letters and Remains*, sigs. 2F3r–2G1r.

141 See Patricia Brewerton, "Locker, John (1693–1760)," *ODNB*.

142 See Bacon, *Letters and Remains*, 232, for a note by Locker recording the point at which Stephens' work on the volume ceased, and his own began.

143 Two contemporary presentation manuscripts that might meet this description are known to survive. One is now in the Huntington Library, MS EL 1721 (*CELM*, BcF 132), which Locker would have had to have seen in the collection of the earls of Bridgewater, where it then was. The other, National Library of Ireland, MS 2582 (*CELM*, BcF 134), entered that collection in May 1954 from W. Corran, of 137 Cavehill Road, Belfast; its life before then is unknown to me.

144 John Locker, "Postscript," to *Letters and Remains*, 518; [Thomas Birch?], "Advertisement," to *The Works of Francis Bacon*, 5 vols (London, 1765), sig. π3r.

By this point, however, the only substantial cache of Bacon's manuscripts that remained unpublished was the volume of letters that is now Lambeth MS 936. It was left to the capable Thomas Birch at last to print these, in 1763, in his volume of *Letters, Speeches, Charges, Remains*. This volume derived from the access that Birch had been granted to the Gibson manuscripts by Archbishop Thomas Secker through the intermediation of Viscount Royston.[145] With certain exceptions, Birch admitted into this volume no manuscript that had been published before.[146] Once his edition had been published, no really significant document that survived from Bacon's own posthumous archive remained to be printed. The next major contribution to the study of Baconian manuscripts was made in the mid-nineteenth century by the indefatigable James Spedding, who would make his way through incalculable quantities of then-uncatalogued State Papers in the Rolls Office at Chancery Lane, looking for documents containing Bacon's name or handwriting—in which endeavour he was conspicuously successful.[147]

Thus with Stephens' and Locker's publication of most of the treatises from the Meautys portion, and with Birch's publication of most of its letters, the three separate streams of Bacon's archival afterlife had run their course. They had all either seen their way to the press, or perished, or both. Of the Boswell portion, some had been printed, and all had perished; of the Rawley portion much had been printed, and most had perished; whereas the Meautys portion had mostly been printed, and yet had mostly survived (Fig. 1.1).

6 Conclusions

The drama of Francis Bacon's archival afterlife is therefore not entirely tragic. Though the majority of his posthumous manuscripts are lost, the loss of most of them came in the course of reaching print. Their texts, therefore, if not the original documents that contained the texts, survive. Hence one important conclusion of this inquiry is that while there are a few attested works that have

145 Thomas Birch, "To the Honourable Charles Yorke," in Bacon, *Letters, Speeches, Charges, Advices*, sig. A2v.

146 Birch, "Preface," sig. A7v.

147 Lisa Jardine and Alan Stewart, "Editing a Hero of Modern Science," in *Books and the Science in History*, ed. Marina Frasca-Spada and Nick Jardine (Cambridge: Cambridge University Press, 1996), 345–68, offer an account of Spedding's editorial labours.

disappeared completely, the great majority of his unpublished manuscript writings that Bacon had retained to his death does seem to have been preserved in one form or another.[148]

The significance of Bacon's archival afterlife is therefore both representative and exceptional at the same time. It is representative insofar as it tells an unusually clearly-defined story about the changing value accorded across our period to the manuscripts of a modern author. Until the turn of the seventeenth century, the print-shop meant destruction for holograph manuscripts. Thereafter, however, the emergence of the large-scale collector—in this case the Harleys—who valued manuscripts for their authorial associations as well as for their content, meant that an autograph document could be published without perishing. In Bacon's case, that tipping-point occurred across the period between two volumes published in 1679 and 1734. In the fifty years between Tenison's *Baconiana* (1679) and Stephens' *Letters and Remains* (1734), a new respect for the manuscript as a document of interest in its own right had demonstrably come into being. When printing manuscripts from the noble collections of the earls of Oxford, or of his Grace the Lord Archbishop of Canterbury, it was no longer appropriate to destroy them in the process.

But the story of Bacon's archival afterlife is also exceptional. It is exceptional for the intensity with which his manuscripts were sought after; for the effort that went into publishing them; and for the size of the readership that made collections of his remains marketable. Following his death, Bacon relatively quickly became a modern classic—a writer, in Thomas Birch's words, "equal to the most valuable of the ancients."[149] It was this reputation, above all, that animated the interest of archbishops and future archbishops, and which facilitated the preservation and the publication of his manuscripts.

But why was the seventeenth-century print-shop so destructive? A number of reasons suggest themselves. At a practical level, the print-shop destroyed manuscripts by virtue of being cast-off, composed, and used for correction. The notoriously low proportion of surviving manuscripts that bear witness to these processes demonstrate that printers were not in the habit of returning

---

148 Among the attested but unlocated manuscript works by Bacon are his "De arthritide" (n. 47, above); his "Ornamenta Rationalia" (n. 150, below); and an early treatise entitled "Of Active Knowledge," discussed in Richard Serjeantson, "Francis Bacon's *Valerius Terminus* and the Voyage to the *Great Instauration*," *Journal of the History of Ideas* 78 (2017), 341–68.

149 Birch, "Preface," sig. A8r.

copy to its authors or editors.[150] Given this, however, we must also assume that editors of posthumous works did not expect to have the manuscript they submitted to the print-shop returned to them. Yet the loss, without exception, of the documents that were published by Rawley and Tenison strongly implies that they sent originals, and not copies, to the press. Tenison bitterly laments the loss of the manuscript of Bacon's *Ornamenta Rationalia*, yet one infers that had he found it, and printed it, the document itself would have been freely discarded.[151] Hence, for these seventeenth-century editors, it appears that the value of a manuscript lay in the contents of its text, and that once that text was published, it superseded any specific interest the document itself possessed. Tenison valued the presence of Bacon's handwriting, but only insofar as it bore witness to the authenticity of the text it offered.

The matter of authenticity is therefore also crucial to the story of Bacon's archival afterlife. Every one of his posthumous editors provided a preface asserting the authoritative source of the documents they are printing. Gruter's first sentence stated that what he had to offer derived from copies made "from manuscript volumes which the author had accurately reviewed and variously altered."[152] In the *Resuscitatio* Rawley reminded the reader of his own role as Bacon's amanuensis, and spoke of setting forth his "true, and Genuine, writings themselves."[153] With the publication of the *Opuscula* a year later he noted that no more of Bacon's writings remained in his hands.[154] Tenison insisted that he was publishing nothing in his *Baconiana* that was not in either Bacon's or Rawley's handwriting. Again, therefore, the presence of an author's hand guaranteed authenticity; but it was an authenticity of the text rather than of the document. In no case did an editor before 1734 feel the need to verify their own good faith by appealing to the availability of documents; Tenison, rather, referred anyone "who doubteth of my Veracity in this matter" to a person: his "worthy Friend Mr. *John Rawley*," William's executor and son.[155]

---

150 Some such documents which do survive are accounted for in J. K. Moore, *Primary Materials Relating to Copy and Print in English Books of the Sixteenth and Seventeenth Centuries* (Oxford: Oxford Bibliographical Society, 1992).

151 Tenison, "An Account," 89–90. This document does indeed appear to be lost.

152 Gruter, "Lectori," sig. *5r: "ex Manuscriptis Codicibus, quos accurate recensuerat & varie emendarat author."

153 William Rawley, "To the Reader," sig. (a)4r–v.

154 Rawley, "Lectori S*alutem*," in Bacon, *Opuscula*, sig. *2r: "neque plura mihi in manibus esse."

155 Tenison, "An Account," 79.

One reason why early editors of letters, in particular, might have been coy about the location of original documents, was the persistent suspicion that the publishing of an individual's private correspondence was, as Robert Stephens put it in 1702, a "great ... violation of the secrecy that is due to them."[156] It was precisely "Cabalistick mysteries of State," after all, that was the attraction of printing the circulating volumes of manuscript letters that appeared in Bacon's *Remaines* (1648), and in the *Cabala* (1663).[157] In 1763, by contrast, Thomas Birch was confident that in his own time "the rights of historical truth" ought not to be prejudiced by a misplaced "delicacy."[158] In a parallel development, Birch now also begins his edition by seeking to establish the provenance of the manuscripts he is printing.[159]

But the most far-reaching development that this story exposes is the changing character of manuscript collections across the seventeenth century. Bacon (*d.* 1626), had a substantial archive; but it was divided among heirs he had not chosen, and its contents are now largely lost. Boswell (*d.* 1650), had a notable manuscript collection for his time; but it was lost following his death by someone—Isaac Gruter (*d.* 1680)—who should have appreciated it. Thomas Meautys' (*d.* 1649) box survived only by a "strange providence." Thomas Tenison (*d.* 1715), by contrast, was a serious clerical collector, who went so far as to found his own parochial library, and whose interests and energies united the surviving Bacon manuscripts along with many other documents. Yet he deliberately divided his own collection at his death; and the existence of both his own library at St Martins-in-the-Fields, and that of the archbishops of Canterbury at Lambeth, was not enough to prevent his choicest Baconian documents from passing, silently, into the endlessly receptive and guinea-laden hands of Robert (1661–1724) or Edward (1689–1741) Harley. It was only once Bacon's manuscripts arrived in their collection that their travels finally ended. Moreover, their permanent safety was only really assured when a growing sense of national patrimony, together with a growing national wealth, allowed that extraordinary collection to be purchased for the nation in 1753 and placed

156 Stephens, "The Preface," to Bacon, *Letters* (1702), sig. A3r. In his later "Preface" to Bacon, *Letters and Remains*, v, Stephens is similarly concerned "for his Lordship's honour."

157 G. Bedell and T. Collins, "The Stationers to the Reader," in *Cabala, sive Scrinia Sacra, Mysteries of State and Government: in Letters of Illustrious Persons and Great Ministers of State* (London, 1663), sigs. A3v–A4v.

158 Birch, "Preface," sig. A6v.

159 Birch, "Preface," sigs. A6v–A7v.

in the newly-founded British Museum "to be consulted by the curious, and for public Use to all Posterity."[160] So far, at least, that noble goal remains intact.

This story illustrates, finally, that two kinds of person, in particular, tend to make lists of an author's writings: biographers, such as John Aubrey; and editors, such as Rawley or Tenison. Seventeenth-century editors used lists to master a multiplicity of writings which, in the case of Francis Bacon, are uniquely numerous, various, and complicated. The catalogues they created have a general historical interest for what they tell us about attitudes to manuscripts and their circulation. But for any modern editor of Bacon's writings these lists now also have a specific functional necessity: they help establish what papers once existed; from what sources texts were printed; and where to look for any documents that might still exist to be discovered.

But the early-modern manuscript-list is not ultimately a merely banausic document. For there was also a further kind of person who made and collected such catalogues in the seventeenth century. These were the intelligencers, the collectors, the readers, and the thinkers, who were excited (and sometimes also disappointed) by the manuscripts they found, and who laid plans to use them.[161] For these people, the manuscript-list was a cousin to the *desiderata* or wish-list—a genre that Francis Bacon helped bequeath to the modern world.[162] For figures such as John Sparrow, Justinian Isham, or Samuel Hartlib, manuscript-lists served as tantalising promises of knowledge yet to be acquired. Even now, they retain something of that quality.[163]

---

160 The words are those of the Act authorizing the sale, quoted in *A Catalogue of the Harleian Manuscripts in the British Museum* (London, 1808), 1:1.

161 See e.g. Justinian Isham to Samuel Hartlib, 20 Jan. 1640 (*Hartlib Papers*, SUL, HP 44/2/9A): "I doubt not but to obtaine *the* Lord Veru*lams* MS. againe, to compare w*i*th that you shall send."

162 Vera Keller, *Knowledge and the Public Interest, 1575–1725* (Cambridge: Cambridge University Press, 2016), 127–66.

163 I am grateful to Angus Vine for the expertise that he has contributed to this study; to Noel Malcolm for remarks that helped stimulate its conclusions; and to the editors and anonymous referee of this volume for their judicious advice. The research leading to these results has received funding from the European Research Council under the European Union's Seventh Framework Programme (FP7/2007–2013)/ERC grant agreement no. 617391.

Appendix

Some Seventeenth-Century Lists of Bacon's Writings

1625	FRANCIS BACON, "Epistola ad Fulgentium." An account of his own printed writings.[164]
after 1626	WILLIAM RAWLEY, "Lord St Albans works printed."[165]
by 1639	WILLIAM RAWLEY/JOHN SPARROW, "A Catalogue of the extant Manuscripts of Bacon, Lord Verulam viscount St Alban."[166]
by 1639	WILLIAM RAWLEY, "a full Catalogue of Verulams MS. which Dr Raleigh hase."[167]
1652	ISAAC GRUTER sends William Rawley a catalogue (*Indeterminate*) of William Boswell's Bacon manuscripts (now lost).[168]
1655	WILLIAM RAWLEY sends Gruter a "little catalogue" (*indiculum*) of his own holdings (also lost).[169]
1657	WILLIAM RAWLEY, "A Perfect List of his Lordships true Works, both in English, and Latin." Lists works printed and about-to-be printed.[170]
after 1657	JOHN AUBREY, "His Writings." Lists Bacon's printed writings.[171]
1660	THOMAS BUSHELL, "Post-script to the Judicious Reader." Lists some of Bacon's printed writings.[172]

---

164 Bacon to Fulgenzio Micanzio [*c.*1625], in Francis Bacon, *Opuscula* (London, 1658), 172–4.

165 LPL, MS 2086, fol. 29v.

166 *Hartlib Papers*, Ephemerides, 1639 (SUL, HP 30/4/5A). The same list, not in Hartlib's hand and with very minor differences, is at BL, MS Sloane 429, fol. 243r (papers of Daniel Foote). Hartlib sent this "Catalouge" of Bacon's manuscripts to Justinian Isham in 1639 (Isham to Hartlib, 2 Sept. 1639; *Hartlib Papers*, YALE/16 [*i.e.* Beinecke Library, Osborn Collection, MS File 16792]).

167 Copy at BL, MS Sloane 629, fols. 244r–245v (papers of Daniel Foote). Mentioned in *Hartlib Papers*, Ephemerides, 1639 (SUL, HP 30/4/27B). Hartlib also sent John, Baron Robartes, a "Register of my Lord Verulams M.S." on 10 March 1640, which will have been either this or the previous item (*Hartlib Papers*, SUL, HP 7/55/1A).

168 Tenison, *Baconiana*, 222 (Gruter to Rawley, 29 May 1652): "Mitto indicem eorum, quæ ex *Boswelliani* Musei scriniis chartaceis penes me extant."

169 Tenison, *Baconiana*, 231 (Gruter to Rawley, 20 Mar. 1655): "Et quamvis pauxillum erat quod præter gratias pro indiculo reponerem ..."

170 Francis Bacon, *Resuscitatio*, ed. William Rawley (London, 1657), sig. 3Q4r.

171 OB, MS Aubrey 6, fol. 74, printed in Aubrey, *Brief Lives*, 1:73–74 (life of Bacon).

172 Thomas Bushell, *An Extract ... of his Late Abridgement* (London, 1660), 21.

1679 THOMAS TENISON, "An Account of All the Lord Bacon's Works." Identifies printed, manuscript, and lost works; also mentions possessing "three catalogues" of Bacon's "unpublished Papers" by RAWLEY.[173]

1682 THOMAS TENISON, "A Collection of Tracts, Speeches, & Letters written by Francis Bacon Baron of Verulam & Lord Saint Alban." Records the manuscripts Tenison found in Meautys' chest.[174]

---

173 Thomas Tenison, *Baconiana* (London, 1679), 21–104 (qu. at 89).

174 OB, MS Tanner 35, fols. 145r–146r.

CHAPTER 2

Scarlet Letters: Sir Theodore de Mayerne and the Early Stuart Color World in the Royal Society

Vera Keller

1 Introduction: Archives in Afterlife

Death often stymied the pursuit of long-term research goals in the early modern experimental study of nature. As Dániel Margócsy has argued, in contrast to the impersonal institutions that would later make individual participants replaceable parts of large-scale research projects, in the early modern period, "there was no established method to replace a human correspondent. Death put an end to communication and to scientific collaborations."[1] This was not always the case. Experiment could survive death within the papers that were left behind. *Archival Afterlives* explores the myriad ways that later experimentalists sought to interrogate, collaborate with, or spy upon the deceased through their papers. Margócsy is right, however, to stress the personal nature of the human correspondents making up early modern scientific networks, as well as the methodological pluralism of the time. As Richard Yeo has discussed, this was a period of fertile creativity concerning methods of archival practice, particularly for experimental records.[2] Those investigating the experimental papers of the recent past in the early modern period thus faced peculiar quandaries, as well as opportunities. They had to engage not just with the content of an experiment, but with a personality. The form in which past experimental content appeared might be very different from their own, and it would be perhaps highly inflected by the social world and idiosyncracies of the deceased personality. Finally, in order to enter into and trace the course

1 Dániel Margócsy, "A long history of breakdowns: A historiographical review," *Social Studies of Science*, 47, 3 (2017): 307–325 (on p. 315). I would like to thank Rupert Baker and Katherine Marshall of the Royal Society Archive for assistance with manuscripts and permission to reproduce images. I would also like to thank Michael Hunter for advice on this chapter, as well as for pointing me toward the additional Mayerne manuscript transcriptions in the Hooke papers, and my fellow editors, Anna Marie Roos and Elizabeth Yale.

2 Richard Yeo, *Notebooks, English Virtuosi, and Early Modern Science* (Chicago: University of Chicago Press, 2014).

 | DOI 10.1163/9789004324305_004

of previous research, the reader of an archive would have to, as it were, reanimate dead social networks, following the threads of past conversations and perhaps picking them up again in the present. The personal nature of early modern scientific networks made passage through death challenging. It also allowed archives a robust afterlife, one rich not just in information, but in the personalities, aesthetics, and diverse methodologies of the past.

Scholars of material culture have long been interested in afterlives, from the "fascination with ruins" to the study of wear and tear. Textiles, in particular, because they "often function in intimate contact with the human body—garments, coverings, bandages, shrouds" are "exemplary 'biographical' objects," telling poignant stories of "people, events, and passing time in their physical dissolution."[3] By contrast, archives in their very nature aim for pristine preservation, legibility, and credence. The impersonal nature of government documents and large institutions today do not make archives obvious candidates for bringing us in contact with lives once lived. Some exceptional archives contain biographical objects due to their peculiarly personal, emotional or bodily contents, such as autographs, orphans' tokens, or the Leigh Hunt hair collection at the Ransom Center.[4] Still rarer are archival files that themselves serve as evocative material objects.[5] Recently, however, apparent registers of fact in the early modern period, such as account and commonplace books, have been reconsidered as more akin to traditional egodocuments.[6] The very evocation of personality and the passage of time in an archive, complete with idiosyncratic methods of organization, could serve a function for later users of the archive, above and beyond the information that it contained.

The archival afterlife explored in this chapter offers, I argue, a case in point. This chapter identifies a previously unknown cache of papers on inks and dyes dating from 1618 to 1650 (with some undated pieces likely older),

3 Victoria Kelley, "Time, Wear and Maintenance: The Afterlife of Things," *Writing Material Culture History*, Anne Gerritsen and Giorgio Riello, eds. (London: Bloomsbury Academic, 2015), 191–7.

4 John Styles, "Objects of Emotion: The London Foundling Hospital Tokens, 1740–60," *Writing Material Culture History*, Anne Gerritsen and Giorgio Riello, eds. (London: Bloomsbury Academic, 2015), 165–172.

5 A rare example is the Archive Gallery at the Haus der Kunst in Munich, which opened in 2014.

6 Adam Smyth, *Autobiography in early modern England* (Cambridge: Cambridge University Press, 2010) and Jason Scott-Warren, "Early Modern Bookkeeping and Life-Writing Revisited: Accounting for Richard Stonley," *Past & Present*, 230, Issue Supplement 11 (2016): 151–170.

in the archives of the Royal Society originally collected by Sir Theodore de Mayerne (1573–1655), the Huguenot royal physician in the early Stuart court (Fig. 2.1).[7] The papers were brought to the Royal Society by Mayerne's godson and fellow Huguenot royal physician, Sir Theodore de Vaux (1628–1694), a year after his election to the Society in 1665. The story of the transmission of the Mayerne papers offers a veritable palimpsest of archival afterlives, since Mayerne's papers themselves contain papers drawn from a wide range of other informants, some of them holographs in French, English, German and Latin.[8] This group of papers offers a more immediate view of Mayerne's interaction with his informants and of his own experimental method than do the formal volumes he drew up, with the help of amanuenses, on the basis of such documents.[9] Mayerne's experimental method is self-consciously innovative and idiosyncratic, everywhere marked with his possessive monogram claiming credit for his ideas. He saturated his papers with the color inks that were themselves the objects of his research. As the remains of a practice Mayerne doggedly pursued on a daily basis for years within a domestic setting, his papers, continually re-read and cross-referenced by him, can be compared to a favorite piece of clothing interacting closely with the body of its wearer. The personal nature of the archive, I argue, at times served as an obstacle but was also part of its allure to later experimentalists.

Despite Mayerne's massive collections and their prestigious resting places in the collections of Sloane, the Royal Society, Oxford, Cambridge, and the Royal College of Physicians, the significance of his papers to later naturalists has

---

7 Royal Society, ClP/3i/27, ClP/3i/28, ClP/3i/29, ClP/3i/30, ClP/3i/31, ClP/3i/33, ClP/3i/34, ClP/3i/35,ClP/3i/36, ClP/3i/37, ClP/3i/38, ClP/3i/39, ClP/3i/40. ClP/3i/32, on treating mouse skins, doesn't seem to be related to dyes. It was a technique (in German) taught to the original owner of the paper by Johann Friedrich Fuchs of Hagenau on June 7, 1618. See also ClP/24/80, ClP/24/81 and ClP/24/82, and thanks to Michael Hunter for pointing out the materials in volume 24.

8 All translations are my own. Transcriptions are not modernized and follow the source's orthography.

9 Unlike the more formal compilations of the many Mayerne volumes now in the Sloane collection (on these, see Alison Walker's chapter in this volume), the papers in the Royal Society were collected by Mayerne at different times, and not originally as part of a specific collection devoted to color research. ClP/3i/34 mixes an autograph manuscript on wine preservation from "M. Peter Bodinus Germanus" with the dyes of Jan Davidszoon and de La Noy. Bodin's more relevant recipes for cinnabar and whitewash are now ClP/3i/33. Mayerne's personal correspondence was largely lost or destroyed. Hugh Trevor-Roper, *Europe's Physician: The Various Life of Sir Theodore de Mayerne*, Blair Worden, ed. (New Haven: Yale University Press, 2006), 372–4.

FIGURE 2.1 Sir Theodore Turquet de Mayerne. *Attributed to Paul van Somer, probably after 1625.*

faded from recent historiography of London science. Mayerne, for instance, is only mentioned once in passing in Harkness's *Jewel House,* an omission all the stranger given that Paul L'Obel (son of Matthias de l'Obel, one of Harkness's central figures) married Mayerne's sister.[10] Meanwhile, the significance of the

10 Although many individual figures are well known, the culture of early Stuart experimentalism remains underexplored even today. Mordechai Feingold regretted in *The Mathematician's Apprenticeship: Science, Universities and Society in England, 1560–1640* (Cambridge: Cambridge University Press 1984), 198 that "no study of James's interest in, and patronage of, science" exists, noting that "the pursuit of the mechanically curious reached new heights" under the Stuarts. This remains the case. Stephen Pumfrey and Frances Dawbarn attempted a partial survey in "Science and Patronage in England, 1570–1625: A Preliminary Study," *History of Science* 42 (2004): 137–188, with unduly pessimistic findings. Harkness, notably, skipped over the early Stuart period to argue that Elizabethan science formed the basis for Restoration experimentalism. Deborah Harkness, *The Jewel House of Art and Nature: Elizabethan London and the Social Foundations of the Scientific Revolution* (New Haven, Conn.: Yale University Press, 2007). Recently, this orientation

similar project of Samuel Hartlib for many early Fellows of the Royal Society has attracted increasing attention since Hugh Trevor-Roper's 1960 essay, "The Three Foreigners" and Charles Webster's 1976 *Great Instauration*.[11] In his magisterial biography of Mayerne (published posthumously in 2006), Hugh Trevor-Roper ends on a melancholic note, seeing Mayerne's "Paracelsian and Hermetic ideas" "discredited" in the "scientific revolution," and concluding that the "fate of Mayerne's writings is likewise a melancholy tale," and "Mayerne's mental world had passed."[12] In his 1960 essay, "The Three Foreigners," Trevor-Roper likewise cast the efforts of the similar (although rival) figure of Samuel Hartlib as evaporating in the advent of the Royal Society.[13] Since then Hartlib, and his contributions to the aims and practices of the nascent Royal Society have received ever more attention. Oddly, the same has not been true for Mayerne. It was in part, I argue, because Mayerne's papers did point to a world that had passed, that is, to the experimental culture of the early Stuart courtier, that they became a precious resource to the Royal Society, with an afterlife in the Society's archive.

This chapter begins with exploring the reasons behind Mayerne's assembly of his collections. It then turns to the retrospective valences of early Stuart color research among Hartlib's associates during the Interregnum and following the Restoration. It compares Mayerne's project to the History of Trades program and sets the story of de Vaux's gift of the Mayerne manuscripts in the context of the Royal Society's turn toward artistic techniques and Robert Hooke's investigations in the History of Trades. Finally, it will suggest that the value of Mayerne's manuscripts was threefold; they offered access to the techniques of artisans, to Mayerne's own practices of experimental note-taking and speculation, and to the lost color world visibly suffusing his papers.[14]

---

toward the Tudors has begun to shift. Eric Ash, whose prior work stressed Elizabethan expertise, has turned to early Stuart projecting. Eric Ash, *Power, Knowledge and Expertise in Elizabethan England* (Baltimore: Johns Hopkins University Press, 2004). Eric Ash, *The Draining of the Fens: Projectors, Popular Politics and State Building in Early Modern England* (Baltimore: Johns Hopkins University Press, 2017). See also William Cavert, *The Smoke of London: Energy and Environment in the Early Modern City* (Cambridge: Cambridge University Press, 2016). I hope to contribute to this trend in a monograph tentatively entitled *Interlopers: Cornelis Drebbel (1572–1633) and Early Stuart Science on the World Stage.*

11 Mark Greengrass, "Three Foreigners: The Philosophers of the Puritan Revolution," in *Hugh Trevor-Roper: The Historian*, Blair Worden, ed. (London: Tauris, 2016), 85–98.

12 Trevor-Roper, *Europe's Physician*, 367–8.

13 Trevor-Roper points out the similarities between Mayerne and Hartlib in *Europe's Physician*, 338.

14 For the term "color world," see Tawrin Baker, Sven Dupré, Sachiko Kusukawa and Karin Leonhard, "Introduction. Early Modern Color Worlds," *Early Science and Medicine* 20 (2015): 289–307.

2 Mayerne and the Social Status of Craft

Among the many manuscripts Mayerne left on his death in 1655, the best known today is indubitably Sloane 2052, "*Pictoria, Sculptoria & quae subalternarum artium*," the so-called "Mayerne manuscript," published in multiple versions.[15] Art historians have long mined this manuscript for techniques of the old masters that would otherwise be unknown. The fact that this manuscript still serves as such a resource suggests the unprecedented degree of access to artistic techniques Mayerne succeeded in gaining, and his success in preserving those techniques in a lasting and accessible form. The fame of "the Mayerne manuscript," however, has obscured the importance of the many other scattered Mayerne manuscripts on pigments, inks and dyes, such as Sloane 1990, Sloane 2079 and Sloane 3423, in addition to Mayerne's numerous medical and chymical collections.[16] As Trevor-Roper wrote, although "art historians mostly confine themselves to the 'Mayerne manuscript,' it is only the most striking of the many papers of Mayerne that reveal his artistic interests."[17] These include the Mayerne papers in the Royal Society, which have not been identified before and were unknown to Trevor-Roper himself. Several of the informants of the Mayerne papers in the Royal Society, such as Nicolas Briot, Jean Petitot, Louis le Myre, Pierre Antoine Bourdin, Mark Anthony of Brussels, and Nicholas Lanier, were also informants of the famous "Mayerne manuscript," Sloane 2052.

The reasons why Mayerne collected artisanal techniques so methodically in the first place have eluded Mayerne scholars. Both Trevor-Roper and Karin Leonhard raised this question, suggesting that perhaps the answer might lie in Mayerne's chymical interests, since such techniques also supplied him with materials for chymical processes.[18] Romana Sammern suggested that art writing by physicians of Mayerne's generation served to

15 J. A. van de Graaf, ed., *Het de Mayerne Manuscript als bron voor de schildertechniek van de barok. British Museum. Sloane 2052* (Mijdrecht: Verweij, 1968). Donald C. Fels, trans., *Lost Secrets of Flemish Painting: Including the First Complete English Translation of the De Mayerne Manuscrip*t, B.M. Sloane 2052 (Hillsville, Va.: Alchemist, 2001).

16 On Sloane 1990, see A. E. Werner, "A 'New' de Mayerne Manuscript," *Studies in Conservation*, 9,4 (1964): 130–134.

17 Trevor-Roper, *Europe's Physician*, 346.

18 Karin Leonhard, "Painted Gems. The Color Worlds of Portrait Miniature Painting in Sixteenth- and Seventeenth- Century Britain," *Early Science and Medicine* 20 (2015): 428–457.

distinguish them as virtuosi from medical empirics.[19] While both these suggestions are doubtless true, neither provides a full explanation for the extent and duration of Mayerne's project. Artistic and alchemical techniques often mingled in books of secrets, but Mayerne's artistic collections far surpassed the norm.[20] Someone of Mayerne's erudition and family background did not require such extensive collections to confirm his status. Furthermore, Mayerne's interests encompassed not only techniques that we recognize as artistic today, but also practical skills that may not have served similarly as a marker of social elevation.

Trevor-Roper speculated that a Vasari translation by Mayerne's father, the political writer and would-be reformer Louis Turquet (1533/4-1618), may have stimulated Mayerne's artistic interests. It may also point to the notions of socioeconomic reform animating the craft research of both Louis Turquet and Mayerne. Louis Turquet did not translate Vasari's artistic biographies, but his introductory material dealing with materials and craft techniques (the draft, with Theodore's marginalia, is now Sloane 2057; the presentation copy is Sloane 2001).[21] In his dedicatory letter to François de Roaldès, Louis Turquet defended the political benefits of the arts, which taught man how "to live well, comfortably, and happily."[22] Like others of the time, Louis Turquet championed the nobility of the mechanical arts and castigated those who considered the once honorable title, "mechanical," as the "most vile and abject epithet one could imagine."[23] Turquet was among those writing on the reason of state and "good police" (and one of the first to employ the term "political economy"). He also published works emphasizing manufactures within the well-run polity and promoted in particular silk manufacture, an industry that Louis Turquet's father, Étienne Turquet, had established in Lyon.[24]

19 Romana Sammern, "Red, White and Black: Colors of Beauty, Tints of Health and Cosmetic Materials in Early Modern English Art Writing," *Early Science and Medicine* 20 (2015): 397–427.

20 Compare, for instance, Mayerne's extensive experimentation on dyes in Sloane 3423 with the more typical collection of medical, chymical, and artisanal receipts in Sloane 3426, 1-28v (excerpted by Mayerne in London, November, 1626, "ex Miscellaneis D. [Nicholas] Briot") or with the receipts of Sloane 2079.

21 Trevor-Roper, *Europe's Physician,* 18 and 347.

22 Sloane 2057. "de l'homme a bien, commodement, et heuresement vivre."

23 Ibid. "aujourdhuy attribué pour le plus vil et abjet epithete qu'on scauroit penser." For similar defenses of the nobility of the mechanical arts by Agricola and Guidobaldo del Monte, see Paolo Rossi, *I filosofi e le macchine, 1400–1700* (Milan: Feltrinelli, 2002), 72–75.

24 Mark Greengrass, "The Calvinist and the Chancellor: The Mental World of Louis Turquet de Mayerne," *Francia-Forschungen zur westeuropäischen Geschichte* 34,2

Mayerne's interests in even the lowliest of the mechanical crafts, such as tallow chandling, are perhaps not as difficult to explain as has been supposed, given period attempts to elevate the status of the mechanical arts for the benefit of the polity.[25] Mayerne may have left us a signally methodical collection, yet he was far from alone among the early Stuart courtiers developing profitable consumer trades.[26] Trevor-Roper himself pointed out Mayerne's many entrepreneurial projects, although he interpreted Mayerne's experimentation as a retreat into a private world, following a period of disappointment with the possibilities for international Calvinism ca. 1620. Given the views current in Mayerne's context, including those voiced by his father, concerning the political importance of fostering profitable industries, it is possible to see his experiments rather as public-minded and part of a wider culture of elite projectors.[27]

3 Mayerne and the Early Stuart Fashion for Experiment

Mayerne participated in a generation of methodologically creative early Stuart experimentalists that has been occluded by sociopolitical disorder, the later dominance of Sir Francis Bacon, and by changing mores in the early Royal Society regarding the long-standing relationship between courtly experimentation and craft knowledge. Like other social elites of his generation, Mayerne did not exhibit the same squeamishness regarding profitable crafts as some Fellows of the Royal Society later would. In his published discussion of colors, Robert Boyle purposefully omitted "the Lucriferous practise of Trades-men about colours; as the ways of making Pigments, of Bleanching wax, of dying Scarlet, &c. though to divers of them I be not a stranger, and of some I have myself made Tryall."[28] William Petty likewise affected to leave the "Fixing of colours for Use" to "more

(2007): 1–23. Germano Maifreda, *From Oikonomia to Political Economy: Constructing Economic Knowledge from the Renaissance to the Scientific Revolution* (London: Routledge, 2016), 179.

25 For tallow, ClP/3i/26.

26 Linda Levy Peck, *Consuming Splendor: Society and Culture in Seventeenth-century England* (Cambridge: Cambridge University Press, 2005).

27 Trevor-Roper, *Europe's Physician*, 348. Vera Keller and Ted McCormick, "Towards a History of Projects," *Early Science and Medicine*, 21, 5 (2016): 423–444.

28 Robert Boyle, *The Works of Robert Boyle; Vol. 4, Colours* and *Cold, 1664–5*, ed. Michael Hunter and Edward B. Davis (London: Pickering & Chatto, 1999), 7–8.

experienced persons."[29] In manuscript notes to his *Experimental Philosophy*, Henry Power defended his color research, noting, "Though our designe here be the Employment of Knowledge & not of Trades, yet these Experiments of Colours, may bee not only of Speculative but practicall use & may not only advantage the Contemplative Naturalist, but also ennoble the Painters art & dyers Trade. Especially if we understand the Nature of Sulphureous Acid & Alkalizate Salts or Liquors."[30]

These later views have made Mayerne's pursuit of profitable trades puzzling in retrospect. Mayerne's motivations for experimenting with dyes in particular baffled Trevor-Roper since Mayerne "was not himself an artist or a craftsman: he did not intend to exercise the arts which he studies. It is difficult to detect an economic motive in this case…."[31] A few of Mayerne's experiments were destined to serve as gifts, as in a pair of stockings dyed for "la petite Colladon (presumably an offspring of Mayerne's protegé, Jean Colladon who had married his favorite niece, Aimée)."[32] Most, though, aimed at the most lucrative dyes. Despite his wealth and social standing, Mayerne pursued profit unabashedly, a phenomenon only in need of explanation from the later perspective of the Royal Society.

Mayerne moved easily between multiple domains of knowledge production concerning color, including dyers, painters, and heralds, who recorded the significance of and recipes for inks, dyes and pigments.[33] The rich access to multi-lingual and international artisans his papers offered would have made them a precious resource to the Royal Society's History of Trades program, which struggled to collect such firsthand accounts from practitioners.[34] Mayerne also drew upon a world of early Stuart gentlemen

29 William Petty, "Some Observations, Touching Colours, in Order to the Increase of Dyes, and the Fixation of Colours," *Philosophical Transactions* 6 (1671): 2132–2136 (on p. 2134).

30 Henry Power, *Experimental Philosophy* (London: Martin, 1664), 74, ms. notes by the author, British Library, General Reference Collection 537.h.1. Power's notes on the various colors produced by calcining metals on glass are at p. 45.

31 Trevor-Roper, *Europe's Physician*, 347.

32 Sloane 3423, 15r.

33 eg. British Library Stowe 680 and Additional MS 6284.

34 Early Fellows of the Royal Society such as Robert Boyle and Nehemiah Grew purposefully sought access to the color world of dyers, using their processes as a central platform from which to analyse the saline chymistry of color. Anna Marie Roos, "The Saline Chymistry of Color in Seventeenth-Century English Natural History," *Early Science and Medicine* 20 (2015): 562–588. Michael Hunter, *Science and Society*

experimentalists to which many of the Royal Society Fellows were intimately related but which, like the color world of artisanal informants, had become difficult of access in the Restoration.[35] He offered a firsthand view of the culture of early Stuart curious gentlemen (known variously as *virtuosi, liefhebbers,* or *amateurs*) encouraged by Henry Peacham (1578–1644?), who urged the student of heraldry "to undertake more Philosophicall and particular examination of the causes of colours."[36] While Mayerne's papers on coloration do not preserve material relics like the fabled Garter jewel Charles I passed off on the scaffold, they do contain a paper, discussed further below, with the once famous and otherwise lost technique of staining images into artificial agates developed by Sir Edmund Bacon (1566–1649). The paper, complete with Mayerne's interview with Bacon and further speculations on the technique, represents a remnant from a once vivid circle of courtier-researchers into color techniques, a circle later lyrically recalled by figures such as John Beale (1608–1683), a future fellow of the Royal Society.

This view of early Stuart *virtuoso* experimenters, made distant not so much through passing time as by intervening political fortune, afforded Restoration gentlemen philosophers elite models of experimental practice above and beyond the papers' role as a repository of matters of fact. Mayerne's papers explicitly served as one possible model for experiment in Royal Society discussions. His literary techniques and apparatus were carefully copied in the

---

in Restoration England (Cambridge: Cambridge University Press, 1981), 87–112. Kathleen H. Ochs, "The Royal Society of London's History of Trades Programme: An Early Episode in Applied Science," *Notes and Records of the Royal Society of London* 39, 2 (1985): 129–158. William Eamon, *Science and the Secrets of Nature: Books of Secrets in Medieval and Early Modern Culture* (Princeton University Press: Princeton, 1994), 343–6.

35 Harkness, *The Jewel House*. Mayerne does not appear at all in Yeo, *Notebooks, English Virtuosi, and Early Modern Science*. Charles Webster makes slight references to him in *The Great Instauration: Science, Medicine and Reform, 1626–1660* (Oxford: Peter Lang, 2002), 254, 273, and 317, although Webster does discuss the research into dyes of Drebbel, the Küfflers, Petty, Boyle, and Beale (at 388–389). Despite his importance for Hooke, Mayerne does not appear in, for instance, Matthew Hunter, *Wicked Intelligence: Visual Art and the Science of Experiment in Restoration London* (Chicago: University of Chicago Press, 2013). Trevor-Roper discusses the fortunes of Mayerne's papers in "Appendix C: Mayerne's Papers," *Europe's Physician*, 372–4.

36 Anne Geoffroy, "English Perceptions and Representations of Venetian Chromatic Variations," *E-rea* [online], 12.2 (2015): URL: http://erea.revues.org/4509; DOI: 10.4000/erea.4509.

English translations of these papers ordered by the Society, down to Mayerne's monograms (marking his own ideas or innovations), manicules, and marginalia. These papers were prized for Mayerne's studied techniques and majestic style that shaped the evanescent particulars of experimental practice into an enduring and regal body of scientific knowledge.

Mayerne lent stature to his manuscripts through their physical form. Sloane 3423, for instance, appears in a striking array of brilliant purple, orange, and red inks, which he was also producing himself in concert with his research on dyes. Mayerne's other manuscripts as well as his papers in the Royal Society include several ink recipes and discussions.[37] In Mayerne's practice the relationship between dye and ink was particularly close, since Mayerne often tested colors on paper.[38] Mayerne's colors suffused all of his records; in her chapter in this volume, Walker also notes Mayerne's use of colorful ink in his annotations in printed volumes in the Sloane collection. Thus, while Mayerne's experimental records were functional, as in the case of Samuel Hartlib's papers discussed by Carol Pal in this volume, they were a far cry from Hartlib's monotone, hurried notes. In his careful use of color and layout, Mayerne collects and memorializes his experimental practice both for immediate use and for long-term preservation. Decades later, fellows of the Royal Society would note the form, as well as the content, of Mayerne's papers.

This argument, that early Restoration gentlemen philosophers would have found the style of early Stuart experimentation of interest, might appear to parrot the arguments made by Shapin and Schaffer in *Leviathan and the Air-pump* concerning credibility and social status, but it differs in important respects. First, the disinterestedness granted by social status according to Shapin and Schaffer inherently depends on a denial of financial interest in the pursuit of knowledge. This was not the case for Mayerne nor for many of his fellow early Stuart experimenters. Early Stuart experimenters belonged to a notoriously corrupt profiteering court culture, and association with that

37 eg. Sloane 1512, 187v. "Pour faire l'encre rouge," with brasilwood and alum, "ils sera fort rouge et luisant & sanguin, et fort agreeable à voir."

38 For example, in his *Experimenta tinctoria*, a now lost volume transcribed by Robert Hooke in ClP/24/81, Mayerne experiments with making colors from poppies. He notes that "the liquor spread upon paper made" a very glorious purple, like that which comes from logwood ("la liquor estendue sur du papier a faict un pourpres tres glorieux, comme celuy qui vient de Logwood"). Another poppy color "was a dark red, which when laid on paper with a brush was violet and not at all red (La couleur a este dun rouge obscur laquelle couchée sur du papier avec un pinceau a este violette nullement rouge)."

culture was no guarantee of propriety or disinterest. My account of the personal nature of the experimental archive and the social networks it can reanimate serves to emphasize the role of courtly magnificence, social animus and rivalries in experimental practice rather than the qualities of the ideal gentleman-philosopher according to Shapin and Schaffer, namely, modesty and credibility. As a result of such differing ideals and mores, the aesthetics highlighted here differ too. In line with the resplendent fashions of the early Stuart court, Mayerne's experimental notes are colorful and dashing. By contrast, Shapin and Schaffer argue that the aesthetic Boyle aimed for, in his published engravings such as that of the air-pump, was one of unadorned "virtual witnessing," that is, an aesthetic that aims to erase its own existence as a visual representation, akin to Boyle's "professedly 'naked way of writing.'"[39]

More recently, Michael Hunter has emphasized how individuals such as John Evelyn forefronted the artistry of the Society.[40] It was the latter interest that spurred the Society's investigation into Mayerne's papers on color. In a society already attuned to reading symbolic and heraldic codes, color became highly politicized leading up to, during, and after the Civil War.[41] Mayerne's brightly hued courtly world was swept away by the duller tones of the Interregnum. In later memory, the former may have shone all the more brightly, since nostalgia so heavily tinged Restoration elite memories of the time before the troubles. As Edward Hyde, first Earl of Clarendon (1609–1674) wrote in his influential *History of the Rebellion*, the reign of King Charles before the long Parliament (1640–1660) enjoyed "the fullest measure of Felicity, that any People in any Age, for so long time together have been bless'd with; to the wonder, and envy of all the other parts of Christendom." The happiness of his reign supposedly surpassed that of either Elizabeth or James. The "Kingdoms we now lament, were alone look'd upon as the Garden of the World ..." according to Hyde.[42]

39 Steven Shapin and Simon Schaffer do not mention archives in *Leviathan and the Air-pump: Hobbes, Boyle, and the Experimental Life* (Princeton: Princeton University Press, 1985), p. 66, discussed also in Steven Shapin, "Pump and Circumstance: Robert Boyle's Literary Technology," *Social Studies of Science* 14,4 (1984): 481–520.

40 Michael Hunter, *The Image of Restoration Science: The Frontispiece to Thomas Sprat's* History of the Royal Society *(1667)* (London: Routledge, 2017). Image production in the Royal Society has garnered great interest in recent years, notably in the current project, *Making Visible: The Visual and Graphic Practices of the Early Royal Society*, at the Centre for Research in the Arts, Social Sciences, and Humanities, Cambridge.

41 Sophie Chiari, "General Introduction: '*Chamelion* like' England," *E-rea* [online], 12.2 (2015), URL: http://erea.revues.org/4331.

42 Edward Hyde, *The History of the Rebellion and Civil Wars in England, Begun in the Year 1641*, Vol. 1, Part 1 (Oxford: Sheldonian Theater, 1707), 74–75.

Mayerne's papers brought back to life a lost early Stuart color world and a manner of courtly experimentation more generally to a generation eager to dress experimental philosophy in the majestic and enduring garb befitting its status.

Sloane 3423, *Experiments & operations en matiere de teincture faites par moy* [monogram] *1639*, a series of dated experiments Mayerne performed from 1639–1650 with the assistance of practitioners, Thomas Fletcher and his wife, as well as Mayerne's Flemish chambermaid Elisabeth, illustrates the extent, duration and social setting of Mayerne's quests. In 1639, Mayerne had moved to a small riverfront property in Chelsea that had been the old farmhouse on the much grander estate of the royal favorite and arbiter of fashion, George Villiers, the Duke of Buckingham (1592–1628). Buckingham's daughter, Mary Villiers, Duchess of Richmond and Lennox (1622–1685), at first received permission from Parliament to reside in the main estate in order to be treated by Mayerne. Buckingham House was then confiscated by Parliament in 1648 and granted to Bulstrode Whitelocke, the Commonwealth's Librarian and Commissioner of the Great Seal, who was friendly with Mayerne despite the latter's royal appointment.[43] From this elite western suburb, Mayerne surreptitiously investigated the lucrative techniques deployed in a very different social setting to the east of London by the Küffler brothers of Cologne.

Residents of Stratford-Langton, the Küffler brothers were associates and sons-in-law of Cornelis Drebbel (1572–1633), Chief Engineer of the Ordnance Office before his death. The Küfflers established a dyework for the famous scarlet dye (known as Bow-dye after Stratford-Bow) invented by Drebbel. Like other lucrative dyes Mayerne researched, such as calico, scarlet was a notoriously difficult dye to reproduce, especially for silk.[44] Variations of water supply and natural ingredients, proportions of mordant and cochineal, and various options for heating and timing of the operation made dyeing silk with cochineal "a classic example of the problem of controlling a dynamic system with many variables."[45] Of the four Küffler brothers (Abraham, Dr. Johann Sibbert, Aegidius or Gilles, and Jakob), the first three were involved in the scarlet dye business, and the first two were also inventors and chymists.[46]

---

43 "Key to Kip's View," *Survey of London: Vol. 4, Chelsea, Pt. II* (London: County Council, 1913), 18–27. Trevor-Roper, *Europe's Physician*, 420.

44 On the obstacles to achieving calico printing in England, see P. C. Floud, "The Origins of English Calico Printing," *Journal of the Society of Dyers and Colourists* 5 (1960): 275–281.

45 Valery Golikov, "The Technology of Silk Dyeing by Cochineal. II. The Experimental Investigation of the Influences of Types and Concentrations of Cations," *Dyes in History and Archaeology* 16/17 (2001): 10–20; 11–12.

46 Francis Mauritius Jaeger, *Cornelis Drebbel en zijne Tijdgenooten* (Groningen: Noordhoff, 1922), 50.

Mayerne's informants did not blush to query both Abraham and his brother, the physician Johann Sibbert, about their profits, as in:

> Kefl. dyes a piece of cloth for 25 lb; his brother said 16 lb with Cocheneel.
> The ordinary dyers dye a piece of cloth with powder of graine of kermes for 35 lb.
> He makes profit in every piece of cloath, 10 lb sterl.
> The pound of scarlet grains costs him 6 shill. To 8 for ye best.[47]

Mayerne proved a connoisseur of color, continually comparing his successes to the Küfflers' products, always with an eye to their market value, as in, "the scarlet being dried its color was very rich, and according to the merchant better and of a greater value than that of Keffler [Küffler]," and "A scarlet as beautiful as that of Keffler [Küffler], the comparison having been made."[48] One "incomparable crimson" Mayerne achieved "kept its luster" after the silk dried, and according to Fletcher, was worth 30 shillings for a pound of the dye alone.[49]

Dr. Küffler, or Keffler as Mayerne called him, figures centrally throughout Mayerne's writing on dyes.[50] Mayerne sought out informants who could follow his trail through London. He noted how Nicolas Briot (1579–1646), the alchemist and chief engraver to the Mint, had seen a number of glass retorts in the glass shop, where the glass-man informed him "that Keffler used 5 or 6 of them every week for his necessarys about the tincture."[51] The factor of a merchant who employed Küffler every year to dye several pieces of scarlet said that after Küffler "had dyed it as well as he could, he put something on the cloth that gave it that glorious yellow appearance so desired by everyone (... cette oeil glorieuse jaune tant recherché par tout le mond)."[52] As he typically did, Mayerne added his own speculation to this information, surmising that Küffler's additive was arsenic and offering as evidence the fact that gold or silver brocade

47 ClP/3i/30.

48 Sloane 3423, 1 and 9v. "Estant [Escarlate] seiche la couleur a esté riche, & au rapport des marchande meilleur et de plus grand prix que celle du Keffler " and "La couleur est venue suffisament haulte. Escarlatte aussi belle que celle du Keffler, comparaisan en ayant este faitte. Nacarat, coulur de feu [underlined with an NB in pencil]."

49 ClP/24/82 at # 79.

50 For instance, Mayerne refers to Küffler at least fifteen times throughout ClP/24/80, 81 and 82.

51 ClP/24/80/175. On Nicolas Briot, see Trevor-Roper, *Europe's Physician*, 63–4.

52 ClP/24/81 at #57.

touching the Küfflers' cloth would turn black. He noted three different ways to try adding arsenic.

The Küffler dyes suggested the commercial profits that could accrue if superior chymical learning were brought to bear on the craft world of dyes. Robert Boyle prized scarlet not as a desirable artisanal secret, but as an example of what individuals who were *not* artisans, like "Cornelius Drebble, who was not bred a dyer, nor other tradesman" could achieve.[53] Mayerne too considered Küffler's successes the result of a more philosophical approach ignored by the common sort of dyer. The scarlet dye was prepared with an innovative tin mordant that rendered cochineal dye redder than the traditional alum mordant. The mordant was also added to the cloth before the dye, which Mayerne believed was an unusual approach. Mayerne noted how Keffler told Mr. Briot "that the secret of his tincture consisted in the preparation of the stuff the which he boyled with some ingredient which changed not the colour but it remained white but by its preparation acquires that force that when you put it in the tinctures it drawes all the colour of the cocheneel in such manner that the water remaining white or rather a little reddish soe that you loose nothing of the tincture but all goes into the cloth."[54] While the "ordinary dyers beat the cochineel in powder and make it boyle in the liquor," "Fletcher [the expert dyer who assisted Mayerne] believes that it is better to draw all the tincture of the cochineel intire," and thus "dye the stuff prepared," "beleiving Keffler does soe." Mayerne approved of this manner of dyeing as "most philosophicall & proper."[55] Mayerne also appreciated Küffler's thriftiness. Rather than grinding his cochineal, Küffler used it whole. After drawing the "most glorious colour" for his cloth, he saved the water, adding "a little new cocheneel," and used it to dye stockings, sold for 7 shillings.[56]

53 Robert Boyle, "Usefulness of Natural Philosophy, II, 2," *Works of Boyle*, Vol. 6, 1668–71, Michael Hunter, ed. (London: Pickering & Chatto, 2000), 400.

54 ClP/24/80/175. The French academician Pierre-Joseph Macquer believed as late as 1768 that pre-mordanting silk for the scarlet dye was his own discovery and a signal example of how a superior knowledge of nature might improve the arts. Christine Lehman, "L'art de la teinture à l'Académie royale des sciences au XVIIIe siècle," *Methodos* 12 (2012), consulted on 31 January 2017: URL: http://methodos.revues.org/2874; DOI: 10.4000/methodos.2874. Evidently, common dyers were also pre-mordanting, eg Sloane 3292, item #4, "A Booke of Dyers good Receipts," for adding the "fixing liquor" before the pigment.

55 ClP/24/80/177. See also ClP/ 24/80/173, "… Fletcher beleives that Dr. Kefler boyles the cochineel as the Logwood and by the help of the spirits draws the tinctures, and is assured that he doth not proceed after the common manner.…"

56 ClP/24/80/168.

Mayerne certainly felt that his chymical knowledge could trump the advice of experienced dyers. Both his sources of information and his chymical investigations of dyes were in advance of others in England at the time.[57] Yet, unlike Henry Power and other Fellows of the Royal Society, quoted above, he did not use his chymical approach to justify his craft research as epistemically elevated above profitable trades. Nor were his craft experiments merely tributary to his chymical research. Rather, Mayerne deployed his chymical knowledge in the pursuit and further development of valuable dyes.

4 Mayerne's Method

As Trevor-Roper has noted, Mayerne applied the same experimental approach to dyeing and other arts and crafts as he did to his medical and chymical studies. He would begin by collecting valuable intelligence, either directly from informants or indirectly through what we might call craft espionage, as well as from written sources. He often noted the date and address of the source on holograph artisanal papers or on the records of his oral interviews with practitioners. Drawing on his sources, he then speculated as to what other utensils, ingredients, and operations might be relevant. He continually re-tested a wide range of factors (water sources, a wide array of organic and inorganic dyes and mordants, different cloths, utensils made of different metals, and various additives ranging from metal plates to urine) in different combinations and proportions. As his informant Nicolas Briot wrote to him concerning "divers tinctures" of metals and their use in dyes, the "whole way lyes in the tryall of all."[58]

Mayerne called his experiments, "experiments" or "essays," and included dates and locations, notes of those who were present, speculations about what happened, general observations, "censures" of failures, and suggestions for future attempts. He amalgamated his findings into more general principles and systematic catalogs, such as a list of materials to be dyed and vessels to use, a catalog of woods still to be tried as pigments, and a bibliography of forty-four authors writing on color. Throughout his discussions of particular

---

57 Compare Leonard Trengove, "Chemistry at the Royal Society of London in the Eighteenth Century—IV. Dyes," *Annals of Science* 4 (1970): 331–353 and Giorgio Riello, "Asian Knowledge and the Development of Calico Printing in Europe in the Seventeenth and Eighteenth Centuries," *Journal of Global History* 5 (2010): 1–28.

58 ClP/24/80/172.

experiments, he referred to the printed works he included in his bibliography on color; these ranged from books of secrets, herbals, and alchemical works, to studies specifically devoted to inks and dyes and reference works such as dictionaries.[59]

Although Mayerne's approach moves from the particular to the general, it does not attempt to appear disinterested or to remove traces of the self in the process. Throughout, Mayerne indulged in commonplaces on ideal experimental behavior through which he reflected on failure or success and encouraged himself to persevere.[60] These moments of personal introspection also appear in Mayerne's casebooks, a practice that Trevor-Roper has referred to as the inclusion of "some apparently irrelevant verses in Spanish and Latin advising himself to avoid the company of false friends and rely only on his own virtue and divine support."[61] Mayerne also made his papers personal by noting his own ideas and discoveries with his possessive monogram (Fig. 2.2).

5 Mayerne in the Hartlib Papers

While Mayerne deployed his many informants to research the Küffler dye, Samuel Hartlib befriended the brothers, especially Johann Sibbert, and investigated their techniques more directly. Hartlib learned in 1636 that Küffler "hase also a special Art of dying of Scarlet for to save 50. lib. in a peece, as I take it."[62] An undated agreement between one of the Küffler brothers and Hartlib sets out

59 Sloane 3423, 36v-37, "Variae species quae tinctoribus supra Lanam, Sericum, filum &c. laborantibus usui esse possunt." Sloane 3423, 38v, "faist essayer les bois &c. suivant." ClP/3i/40, "Autheurs traitans de la Teincture & Couleurs." For references to works on the list, see Sloane 3423, 5 with a reference to "Plichto sur la teincture" and "les secrets de Birelli & autres;" ClP/3i/31, for a reference to the "illumineer buch;" ClP/81/unpaginated, *Experimenta tinctoria,* for a reference to Johann Rudolph Glauber's *Furni philosophici,* as well as "Alexis Piedemontois, Russelli, Fioraventi, Fallope, Birelli, Caneparius, de atramentis, Isabella cortesse & aultres qui ont escrit des livres de secrets."

60 eg. Sloane 3423, "Canis festinans caecos parit catulos" (3r); "Chi troppo s'assotiglia si scavezza" (7v); "Bené qui latuit, bené vixit" (8r); "Experientia rerum Magistra" (17v); "Per varios casus artem experientia fecit/exemplo monstrante viam" (20r); "Debile principium melior fortuna sequatur" (27r).

61 Hugh Trevor-Roper, "Harriot's Physician: Theodore de Mayerne," *Thomas Harriot, an Elizabethan Man of Science,* Robert Fox, ed. (New York: Routledge, 2017), 48–63 (on p. 57). See Scott-Warren, "Early Modern Book-keeping," for a comparable example of divine or ethical quotations in an account book.

62 Hartlib Papers Online (hereafter Hartli), [29/3/57B].

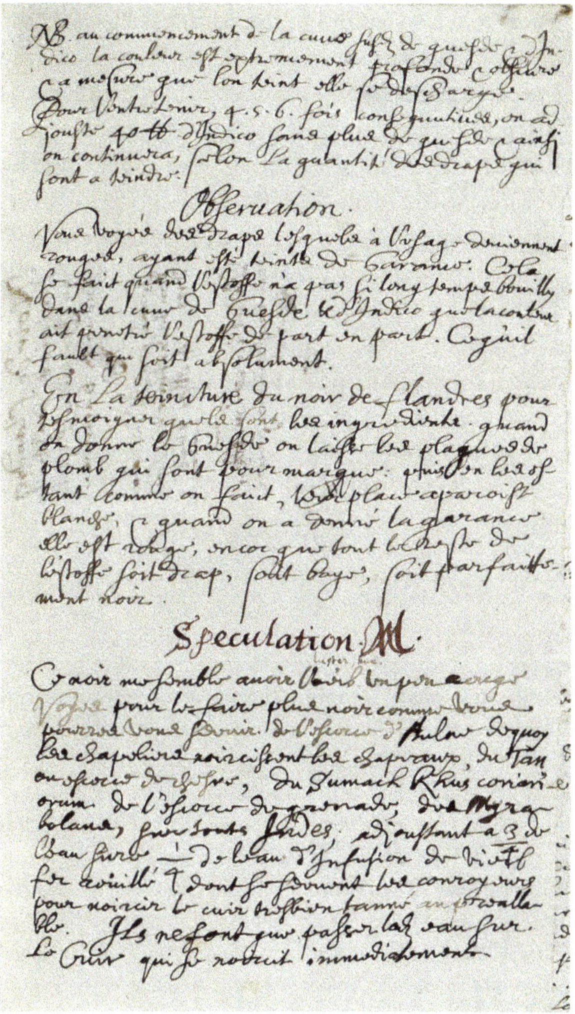

FIGURE 2.2 *Royal Society, ClP/3i/35. Mayerne typically recorded observations, speculations, and experiment. Not the possessive monogram next to Mayerne's speculation.*

terms for teaching Hartlib two of the secrets Drebbel passed on to the Küfflers, the scarlet dye and a self-regulating oven (as well as sundry of Küffler's own chymical techniques).[63] Twenty years later, still apparently pursuing the technique, Hartlib noted that Catherina Küffler, Drebbel's daughter and Dr. Küffler's wife, an "understanding Woman," also "know's the way" of perfecting the scarlet

63 Hartlib, [27/13/7A-B].

dye.[64] Ultimately, the Küfflers would lose control of the scarlet recipe, and the subsequent lowering of the price of scarlet was widely noted.[65] Fortunately, the Küfflers had expanded their repertoire. As Hartlib recorded in 1655, Küffler was also working on "an excellent approoved Way of colouring of furres in black, w*hi*ch may proove very gainful," as well as a way to "fix a greene colour" that he thought "would bee of equal valew with that of scarlet."[66]

Hartlib was also au courant with Mayerne's efforts. In 1639, when Mayerne was just beginning his scarlet dyeing experiments with Fletcher, Hartlib noted, "Dr Mayerne is trying experiments for dying without cutchenel by the herbes growing in England."[67] Hartlib might be referring to Mayerne's experiments with Mistress Fletcher for using safflower, which could produce a rosy pink (*incarnat*), pink, and scarlet.[68] Producing scarlet dye through local ingredients would significantly raise his profits.[69] Hartlib, who also possessed a number of medical receipts copied "ex Mayerni MSS," was also apprised later of Mayerne's experiments on fixing logwood, another South American dye.[70]

---

64 Hartlib, 1656, [29/5/88B].

65 eg. Johann Joachim Becher, *Minera Arenaria Perpetua* (London: Pardoe, 1680), 55. "Quae jam in Hollandia circa scientias expiscandas technae adhibeantur, meum non est, heic loci exponere, notum est quibus technis Kifflero colorem suum scarletinum furati, & imitati sunt."

66 Hartlib, 1655, [29/5/16B]. According to what Ole Borch recorded, Küffler may have succeeded in developing these new dyes. Ole Borch, *Itinerarium 1660–1665*. II. Oct. 1661-May 1663 (London: Brill, 1983), 165–6. Johann Moriaen, Küffler's old business partner and now a septuagenarian, was still dying a black dye as well as a blue that could also be used as a green. Scarlet had become so cheap that Moriaen wasn't producing it anymore. On Moraien and Küffler, see John Young, *Faith, Medical Alchemy, and Natural Philosophy: Johann Moriaen, Reformed Intelligencer and the Hartlib Circle* (Brookfield, Vt.: Ashgate, 1998).

67 Hartlib, 1639, [30/4/9B].

68 eg. Sloane 3423. 24v-25v. cf. Joan Thirsk, who notes Hartlib's comment that Mayerne was seeking scarlet without cochineal, but dates the first proposal for safflower to 1663–4. Joan Thirsk, *Alternative Agriculture: A History from the Black Death to the Present Day* (Oxford: Oxford University Press, 1997), 131.

69 In his later *Faber Fortunae*, John Aubrey, who was also apprized of the dyeing experiments of Hunniades, suggested dyeing scarlet with safflower as a profitable project. Bodleian, MS Aubrey 26, 4r.

70 For the receipts: Hartlib, undated, [30/1/12A-14B]. Hartlib, 1650, [28/1/79A]. "One wrote a Leter of late to Mr Ash a Memb*er* of the C*ouncel* for Trade that hee had found out an infallible Exp*eriment* for the fixing of Lockwood heretofore soe much forbidden. Dr Mayerne and Dr Corydon [probably Colladon] have likewise beene about it." 60/4/216A-228B: 225A-228B also appears to be transcribed from Mayerne manuscripts, although a cipher is used for his name.

Dr. Küffler, unsurprisingly, was no admirer of Dr. Mayerne. Referring to Mayerne's assistant, the Hungarian alchemist Johannes Bánfi Hunyadi, active in London from 1608–1646, Küffler informed Hartlib in 1656 that "Hans Hunniades [was] a very idiot Laborant and one that knew nothing at all, but only was cryed up by Dr. Mayerne, who was no chymist at all. The chiefe thing that was in him was his Art of enamelling or making of Artificial Stones or Jewels."[71] Much later, in his *Numismata,* John Evelyn, FRS, would praise "the Noble *Hunniades* Inventor of the *Scarlet-Bow Die,* since the loss of the antient *Purple,*" perhaps reflecting the competition between Küffler and Mayerne/ Hunyadi by attributing scarlet to the latter; some of Hunniades' recipes are included in the Mayerne papers now in the Royal Society.[72]

6 Lost Colors: Beale, Evelyn, and the Retrospective Meaning of Early Stuart Color Research

The improved scarlet Mayerne sought bore political overtones. As several of the sources Mayerne drew upon discussed, scarlet, the color of senatorial and judicial robes, frequently competed with the lost ancient purple dye extracted from the poison of eastern Mediterranean predatory shellfish. One such source was André de Nesmond, "The Robe or the Purple of Justice," written for the opening of the Parliament of Bordeaux in 1615. Nesmond related modern scarlet robes to ancient imperial purple, drawing on many ancient discussions of color, and discussing the significance of various shades of scarlet and purple.[73] Another was the 1619 work of the learned physician active in Venice, Pietro Maria Canepari, *De atramentis* (on inks), which also related Venetian scarlet to ancient purple. Mayerne himself considered one color extracted from a Scottish snail as the "analog of the purple of the ancients."[74]

---

71 Hartlib, [29/5/91B]. John Aubrey records many details of Hunyadi's dyes in his *Faber Fortunae*, Bodleian, MS Aubrey 26, including [4r], "Mr. hunniades, secret of making the Raven-black dye, with Vitriol of Iron" and [9v], "Mr. huniades can make a Dye not to be washed-out e.g. An Indian way of Dyeing."

72 John Evelyn, *Numismata, a Discourse of Medals, Antient and Modern* (London: Tooke, 1697), 280. See Webster, *Great Instauration,* 389, note 162. A recipe for purple calling for the "water of Hunyadi" appears in Mayerne's volume, *Experimenta tinctoria*, p. 63, copied by Hooke. ClP/24/82.

73 André de Nesmond, "La Robbe ou la Pourpre de Justice" *Remonstrances, ouvertures de Palais, et Arrestz prononcez en Robes Rouges* (Poictiers: Mesnier, 1617), 439–478. ClP/3i/40.

74 "Analogon purpurae antiquorum." ClP/24/81, on p. 6 of Mayerne's *Experimenta tinctoria.*

Restoring purple was a special interest for returning monarchy; Charles II himself asked Robert Boyle to discuss the ancient purple dye with him.[75] When the color was successfully recovered in the Oxford Philosophical Society, it was deployed to stain the Royal Society's motto, "nullius in verba."[76] Throughout the Interregnum, future Fellows of the Royal Society would reflect upon the lost vibrancy of the early Stuart court by reminiscing about past experimental research into colors and engaging in their own collaborative research. It seems to have been difficult for them to obtain, however, the range of resources Mayerne enjoyed.

Consider Canepari's *De atramentis* of 1619. In 1650, Benjamin Worsley told Samuel Hartlib that (future Fellow) Theodore Haak had a copy, and that the work was among the most "select" chymical books.[77] Finally, in 1658, Boyle sent Hartlib a copy of *De atramentis* from Worsley.[78] Hartlib passed the work on to others to read, including (future Fellow) John Beale (1608–1683) in Hereford and his own son-in-law, Frederick Clodius.[79] Beale praised the work's novelty and the clarity of its chymical discussions, and he immediately drew up a treatise of his own, *The Purple of the Ancients*, based upon it.[80] At Beale's request, Hartlib prefaced Beale's work with a dedicatory letter to Robert Boyle and to John Worthington.[81]

Beale highlighted the political resonances of color research and what was at stake in restoring an experimental culture of glorious inks and dyes. Beale hinted to Hartlib that, given the rapidly shifting political climate, the tract was sensitive.[82] In his preface (signed by Hartlib but composed by Beale), Beale

---

75 Boyle, *Colours* and *Cold*, 143. Charles II's name isn't given, but Boyle describes him as "his Majesty." Vera Keller, *Knowledge and the Public Interest* (New York: Cambridge University Press, 2015), 224–5.

76 Robert Theodore Gunther, "The Philosophical Society," *Early Science in Oxford*, (Oxford: Dawson, 1925), vol. 4, 230–1. For more on a swatch of purple William Cole sent to the Royal Society, see Hunter, *Wicked Intelligence*, 135–7.

77 Hartlib, [28/1/76B].

78 Hartlib, [29/7/8A].

79 Hartlib, [51/40B]. Hartlib to Boyle, 25 May 1658, Robert Boyle, *The Correspondence of Robert Boyle, Vol. 1, 1636–1661*, ed. Michael Hunter, Antonio Clericuzio and Lawrence Principe (London: Pickering & Chatto, 2001), 274–5.

80 Hartlib, [51/52A].

81 John Beale, *Purple of the Ancients*, in Hartlib, [51/8A-B and 10A-12B, 51/107A-144B]. Beale requests Hartlib to procure the "shelter" of Boyle and Worthington in Hartlib, [51/13A].

82 In a further letter of August 31, 1658, Beale asks to revise his writings on color to better fit Boyle's interests. Hartlib, [51/9B], "*You* will guesse by the following preface, That it hath something of importance to the present times."

asked the reader to forgive his sharp criticism of the "fickle fashions" of "our English Courte & Gentry," since he thereby endeavored to "repayre the ruines of Antiquity in that pointe, which restores us to the ornaments of Maiesty" and to invest us "with a more durable lustre of authority & honour" through the "powerfull precepts of the best philosophy."[83]

Beale disagreed with Canepari's claim that the Venetian senatorial scarlet was the lost ancient purple of the ancients. He did agree with Canepari's medical views of colors. He too was concerned both about the poisonous metallic ingredients of many inks and their affects when constantly respired. He also claimed that colors worked powerfully upon the passions; "colors are not without a forcible energy to rayse affections, & passions, to settle or to alter complexions & constitutions, & to compose or to endanger health."[84] Both the color and chemical content of ink directly affected the health and mental state of the reader, suggesting a very immediate way for old writings to have a new lease on life. Beale proved nostalgic for the colors of the past, noting that "The Lord Bacon, in his newe Atlantis, seemes to affect yellowe, as the fittest color of parchment" and that "Our ancient Manuscripts, the richest, especially such as are devotionall, & then chiefely, when they containe Songs or Oraysons to the Holy Virgine, are in the leading letters embellished w*i*th curious worke in beautifull coulors."[85]

Given how much time scholars spent viewing books, particular attention should be paid to the content and color of inks and paper used in printing, Beale urged. Black and white were especially poor choices. He pointed out that many cloth dyes could also be used for books; if scarlet "be dissolvd in the white of an Egge or in gumd water, you may painte, adorne bookes, or write w*i*th it."[86] Beale does not mention it, but printing in red had gained a particular significance in recent years, when several accounts of Charles I's execution evoked the spilling of royal blood through the use of red ink.[87] Beale concentrated instead on the medical benefits of scarlet printing. Scarlet ink would particularly aid the health, since cochineal had noted medical properties that might travel in fumes from the ink through the nostrils and the body.[88] Likewise, an

83 Hartlib, [51/12A-B].

84 John Beale, *Purple of the Ancients*, 7, in Hartlib, [51/127A-144B].

85 Ibid, 55–56.

86 John Beale, *Purple of the Ancients*, 46.

87 Helmer J. Helmers, *The Royalist Republic: Literature, Politics, and Religion in the Anglo-Dutch Public Sphere, 1639–1660* (Cambridge: Cambridge University Press, 2015), 136. See eg. Marcus Zuerius Boxhorn, *Metamorphosis Anglorum* (NA: NA, 1653).

88 Mayerne noted the medical benefits of cochineal in Wellcome MS 716, 93v, but he did not discuss it in his writings on color. This volume is currently ascribed to William Ruthven, but

ink "compounded of the best sacke & well-prepared Vitrioll" might "cheere our spirits, as if wee had a more refined, & spirituall way of drinking Wine all the time."[89] He requested the aid of Hartlib's son-in-law, the chymist Frederick Clodius, in developing a more salubrious printing ink.[90] And in another projected work, *A Garden of Pleasure*, he listed several *desiderata* for future research, including the "compounding of colors" in flower breeding, as well as a "search into *th*e powerfull operations effluxions & motions of colors."[91]

The communal reading of Canepari's *De atramentis* during the Interregnum illustrates the color research shared by future fellows of the Royal Society with Hartlib and others.[92] Hartlib wrote back to Beale with the comments of Worthington and of (future Fellow) William Croone (1633–1684).[93] He also arranged a correspondence on color between Beale and Boyle, who had probably begun the experiments on color that would appear in 1664 as *Experiments and Considerations touching Colours* in 1655–6.[94] Beale suggested that color was a formal shared research aim that Boyle shared with others; he refers to "the engagement upon colors" of "the learned club," apparently in reference to the so-called Oxford philosophical club in which Boyle participated.[95] Another

it is in fact a Mayerne work with accounts of collaborations between Mayerne and William Ruthven's son, the physician Patrick Ruthven, Mayerne's neighbor and Van Dyke's father-in-law. On Ruthven, Trevor-Roper, *Europe's Physician,* 343 and R. Ian McCallum, "Patrick Ruthven, Alchemist and Physician," *Proceedings of the Society of Antiquaries of Scotland* 134 (2004): 471–490. Mayerne also dicusses Ruthven in his discourse on vipers, ClP/15i/50.

89 Beale, *Purple of the Ancients,* 52. In a letter to Hartlib, Beale reveals that he had lost his deceased brother's receipt for an ink composed of sack and vitriol. He never used it himself as an ink, but he and others had enjoyed drinking it. Hartlib, [51/13A].

90 Beale, *Purple of the Ancients,* 54.

91 Beale to Hartlib, undated, [25/6/3A-4B].

92 cf. Trengove, "Chemistry at the Royal Society of London," 346. "Natural philosophers in England paid little or no attention to ink-making until Boyle, prompted probably by the publication of Caneparius's book [in 1660], carried out some inconclusive experiments on it."

93 For Worthington and Croone, Hartlib, [51/52A].

94 Hartlib, [51/8A] and [52/61]. Beale responds to a query from Boyle on September 10, 1658. Hartlib, [51/15A-16B]. He further advocates for printing the Bible in green on September 22, 1658, Hartlib, [51/17A]. Boyle, *Colours* and *Cold,* XI. Beale reacts to Boyle's 1664 work on colors in a letter of 25 April 1665, in Robert Boyle, *The Correspondence of Robert Boyle, Vol. 2, 1662–5,* ed. Michael Hunter, Antonio Clericuzio and Lawrence Principe (London: Pickering & Chatto, 2001), 268–271.

95 Beale to Hartlib, May 21, 1658, Royal Society MSS, Boyle Letters 7.9 1A-2B, cited from Hartlib Papers Online. Beale mentioned the "learned club" again on April 9, 1658. Royal Society MSS, Boyle Letters 7.8 1A-3B, cited from Hartlib Papers Online.

participant, (future Fellow) Thomas Willis (1621–1675) of Christ Church, was also known to be writing his own treatise on color.[96]

In writing to Boyle, Beale took the opportunity to reflect back upon the early Stuart color world, a world with which Boyle was also familiar, but which took on a new significance from the viewpoint of a duller Interregnum palette. Beale described himself as a quondam "favorite" of both the provost at Eton, Sir Henry Wotton (1568–1639; Wotton was also provost when Boyle attended the school) and of the magnificent arbiter of fashion and political fortunes at court, George Villiers, the Duke of Buckingham (1592–1628).[97] He thus enjoyed a privileged position for observing the colors of the old court.

> R. Bacons new Atlantis is full of blewes & yellowes. Sir. H. Wotton perpetuated darke-greene with a black trimming to all his Servants, & lov'd to see Nic Oudart in such colors. I know some Courtiers, that stuck fast to gold upon greene. For some yeares before the Wars, soe many of the Court Nobility gave blew & yellow or orenge tawny for liveryes: That their groomes seem'd to me very generally to weare fooles coates. One Gentleman had a peculiar credite to entice K. James & K. Charles his eye, esp. for ornaments on horse-back: For some yeares I have recreated myself & others in scanning the humours of Court & Country in Whithall & Westminsterhall. Wee could see which favourites gave the Law of coulours, & who were their apes.... Some had a harmonious genius to sorte colours in fit ornaments....[98]

This glittering world of powerful fashion quickly faded, however. Fashions grew more and more muted, as the parti-colored early Stuart cloth gave way to an earthy red and then to a light grey in the Interregnum. As Beale continued,

96 Hartlib, 1656, [29/5/102A]. "Dr or Mr Willis a leading and prime man in the Philosophical Club at Ox*ford.* Hee hath written a Tr*eatise* De Fermentatione and of Colours much *com*mended by Mr Aubrey." Willis' color theories appeared in *Diatribae duae medico-philosophicae quarum prior agit de fermentatione ... altera de febribus* (London: Roycroft, 1659).

97 Mayling Stubbs, "John Beale, Philosophical Gardener of Herefordshire Part I. Prelude to the Royal Society (1608–1663)," *Annals of Science* 39 (1982): 463–489. On Buckingham, see Peck, *Consuming Splendor, passim* and Christiane Hille, *Visions of the Courtly Body: The Patronage of George Villiers, First Duke of Buckingham and the Triumph of Painting at the Stuart Court* (Berlin: Akademie, 2012).

98 Royal Society, RB/3/6/4 [accessed via Hartlib Papers Online].

> In garments three yeares difference makes a Monster. The fallowe or red marle color, which was called my Lord Protectors Colour, would have been scorned by a scavenger, as soone as the light-grey came in fashion.[99]

Beale's reflections on this world are full of criticism of its sartorial and sexual mores (eg, "Sir Francis Pyle was famous for his painted Coach, pied-bald horses, oblique buildings & vnnaturall lusts"). They also express a sense of loss and frustration. Nicholas Oudart (?-1681), whom Wotton loved to see wearing dark green and who had introduced Beale to London instrument-makers in his youth, had, as the former amanuensis of Charles I, become particularly inaccessible to Beale. Beale remarked in 1659, "If it were now lawefull to hold any kind of intelligence with Nic Oudart, I would only aske him Sir H Wottons Art of dressing Mushrooms, & I hope that is not High Treasone."[100]

While critical of the use of colored cloth to attract political notice, Beale was proud of the strides made in early Stuart color research. He agreed neither with Scaliger nor with Cardano on the nature of colors. "Both of them," he said, had much to learn from Sir Henry Wotton's researches into color using glasses tinged various shades, which could be used to explore color combinations. Boyle would later publish his own experiments compounding colors through tinged glasses.[101]

Wotton engaged in his color research with Sir Edmund Bacon (1566–1649), who married his niece, and who was nephew to Sir Francis Bacon and brother to the noted painter, Sir Nathaniel Bacon. In a letter to Edmund Bacon, Wotton attributed the technique of testing colors upon glass to one "Francesco." This was most likely Mayerne's brother-in-law, Gian Francesco Biondi (1572–1644), who was also a contributor to the famous Mayerne manuscript, Sloane 2052, and whom Wotton knew from his days as ambassador to Venice. Wotton informed Edmund Bacon how "*Francesco* hath made a proof of that green which you sent me; against which he taketh this exception, That being tryed upon glass, (which he esteemeth the best of tryals) it is not translucent; arguing (as he saith) too much density of the matter, and consequently, less quickness

99 Royal Society, RB/3/6/4 [accessed via Hartlib Papers Online].

100 Nov. 15, 1659, Hartlib, [62/25/2A].

101 Boyle, *Colours and Cold*, 115–6. See Hartlib, 1656, [29/5/71B]. "An Experim*ent* to make Colours out of herbes as hee shewed mee a glasse of color made out of rue. Mr Boyle."

and spirit then in colours of more tenuity."[102] Wotton also admired Edmund Bacon's own color innovations, sending his protegé Oudart, for example, with flints to Bacon to be colored into agates by Bacon's process, praised by Wotton as a "miraculous invention."[103]

Beale was privy to the philosophical conversations of Wotton and Bacon, and he treasured the vanished colored world of Wotton's researches.[104] In a 1659/1660 tract on pearls, Beale fondly recalled how when he was "an Eton Scholar," researching oysters with Wotton, he had been given a large oyster whose "very beautifull mother of pearle" showed "all the colors of the rayne bowe, the grasse greene & bright purple interchanging, like those colors in changeable taffata."[105] Beale informed Hartlib that the "the Dutch man of Embden [Abraham van Linge], who glased the chappell of Ch*ris*t Church, complained that wee had lost *tha*t true ultramarine tincture, w*hi*ch S*ir* H.W. shewd him in a glasse, w*hi*ch is nowe in my custody."[106] The survival of van Linge's "popish" stained glass windows of Christ Church through the Interregnum was itself precarious.[107] Beale's preservation of Wotton's ultramarine glass, passed from Wotton to van Linge to Beale, represented a cherished surviving shard from a broken color world.

Beale himself mostly experimented with color through the even more transient medium of flowers. He transplanted tulips and other flowers to various soils in order to transform their colors, which he described in ethereal terms of the colored glass combinations used by Biondi and Wotton. He was able, for instance, to train one tulip "fro*m* her deepest purple to a fady blewe *tha*t hath a very small inclinatio*n* to a blush, as if a peach colord glasse & violet colord glasse

102 Ibid, 157. On Biondi and Wotton, see Trevor-Roper, *Europe's Physician*, 203, and on his contributions to Sloane 2052, Trevor-Roper, 341. Mayerne also recorded techniques of glass-painting. Sloane 1990, 85r, "Ars pingendi in vitro" and 86, "colores in vitro." See also Sloane 2052, 141, "Artifice pour faire les vitres de taffetas représentants celle de verre que sont aux églises."

103 *Letters of Sir Henry Wotton to Sir Edmund Bacon* (London: Printed, by R.W. for F.T., 1661), 130. "My servant Nicolas and I hope to send you some good Flints to be Agatized by your miraculous invention."

104 In a letter of January 18, 1658, John Beale fondly recalled the "phansicall discourse" he had overheard Sir Edmund Bacon share with Wotton. Hartlib, [51/57A].

105 John Beale, *Discourse On Pearl-Bearing Shellfish* in Hartlib, [25/17/1A-16B], here at [25/17/7A]. Beale goes on to recall ([25/17/15B]), how an oyster and other things might be petrified, as "Sir H Wotton hath shewed to all that saw his Study."

106 21 May 1658, Beale to Hartlib, [52/61B].

107 Julie Spraggen, *Puritan Iconoclasm during the English Civil War* (Woodbridge: Boydell, 2003), 240.

were ioynd together."[108] Extending his color research across the realms of art and nature, he found it a "gentile delight to observe, howe the same tinctures doe sort in *the* severall receptacles of mettalls, iewells, feathers, silke, flowers &c."[109]

After the founding of the Royal Society, the research exchange on inks and dyes previously brokered by Hartlib continued. Notably, however, the Restoration marked an about course in Beale's views on the luxuriousness of color. While during the 1650's Beale had professed to prefer the multi-colored book to richly dyed cloth, in the Restoration, Beale defended luxurious dyes. He denied the Interregnum view that "silken Gentry" were not patriots, since, Beale contended, they added "luster to the nation," promoting emulation, ingenuity, and the improvement of the arts.[110]

Beale re-circulated his work on the purple dye, tried to republish Canepari's work (which was published in London in 1660 by John Martyn), and also circulated "a tract by his nephew Peter Smith on Flemish and French techniques for bleaching linen."[111] He offered the Society accounts of making parchment, vellum, and ink, and even donated to the Society a dozen sample pieces of parchment made by the process he recorded.[112] He wrote to Boyle in 1663 urging the improvement of "drapery" and the "dyers Arte."[113] In 1666, he also corresponded with Boyle about the best material form Boyle's published works should take.[114]

7 De Vaux and the Entrance of Mayerne's Papers to the Archive

Such were the current lines of color research, tenuously drawn forward from the vanished world of early Stuart experiment through the Interregnum

108 21 May 1658. Hartlib, [52/62B]. Nehemiah Grew would later discuss transplanting tulips for color changes. On Grew, Roos, "Saline Chymistry," 586.

109 Hartlib, [52/61B].

110 Beale to Boyle, 21 November 1663, Boyle, *Correspondence*, Vol. 2 (2001), 209.

111 Mayling Stubbs, "John Beale, Philosophical Gardener of Herefordshire Part II. The Improvement of Agriculture and Trade in the Royal Society (1663–1683)," *Annals of Science* 46 (1989): 323–363 (on p. 347). P. M. Canepari, *De atramentis cujuscunque generis* (London: Martyn, 1660).

112 John Beale, "The Art of making Parchment, Vellum, Glue etc." Royal Society, ClP/3i/18. Read to the Society on 25 May 1664. Beale records the techniques of Matthew Willes alias Coxe of Yeoville, who "makes the best Parchment in England. Challengeth the beste in the World." John Beale, "How to make good ink and how to make gunpowder." Royal Society, ClP/17/12.

113 Beale to Boyle, 21 November 1663, Boyle, *Correspondence*, Vol. 2 (2001), 208.

114 Yeo, *Notebooks*, 146.

by Beale, Boyle, and others, when De Vaux was elected to the Society (proposed as a member by John Wilkins) on May 24, 1665. He brought in his first Mayerne paper, on "worms bred in the teeth of men" at the very next meeting of the Society.[115] de Vaux's candidacy had been unusually contentious. He had been proposed as a member first in 1661 but was then "oppos'd & not admitted."[116] Mayerne's papers offered a means for de Vaux to prove his worth. The other Fellows were eager for more Mayerne papers, and he was requested to communicate to the Society, "what other considerable papers he had of Sir Theodore Mayerne for the purpose of the society," which he proceeded to do.[117] He did so, however, in a piecemeal fashion, continually illustrating his usefulness to the society by producing Mayerne manuscripts until shortly before his own death in 1694. Via de Vaux, the deceased Mayerne participated, as it were, in Royal Society meetings for nearly thirty years. He even, via de Vaux, had articles published in the *Philosophical Transactions* nearly forty years after his death.[118]

De Vaux's access to Mayerne manuscripts resurrected a world familiar to many Royal Society Fellows. Many Fellows shared old ties with Mayerne; Sir William Brouncker (1585–1645), the father of the Society's first present, had helped found the Distiller's Company with Mayerne in 1638.[119] The young Walter Charleton (1619–1707), elected to the Society in 1663, had most likely been Mayerne's assistant.[120] Prince Rupert of the Rhine (1619–1682) and

115 Thomas Birch, *The History of the Royal Society*, Vol. 2 (London: Millar, 1756), 42. May 3, 1665, "Sir Theodore de Vaux, knight was proposed candidate by Dr. Wilkins." 45. May 10, 1665, "Sir Theodore de Vaux was elected." 49. May 24, 1665, "Sir Theodore de Vaux was admitted."

116 Hunter, *The Royal Society and Its Fellows, 1660–1700*, 7 and 172.

117 Ibid, 52. May 31, 1665, "Sir Theodore de Vaux produced a Latin paper of Sir Theodore Mayerne, concerning worms bred in the teeth of men." Ibid, 60. On June 28, 1665, de Vaux presented a Mayerne paper on preserving timber from being worm eaten, and on preserving beef. "Which paper was ordered to be filed up." Royal Society, ClP/3i/24.

118 Theodore de Mayerne and Theodore de Vaux, "An Account of the Diseases of Doggs, and Several Receipts for the Cure of their Madness, and of those Bitten by them. Extracted from the Papers of Sir Theodore Mayern, and Communicated to the Royal Society," *Philosophical Transactions* 16 (1686): 408–410, and Theodore de Mayerne and Theodore de Vaux, "A Discourse of the Viper, and Some Other Poysons, Wrote by Sr. Theodore de Mayerne, after Discoursing with Mr. Pontaeus. Communicated by Sir Theodore de Vaux, M. D. and S. R. S," *Philosophical Transactions* 18 (1694): 162–166.

119 Webster, *Great Instauration*, 254.

120 Emily Booth, *"A subtle and mysterious machine": The Medical World of Walter Charleton (1619–1707)* (Dordrecht: Springer, 2005), 10.

Sir Nicholas Crisp (?–1666) were among Mayerne's personal informants.[121] Individually, Fellows also sought Mayerne manuscripts. Sir Robert Paston (1631–1683) acquired a Mayerne manuscript sometime before 1668.[122] Martin Lister, raised by Mayerne's friend and colleague, his uncle Sir Matthew Lister, later sought out Mayerne manuscripts in Paris in 1692.[123]

Mayerne's research had also been invoked from the early meetings of the Society. While the group was still meeting in Gresham College, Kenelm Digby would recall the delight he and Mayerne used to share within that very building, when together they observed Hunyadi's chymical precipitations that "designed" plants "more exactly" than "any Painter" could.[124] Just the previous week, Jonathan Goddard, who collected Hunyadi's chymical research (including into the color purple), read a paper to the fledgling group on the production of colors through the mixture of "liquors."[125] A living link to Mayerne's color world was restored when De Vaux successfully proposed Nicholas Oudart (what had been so out of reach to Beale in 1659) as a Fellow in 1667.

Mayerne's papers on color would have been highly desirable, as Fellows of the Royal Society sought and lacked access to materials on color. John Evelyn's early volume on mechanical trades is mostly empty; his entry for "dier" is blank, and "painter" has one and a half pages of recipes.[126] William Petty, the son of a clothier, is often cited as contributing importantly to dyeing as part of the History of Trades program. His 1662 "Apparatus to the History of the Common Practices of Dying" was included in Thomas Sprat's *History of the*

---

121 Trevor-Roper, *Europe's Physician,* 346 and 215.

122 Donald Dickson, "Thomas Henshaw and Sir Robert Paston's Pursuit of the Red Elixir: An early Collaboration between Fellows of the Royal Society," *Notes and Records of the Royal Society of London* 51,1 (1997): 57–76 (on pp. 61–2). Thanks to Michael Hunter for this reference.

123 Trevor-Roper, *Europe's Physician,* 373.

124 Kenelm Digby, *A discourse concerning the vegetation of plants. Spoken by Sir Kenelme Digby, at Gresham College, on the 23. of January, 1660. At a Meeting of the Society for promoting Philosophical Knowledge by Experiments* (London: Dakins, 1661), 77–78.

125 Jonathan Goddard, "A brief experimentall Account of the production of some Colours by mixture of severall liquors either having little or no colour or being of different colours from those produced." Read to the Royal Society on 16 January 1660/1. Royal Society, ClP/2/25. Special thanks to Sietske Fransen. For Hunyadi's research in a volume owned by Jonathan Goddard, see eg BL Sloane 1139, 170 to 171 verso.

126 British Library Additional MS 78339. On this volume, see Michael Hunter, "John Evelyn in the 1650s: A Virtuoso in Quest of a Role," *Science and the Shape of Orthodoxy: Intellectual Change in Late Seventeenth-Century Britain* (Woodbridge: Boydell, 1995), 67–98. A later collection, Additional MS 78340, is much fuller.

Royal Society.[127] Yet even Petty appeared to lack desired sources. As he confessed, "of Cummin-seed, Fenugreek-seed, Senna, and Agarick, I have as yet no satisfactory accompt."[128] He was not able to access any Indian information about calico, that "gainfull Mystery."[129] The materials used in Indian painting and dyeing remained on the Society's query list for the East Indies.[130]

Yet de Vaux did not immediately bring in Mayerne's papers on pigments. He first submitted Mayerne's papers on preserving timber and beef, tallow-chandling, and wax candles.[131] It was Thomas Povey who seems to have precipitated de Vaux's hunt for Mayerne's color research in particular, when, during the meeting of April 18, 1666, Povey discussed the painter John Streeter's technique for preventing glare upon his paintings.[132] Soon thereafter, De Vaux located a series on colors in Mayerne's papers. On May 23, 1666, he "produced some papers about coloration." When the Society received this file, it was ordered that de Vaux and seven other individuals, the "rest of the physicians of the Society" as well as Daniel Coxe, Robert Hooke and Mr. Oldenburg (or any two or more of them), form "a committee to consider of the said papers, and to cause them to be translated into English from the French, that so they might be the better digested afterwards." As Felicity Henderson has pointed out, this reference to digestion suggests a desire to make active use of the contents of the papers in the future.[133] The committee was to meet the next Monday at the home of Sir George Ent.

Strangely, however, Mayerne's name was not mentioned. Henry Oldenburg excitedly informed Robert Boyle that "Sir Theodore de Vaux produced some papers about coloration," which were "Drawn up by a very famous and curious physician from the mouth of the most knowing and experienced dyers of

---

127 Royal Society, ClP/3i/6. "An apparatus to the History of Common Practises in Dying," read to the Royal Society on 7 May 1662. Discussed in Theodore McCormick, *William Petty and the Ambitions of Political Arithmetic* (Oxford: Oxford University Press, 2010), 149, note 123. See also British Library MS Additional 72897, 1–37, with recipes in the Petty papers, partially in French, for scarlet, red crimson, crimson tawny, crimson violet, orange crimson, purple, and tawny.

128 Thomas Sprat, The *History of the Royal-Society of London* (London: Martyn, 1667), 294–5.

129 Hartlib, [71/9/1A].

130 Daniel Colwall, John Hoskins, and Henry Oldenburg, "Enquiries for the East Indies," RBO/1/50.

131 ClP/3i/24–26 on preserving wood, salting beef, making fire from coalballs, and tallow candles, respectively. On sugar and soap, see below.

132 Birch, *History*, Vol. 2, 84.

133 Felicity Henderson, "Faithful Interpreters? Translation Theory and Practice at the early Royal Society," *Notes and Records of the Royal Society* 67,2 (2013): 101–122.

Holland and England in his time," but he did not identify the physician.[134] Even though his distinctive monogram can be found all over his papers, nowhere is Mayerne's authorship noted in the surviving papers on color in the Royal Society. Such silence might be connected to the lawsuit in which Colladon and de Vaux were engaged from 1664 to 1680 regarding finances as well as the rightful ownership of Mayerne's papers. Their battle continued for so long that, after de Vaux had a two volume edition of Mayerne's medical cases printed in 1690 and 1696, the Colladons brought out a rival volume in 1700, drawn from their own cache of Mayerne manuscripts, and in which they lambasted de Vaux after his death.[135] While earlier Mayerne papers in the Royal Society were attributed to Mayerne (eg, the paper on worms bred in the teeth of men noted above), it is notable that Mayerne papers were not published in the *Philosophical Transactions* until after the conclusion of the lawsuit (in 1686 and 1694).

8 Digesting the Papers

Mayerne's papers on dyeing examined by the committee were originally enclosed in a paper wrapper, labeled on the front, "Sr. Tho. de Vaux/ several papers about dying/ See June 13. 66," and on the back, "These Papers about Dying were set here in one Parcel by a former Committee." When formerly filed up by Mayerne, the papers had been folded in thirds with a label on the exterior of each grouping.[136] Mayerne's more formal volumes now in the British Library, such as Sloane 3423, also have bound into them some original papers once folded and labeled in the same manner, ghosts of Mayerne's past archiving techniques.[137]

---

134 Birch, Vol. 2, 93. F. W. Gibbs identified Oldenburg's reference as Sloane 1990, another manuscript of Mayerne's on inks and dyes, but Oldenburg certainly referred to the newly identified cache of papers in Classified Papers 3i. F. W. Gibbs, "An account of a manuscript entitled 'saponis artificium,'" *Journal of the Society of Chemical Industry* 57,37 (1938): 877–8.

135 Theodore de Mayerne, *Praxeos mayernianae in morbis internis praecipue gravioribus & chronicis syntagma, ex adversariis, consiliis ac epistolis ejus, summâ curâ ac diligentiâ concinnatum* (London: Smith, 1690 and 1696). Theodore de Mayerne, *Opera Medica* (London: Browne, 1700).

136 For example, ClP/3i/31 is labelled on the exterior, "Fletcher 13 September 1639/ Escarlatte sur soye avec Safflor et Annatto./ Pourpe avec Long Cochenill & cognee d'huistres/ Tanné/ Aurange/ Speculation sur la teincture de la soye en Escarlatte avec Cochenille./ 11."

137 Sloane 3423, 40v, labelled "Noirs avec bois violet pour teindre les peaux et chapeler francois … 1640" and Sloane 3423, 42, "Reiteration de l'operation sur la Cochenille Longue etc. faicte tres exactement a Chelsay par Mr. Fletcher 17 Januier 1640."

The committee set up to review the papers composed a thorough table of contents, lettered A-V (see Appendix), accounting for each sheet and assigning individual items to members of the committee for translation and examination (although Oldenburg and Colwall seemed not to have been burdened with any).[138] They gave each paper a title, sometimes drawing on Mayerne's own labels. The current file includes both original papers and some translations. Some papers seem to have been included by mistake, such as one on making bullets for an arquebus from [Gedéon] Chabray.[139] Many of the papers in the group deal with the Küffler scarlet. They include a list of 6 inventions "not previously practiced in England," dated to 1639, all pertaining to scarlet; "no man in this land hath the arte," notes Mayerne. "What he hath seen and heard of M. K" is apparently an informant's description of the Küffler dye works (Fig. 2.3).[140]

The translations of the papers assigned to de Vaux and to Ent now remain in ClP/3i/30 (see Appendix). de Vaux checked these translations with the original again; on June 13, 1666, "Sir George Ent brought in the translation of those papers communicated by Sir Theodore de Vaux about coloration, which had been committed to Sir George's care; and Sir Theodore de Vaux took them with him again, in order to compare them with the original book."[141] Yet the next year, the originals were out of de Vaux's hands again and apparently some were in the meantime lost. After a break of some time, perhaps due to the intervening Great Fire, on October 10, 1667, "Sir Theodore desiring, that the papers formerly brought in by him about coloration might be called for, and the members, who had undertaken the translation of them into English, spoken to about it, Mr. Oldenburg said, that some of those papers were already translated; but that those, which were committed to the care of Dr. Quatremain and Mr. Daniel Coxe, were not yet accounted for; and that Dr. Quatremain being since dead, those persons, who knew how his effects were disposed of, might be desired to inquire after that part of the said papers, which was referred to him."[142] Indeed, items listed in the index to the file, such as "Le Pape his

138 Michael Hunter notes this committee and its index in *Establishing the New Science: The Experience of the early Royal Society* (Woodbridge: Boydell, 1989), 95.

139 Included within ClP/3i/31. The illustrated instruments depict a perforated piece of cardboard, and indeed, such a perforated piece is included, unbound, within the volume of Classified Papers 3i. Chabray was Mayerne's amanuensis and assistant. Trevor-Roper, *Europe's Physician*, 401.

140 ClP/3i/30.

141 Birch, Vol. 2, 97.

142 Birch, Vol. 2, 199.

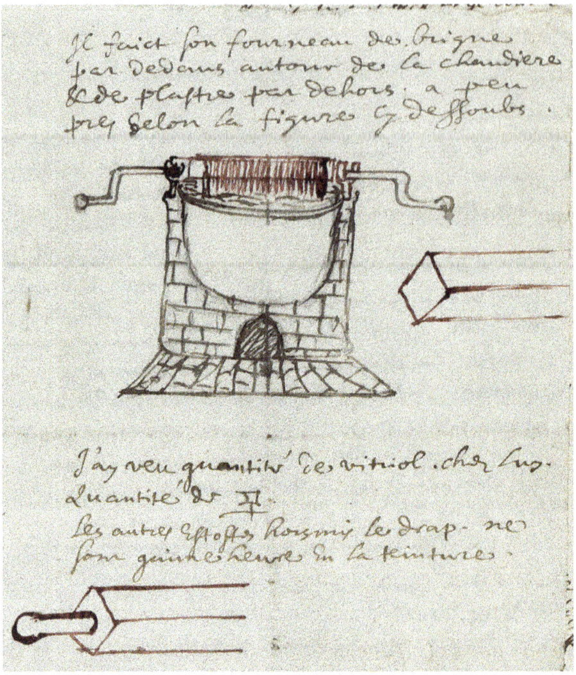

FIGURE 2.3 *Royal Society, ClP/3i/34. A report on scarlet dyeworks.*

paper," assigned to Dr. Quatremain, and "Dr. Cherevix propriò ore," assigned to Dr. Coxe, appear now to be lost.

Robert Hooke apparently also did not return the translations committed to him, or else he retrieved them from the Mayerne bundle at a later date. Further translations of Mayerne papers remain now among Hooke's papers, as Michael Hunter has pointed out to me. Classified Papers 24/80 include copies and translations of many items discussed by the 1666 dyeing committee.[143] As was the case for the translations remaining in ClP/3i/30, those by Hooke also paid close attention to the form of the original, retaining, for instance, Mayerne's monogram, manicules, and marginalia (Fig. 2.4). This was a far closer translation than was standard, according to Henderson's account of early Royal Society translations.

143 The relationship goes both ways; ClP/3i/30 also contains a brief resume of the contents of ClP/24/80. For instance, in ClP/31/30, "To dye Spanish skins red and yellow. Tincture of Stockings" is noted. The recipes for those items can be found at ClP/24/80/168.

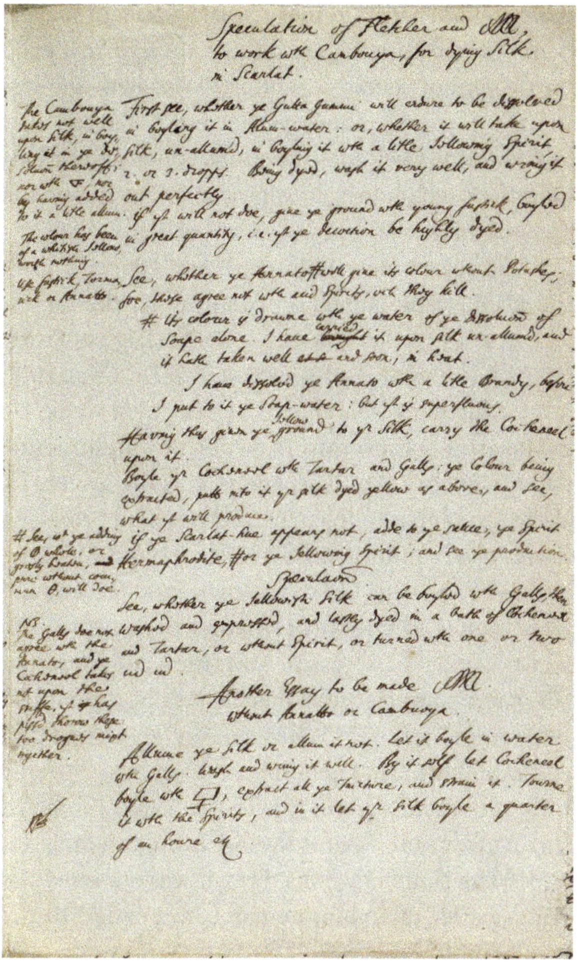

FIGURE 2.4 *Royal Society, ClP/3i/38. The English translation includes Mayerne's monogram, manicule, and marginal comments. In the original, the first paragraph was crossed out in purple ink, but no sign of the crossing out appears in the translation.*

9 De Vaux and Povey: Personal Rivalry and Archival Buildup

Given the bitter relations between de Vaux and the Colladons, it must have seemed quite a threat to de Vaux when Thomas Povey proposed that Jean Colladon be elected to the Society on April 2, 1668. Colladon was not elected.[144]

144 Birch, Vol. 2, 261. April 2, 1668. "Sir John Colladon was proposed candidate by Mr. Povey." Colladon was not elected. Michael Hunter, *The Royal Society and Its Fellows, 1660–1700*, 12 and 59.

Povey's promotion of Colladon might indicate that the two spurs of artistic research of Povey and de Vaux were competing within the Society. Povey too was nostalgic for a lost courtly connoisseurship, writing "lyrically of the lost collections of Charles I" in a 1667 report to the Royal Society.[145] We can only speculate about this competition, but Povey and De Vaux's various suggestions for recovering knowledge of artistic techniques do appear to follow hot on the heels of each other.

Povey had first discussed Streeter's new paint in April 1666. In May, de Vaux brought in the file of Mayerne papers on color, by far the largest cache of Mayerne papers he ever introduced at one time. In August, Povey brought in a trial of Streeter's paint, which was assigned to Dr. Charleton, Mayerne's former assistant, for further research.[146] The very next week, De Vaux brought in another Mayerne manuscript, "a paper of enlumineure, which was recommended to Mr. Evelyn to peruse, and to give the society an account of. He mentioned the art of enameling of Mr. Petitot, and promised to use his interest to procure an account of it for the society."[147] Mayerne had worked with Jean Petitot (1607–1691) in developing new red enamel paints that could depict flesh tones in miniature enamel portraiture. The Colladons, rather than the De Vaux, had access to Mayerne's collaboration with Petitot on enamel. de Vaux suggests here that he would attempt to obtain the technique directly from Petitot.[148] When the Society issued personal *desiderata* lists to its members, Sir Theodore de Vaux was ordered by the Society to fulfill only one *desideratum*: "Sr. Th. de Vaux. Promised to use his interest in procuring from Mr. Petitot the Art of Enamelling, for the Society. Aug. 22. 1666."[149] The same day, John Evelyn was assigned the *desideratum*: "Sr. Th. de Vaux paper of Enlumineur recommended to the perusal of Mr. Evelyn. Aug. 22. 1666."[150]

---

145 Peck, *Consuming Splendor*, 337.

146 Birch, Vol. 2, 107. Charleton's examination of Povey's technique was also on the *desiderata* list assigned to him by the Society. DM/5/72.

147 Birch, Vol. 2, 111. See Sloane 2052, 40r, "Enlumineure," from Edward Norgate. Trevor-Roper, *Europe's Physician*, 338, describes a Mayerne manuscript recording a " 'curious method of using enamel and colours in illumination,' 'as it is done in Limoges'.... This method, say the writers, 'is private to ourselves and must not be indulged.' " Trevor-Roper also described Mayerne's notes on Petitot's "secret" of enamel which is likewise "not to be divulged [ne doibt estre divulguée]."

148 Ibid, 344, for Mayerne's work with Petitot on enamel.

149 DM/5/74B.

150 DM/5/74A.

de Vaux's presentation of the method of illumination immediately then "gave occasion" for Povey to mention that the painters of his acquaintance, Lely, Cooper, and Streeter, would be not be "unwilling to communicate to them the several curiosities and varieties of painting." A committee of nine members was ordered, including both de Vaux and Povey, to consider "what particulars were fit to inquire into, and therupon to discourse with the said masters concerning the same."[151] This attempt to bring de Vaux and Povey together with others for an inquiry into painting techniques does not appear to have progressed.[152] On December 12, 1667 Povey again advised that a further enquiry might be made into "other sorts and ways of paint in oil, distemper or dry colourings ... that it may not be said by the malicious, that you discourse and make flourishes, and subsist chiefly upon what is delivered to you by them, that lived before you." Could this be a reference to the Mayerne papers? Povey suggested that the Society collectively compose a single volume on the history and progress of painting, including all possible ways of coloration, among which painting would be "chief and sovereign," yet which would also include, in an appendix, "several sorts of vernice; browning; staining; graving; etching, and perhaps some other necessary curiosities not foreign to this great subject."[153]

The competition between Povey and the ghost of Mayerne as resuscitated by de Vaux seems to have continued for years. For instance, On July 17, 1679, "Mr. Povey promised to bring in a receipt of making very pleasant, wholesome and strong ale, and as good as any. Thereupon Sir Theodore de Vaux promised to bring in a dozen receipts, which had been experimented by Sir Theodore Mayerne for making of ale of several sorts."[154]

For three decades, de Vaux continually brought piecemeal Mayerniana to the attention of the society. At first, these were medical in nature, yet they quickly came to reflect Mayerne's interest in crafts. They thus informed the Society's History of Trades program at its height. Even as De Vaux brought in Mayerne's papers on crafts, he continued to bring in papers on medicine, chemistry, and natural history, and after the decline of the History of Trades program, these remained his main focus.[155] In a letter of 16 April, 1694 to Robert Southwell,

---

151 Birch, Vol. 2, 111.

152 Peck, *Consuming Splendor*, 338.

153 Birch, Vol. 2, 227–8. Thomas Povey, ClP/2/24, "An Account of a Secrett in the use of Painting In answere to the Commands of the Royal Societie," 17–18.

154 Birch, Vol. 3, 496.

155 On August 29, 1666, de Vaux brought in a paper from Mayerne on the "nature of crawfish; which was ordered to be copied, and filed up, after the copy had been perused by Sir George Ent for his animadversions upon it, and additions to it." Birch, Vol. 2, 113. In

de Vaux was still promising "curious cases from the Collections of Sir Theodore de Mayerne."[156] A new batch of Mayerne papers on poisons was quickly forthcoming. Three papers from Mayerne were read to the Society on 18 April 1694, the last of which was a paper on hydropholica "communicated to Hans Sloane by Thomas [sic] De Vaux."[157] Perhaps it was de Vaux who alerted Sloane to the significance of Mayerne's collections, leading to Sloane's purchase of so many volumes from the Colladon sale in 1712. In May 1694, the last month of de Vaux's life, de Vaux brought in Mayerne's recipe for viper wine and account of viper bites (printed in the *Philosophical Transactions*), Mayerne's "History of serpents in the East and the West Indies," and Mayerne's "History of serpents and crocodiles."[158] De Vaux died on 26 May 1694, concluding a thirty-year career of bringing Mayerne's papers to the Royal Society's attention.[159]

10 Mayerne and Hooke

Individuals Fellows related to Mayerne's papers differently; Hooke engaged with them particularly intensively. de Vaux was ordered to transmit several Mayerne papers directly to Hooke (who had also been a member of the committee examining the papers on dyes). At the meeting of October 10, 1667,

the spring of 1667 he brought in papers on sugar, soap, salt, and coal-ball production. On March 20, 1669, 372, "Sir Theodore de Vaux produced out of Sir Theodore Mayerne's collection a paper dated June 17, 1647, containing an account of an accident, which happened to one Mr. John Stevenson, who swallowed a bodkin, and after keeping it six weeks in his stomach voided it by the anus. It was read, and ordered to be kept on the file." In March and July, 1668, he brough in recipes for sealing wax, and both a particular Mayerne receipt for copperas as well as another collection of papers "about chemistry, which he had in his hands from Sir Theodore Mayerne. It was ordered, that they should be referred to the committee for chemistry." Royal Society, ClP/11i/9. Birch, Vol. 2, 311. July 3, 1668. I have not as yet identified Mayerne's papers on chemistry. Michael Hunter describes the Royal Society's eight standing committees, including the chemical committee, as lasting from 1663–5 in *Establishing the New Science*, 73–121, and not as late as 1668. "Remedies that Theodore Mayerne prescribed (for his children and the rest of his Family) against the biting of a Dog or Cat" was read on 2 November 1687, and a paper on this topic was published in the *Philosophical Transactions* in 1686. Royal Society, RBO/9/10.

156 De Vaux to Sir Robert Southwell. Royal Society, LBO/12/111.

157 Royal Society, RBO/9/9. RBO/9/11. ClP/14i/38. ClP/15i/50.

158 Royal Society, ClP/15i/45, ClP/15i/46, and ClP/15i/49.

159 William Munk, *The Roll of the Royal College of Physicians of London* (London: Royal College of Physicians, 1878), 333.

De Vaux brought in several papers concerning sugar (in French) and soap, which were delivered to Haak and to Hooke for use in writing the history of sugar-works and the history of soap-making.[160] On May 13, 1680, Hooke "produced three papers delivered to him by Sir Theodore de Vaux, being some of Sir Theodore Mayerne's, containing some account of the mixture of metals. They were ordered to be transcribed and the papers to be returned to Sir Theodore."[161] A year later, on May 18, 1681, Hooke also "read a translation, which he had made of a paper of Sir Theodore Mayerne, brought in by Sir Theodore de Vaux, about a method of staining agates." De Vaux was once again desired "to communicate some others of his collections of Sir Theodore Mayerne's papers: which promised to do."[162]

Although they do not include any holograph Mayerne papers, by far the most extensive transcriptions, over one hundred pages, from Mayerne's archive can now be found among the papers of Robert Hooke. Classified Papers 24/80 contains many of the items collectively studied by the committee on dyeing (see Appendix). The first half of Classified Papers 24/81 includes transcriptions of additional papers not discussed by the committee. The second half of Classified Papers 24/81 contains a partial transcription of Sloane 3423, beginning "En presence de Fletcher 11 Septem. 1639 Experiments et operations en matiere de teincture faicte par moy."[163] This is followed by the partial transcription of a now lost 99-page Mayerne manuscript on dyes, *Experimenta tinctoria ... 21 May 1639* (Hooke includes the original pagination). Hooke begins the transcription with a note, "Here the book beginneth" and "again to be transcribed Nov. 24 1669."

In *Experimenta tinctoria*, Mayerne drew on many of the same informants whose original papers were studied by the committee on dyes, such as Mr. and Mrs. Fletcher, de Lanoy, and Peter van der Couter (whose file on calico studied by the committee on dyes has been lost). He also drew on figures known from other Mayerne manuscripts such as the Colladons, the painter Mark Anthony

160 Birch, Vol. 2, 199. For sugar, ClP/19/29, translated from French. Hooke apparently lost the paper on soap. See Archibald Clow and Nan L. Clow, *The Chemical Revolution: A Contribution to Social Technology* (London: Gordon and Breach [1952], 1992), 117. F. W. Gibbs, "An account of a manuscript entitled 'saponis artificium,'" *Journal of the Society of Chemical Industry* 57:37 (1938), 877–8.

161 Birch, Vol. 4, 37. I have not identified these papers.

162 Birch, Vol. 4, 87. ClP/3i/41.

163 While Hooke supplies the pagination from the manuscript he transcribes, this does not match the pagination of Sloane 3423, indicating that he might have been copying another transcription.

of Brussels, and Pierre Antoine Bourdin, Charles I's master of horse, who informed Mayerne of techniques for dyeing horse skin, manes and tails.[164] He also sought out individuals who could inform him about global dyeing practices and materials. From Sir Nicholas Crisp, a future Fellow of the Royal Society who established English trade (including in slaves) on the Gold Coast, he learned of a "red wood of Guinea" used in dyeing black.[165] In 1648, an unnamed English surgeon who had "lived long in Goa" informed him about Indian methods of dyeing with the lac insect. Mayerne learned more about this material in 1650 from Captain Methold, the deputy governor of the English East India Company, who gave him an Indian dyeing receipt calling for five ingredients (gum-lac, the bark called "Lodur," the salt called "Sunchura," lemon, and limewater [calcium hydroxide]). Mayerne returned to his previous notes on lac to add a cross-reference to Methold's account and also speculated at length about the Indian "Sunchura" salt.[166] Mayerne had long been particularly curious about Indian mordants, showing a piece of printed calico to the Dutch dyer Jan Davidszoon in 1639 and asking him to speculate about what mordant had been used.[167]

Hooke, like many other members of the Society, was also interested in calico, and this shows in his decision about what to transcribe from Mayerne's works. Although his transcriptions were partial, he noted the contents that he was omitting, and he attempted to capture the three-dimensional nature of Mayerne's original files. In his transcription of loose Mayerne files, Hooke attempted to record the folding and filing system of the original. For instance, as Hooke noted in one case, one file had originally been labeled "on the outside of the paper" with the monograms of Mayerne and of Jean Colladon and the subjects, "Coccinea. Chameleon." As Hooke also notes, it contained "in the same paper" other material, including a dyed swatch that remains in the file. In his transcription of Mayerne's bound volumes, Hooke also attempted to portray how the volume contained folded files bound within it. Hooke noted within brackets, "a little scedule inserted between the 28 &29 page," and "a small scedule inserted within the former scedule," and again, "a paper lying

164 ClP/24/81. On Bourdin, see Trevor-Roper, *Europe's Physician,* 106 and 326.

165 Ibid, 215. Hunter, *The Royal Society and Its Fellows,* 160. R. Porter, "The Crispe Family and the African Trade in the Seventeenth Century," *The Journal of African History* 9,1 (1968): 57–77. ClP/24/82.

166 ClP/24/82 at #57. "vide in sequentibus circa haec materia a capitaneo Methwell." Captain "Methwell" or "Methuel," as Mayerne refers to him, became deputy governor of the East India Company in 1650 and lived in India for more than fifteen years, Mayerne noted on the third page of the unpaginated Classified Papers 24/82. See also Sloane 3423, 32v-33r.

167 ClP/3i/37.

between the 70 & 71 pages conteind between these marks {}."[168] Notably, however, Hooke's transcriptions did not capture the array of beautifully colored inks Mayerne had employed, for instance, in Sloane 3423. Yet, halfway through *Experimenta tinctoria* (from page 29 to 43 out of 99), Hooke briefly did employ what is now a pale purple ink, perhaps drawn directly from his own experiments with Mayerne's research.

Hooke's transcription of *Experimenta tinctoria* continues in what is now Classified Papers 24/82. Hooke concluded his transcription of this work with the note, "Finished the 6th of December 1669 at 7 at night." He immediately began transcribing another lost Mayerne volume, originally of 32 pages, entitled *Tinctoria Belgica*, in which Mayerne collected the work of a number of Dutch dyers. Hooke followed this with the transcription of a fourth volume, a French recipe collection perhaps originally assembled by Caspar Tomann of Zurich ("S. Gaspar Thoman), *Precepts for dyeing many sorts of things* (Praeceptes pour teindre plusieurs sortes de choses)."[169]

These transcriptions indicate that Hooke returned to the Mayerne material he had previously explored along with the other members of the 1666 committee on dyeing again in 1669 when he was undertaking his own research into dyes. Hooke's research began around October 21, 1669, when he brought to the Society "a piece of stuff stained by a way of his own contrivance which he hoped to perfect." Again on November 11, "Mr. Hooke produced a piece of calico stained after the way contrived by himself."[170] He was desired to proceed

168 ClP/24/82.

169 On Tomann, see Arnold Lätt, "Schweizer in England im 17. Jahrhundert," *Zeitschrift für Schweizerische Geschichte* 11 (1931): 316–352, and William Poole, "Theodoricus Gravius (fl. 1600–1661): Some Biographical Notes on a German Chymist and Scribe Working in Seventeenth-century England," *Ambix* 56,3 (2009): 239–252 (at n. 20). Tomann originally came to England to study in Oxford in 1600 (having previously studied in Zürich, Genf and Montpellier), but settled permanently in London. A chymist, Tomann came into conflict with the College of Physicians; he defended himself by claiming to only practice among foreigners. Margaret Pelling, *Medical Conflicts in early modern London: Patronage, Physicians, and Irregular Practitioners, 1550–1640* (Oxford: Clarendon Press, 2003), 185. Tomann was in touch with many visitors to London, signing the album amicorum of Friedrich Rechlinger in Oxford in 1603 (Bodleian MS Douce 244, 9v), of Francis Segar in London in 1609 (with the alchemical tessera, "Viridis linea omnia gyrat," Huntington Library Ms. 743, 338), and of the "mysteriarch" Joachim Morsius in London in 1620 (Stadtbibliothek Lübeck, Ms. hist. 4° 25,2, 397v). Caspar Tomann communicated the panacea of Mayerne to Ole Worm in 1630. Ole Worm, *Breve fra og til Ole Worm: 1607–1636*, vol. I, (Copenhagen: Munksgaard, 1965), 214 and 236.

170 Birch, Vol. 2, 396 and 401.

in these attempts, and on December 9, he brought to the Society's meeting further specimens of "yellow, red, green, blue and purple colours."[171] Thus, Hooke was actively experimenting with dyes while he busily transcribed Mayerne's collections, perhaps reflected in his unusual and sudden adoption of purple ink in his transcription of Mayerne's *Experimenta tinctoria*. Evidently, he felt that he had progressed in his color production. On January 13, 1669/70, he "brought in two specimens of staining better than those produced by him before." Perhaps drawing from Mayerne's notions about pre-mordanting, for these superior specimens, Hooke had determined "that the preparing of the cloth or stuff to be stained was a main thing in this work…."[172]

11 Mayerne as a Source of Experimental Method

While Mayerne's techniques seemed particularly of interest to Hooke, Fellows of the Royal Society did not only consider Mayerne's papers as a source of raw information for the writing of the history of trades. They also considered its experimental literary technologies. Robert Moray had previously criticized the Society's "promiscuous way" of experimentation and suggested a more organized approach.[173] At the January 25, 1679/80 meeting, while the mixture of metals was being discussed, the question arose again of "the best method of prosecuting experiments;" "Sir Theodore de Vaux queried, whether it would be agreeable to the society to bring in some accounts, which he had of experiments made by Sir Theodore Mayerne, that the society might examine, which of them were considerable and fit to be again examined. Upon which the society desired, that he would produce such papers, and spoke of appointing some persons to examine them."[174]

On February 5, 1679/80, "The method of making experiments was also farther discoursed of," and again, "Sir Theodore de Vaux was desired to bring in such papers, as he should think fit, and order should be taken to have them fairly copied into a book by themselves, that they might be perused; and that the papers, after having been copied, should be safely delivered to him again."[175] As the fellows were considering more methodical ways of prosecuting experiments, they envisioned a "fairly copied" separate book

171 Ibid, 411.

172 Ibid, 414.

173 On the "promiscuous way," Birch, Vol. 2, 132.

174 Birch, Vol. 4, 6.

175 Birch, Vol. 4, 8.

of Mayerne's experiments that they might consider as an experimental agenda. At the meeting of May 25, 1680, those present considered again whether or not Mayerne's papers represented a true experimental method, particularly with reference to Hooke's translation of Mayerne's paper on staining agates.[176]

The original paper "about a method of staining agates" is now Classified Papers 3i/41, and like the dyeing papers, is also not currently identified as Mayerne's. Mayerne's paper records the technique of Sir Edmund Bacon for agatizing flints that Wotton had so admired. Bacon's technique had already been brought to the Society's attention in 1662 by Sir Thomas Browne, who called it a "peculiar art."[177] Were it not for Mayerne's account, this widely celebrated technique would have failed to survive the Interregnum.

Bacon's artistic legacy was notably precarious. For instance, Samuel Hartlib commended Bacon for having "made a most compleate Laboratory in his House there w*i*th all manner of furnaces" but blamed him "for not choosing a right heire to his knowledges and ingenuities wherin hee excelled … Hee made a curious George out of Stone or Pebble, w*hi*ch hee gave to K*ing* Charles, w*hi*ch was much admired for the artifice of it."[178] In his will of 1648, Bacon left his nephew several such artificial stones, including an "achate with Queen Elizabeth's picture in it." The "particular combination of gifts and conditions" of this will, "set by a man who moved in royalist circles during the civil wars" has been seen as "an effort to transmit … the values of monarchy and the Church of England."[179] Both Queen Elizabeth and Saint George were immensely popular subjects of carved "agate" (a term employed for any carved gem in the period). Saint George was worn for daily use by members of the Order of the Garter, and by Charles I, who legendarily handed off his carved onyx George on the scaffold.[180] In passing the agate to the next generation, Bacon attempted not only to transmit an imperiled royalist image, but the courtly art in which it was literally imbedded.

---

176 Birch, Vol. 4, 87–8.

177 Thomas Browne, "An Account of Island, alias Iceland, in the year MDCLXII," *Works*, Simon Wilkin, ed., vol. IV (London: Pickering, 1835), 254–6; 255.

178 Hartlib, [29/5/17A].

179 Margery Kingsley, "Family, Inheritance, and Clarendon's *History of the Rebellion*," in *The Age of Projects*, Maximillian E. Novak, ed. (Toronto: University of Toronto Press, 2008), 29–50, 34. There is an artificially dyed agate portrait of Queen Elizabeth in the Victoria and Albert Museum, 1603–1855.

180 Julia Kagan, *Gem Engraving in Britain from Antiquity to the Present* (Oxford: Archaeopress, 2010).

Mayerne recorded the results of an interview with Bacon about his "*secret merveilleux*." He recalled how Bacon was able to make the image "enter into the stone (*entrer … dans la pierre*);" in one stone, Bacon had made a butterfly appear, and in another a spider, by using a "spirit of salt" without fire and by painting the stone on both sides. Bacon used "common pebbles" or flints, which when polished were as transparent as agates and chalcedony. Mayerne offered his own speculation about what Bacon's liquor might have been, with a reminder to himself to try it with the "*eau des deux frères*" (sal armoniac and saltpeter). Bacon also informed Mayerne that he had painted stones with a solution of a metal that created a landscape the color of the most beautiful silver. Mayerne offered his own "*fantaisie*" for the agatizing process, marked with his monogram. On the verso, Mayerne recorded further details he had learned following his "conference" with Bacon, such as Bacon's belief that nitre was a "great secret," his use of a fatty cheese to control the spread of his acid, of light colors that float on the surface made from different metals such as gold, silver, and copper, and of oyster shells as brushes. He noted that it would take six or seven months for the colors to penetrate in the winter, and much less in the summer. Mayerne also included a recipe in Italian for painting on agates sent to him from Italy by Nicholas Lanier (1588–1666) and further information from Bacon concerning pastes and artificial stones that could also take an impression and were very beautiful when dried and polished.

At the Royal Society's meeting of May 25, 1680, those present "concluded that the ways mentioned in the paper were only conjectures, and not the result of experiments; and that there were other ways of opening the bodies of stones, so as that they might imbibe colours, than by means of spirit of nitre, which would rather corrode a stone than sink into it."[181] Later investigators on staining stones would in fact use a spirit of nitre as the solution and metals for colorants.[182] More significantly, those present at this meeting did not believe that Mayerne's method of speculation and "*fantaisie*" represented a technique to be followed. The list of attendees was not recorded for this meeting, so we cannot know who was of this opinion or whether, for instance, de Vaux was present. Some at the meeting did affirm other aspects of Mayerne's paper, such as the suitability of English flints for the process. Yet, despite this mixed reception for Mayerne's paper on agatization, it is worth noting that at the moment when fellows of the Royal Society

181 Birch, Vol. 4, 87–8.

182 William Lewis, *Commercium philosophico-technicum: or, The philosophical commerce of arts* (London: Willock, 1763), 438.

were reviewing the proper method of prosecuting experiments, they did so in light of the cache of Mayerne manuscripts from which they had been drawing, via de Vaux, for the past sixteen years. Mayerne's papers in the Royal Society archive were considered not only for their content but also for their form.

This point is particularly worth considering given that the Royal Society is so often cast as following a Baconian program. The Society's interest in archives and record-keeping has been seen as intimately tied to that program.[183] Yet, the fact that the Fellows of the Society considered Mayerne as a possible experimental model, even if at this meeting they ended up rejecting him, illustrates that they were not satisfied with Bacon alone. They were aware that the early Stuart period offered other examples of experimental method, and they were eager to investigate those alternatives. They did so by considering the papers that de Vaux had brought into their collections, not only as a register of Baconian particulars, but as examples of Mayernian method.

12 Conclusion

Mayerne's larger-than-life personality saturates his papers, evoking his wide-ranging curiosity and his cosmopolitan networks all up and down the social scale, and offering a model of courtly experimental technique. *Archival Afterlives* offers a useful lens through which to explore De Vaux's career of representing Mayerne in the Society, as it allows us to come to grips with the ways in which archives are deployed to transmit personalities, allowing long-dead figures to continue to play a role in the social lives of those who copied, translated, stole, analyzed, re-filed, and competed with the papers of figures past. Mayerne's papers offer a view of a vanished color world, filtered through the lens of the passage of time and of sociopolitical upheaval. They can serve as a biographical object, allowing Mayerne's personality, social network, and style to have an afterlife in informing the establishment of a new aesthetic and culture of experiment.

If, according to Shapin and Schaffer, Boyle's literary technologies were intended to drain experimental reportage of personal subjectivity and interpersonal conflict, Mayerne's papers suffuse the archive with personality. From

183 eg. Palmira Fontes da Costa, *The Singular and the Making of Knowledge at the Royal Society of London in the Eighteenth Century* (Newcastle: Cambridge Scholars Publishing, 2009), 7.

Mayerne's papers to the Royal Society's system of registering inventions, early modern scientific archival practice is very much about assigning credit, dating discovery, and defending personal fame and profit.[184] It therefore highlights a proprietary self, at the same time as it offers that proprietary self a mechanism for interacting with other selves. [185] Mayerne was highly interested in his research, both financially and in order to preserve credit for his ideas. His possessive monogram was far from an abnegation of the self, just as his majestic style was far from a showcase of diffidence. He pursued lucrative crafts during a period of early Stuart projecting that was notoriously corrupt. Furthermore, the story of how various later individuals, like de Vaux, Povey and Hooke, engaged with his papers is one of conflicting personalities, agendas, and interests.

The story of his papers' afterlives highlights the social animus driving the build-up of the Society's archive, in the contentious and competitive relationships between de Vaux, Povey, and the Colladons. When gentlemen philosophers gathered together to interrogate Mayerne's papers, they did not enter a trustworthy sphere of matters of fact evacuated of subjectivity and interest. Their cognizance of this informed techniques for both deploying and neutralizing interest. On the one hand, De Vaux was desired to use "his interest" in unearthing Petitot's account of enamels for the Society's archive. On the other, assigning papers and trials to multiple individuals for cross-examination, such as Charleton's examination of Povey's trial and the examination of Mayerne's papers by the members of the committee on dyes and by Evelyn, Haak, Hooke, and Sloane, offered a means to remove trust in individuals from the equation. Social skepticism shaped the bureaucratic techniques devised for digesting both Mayerne's archive and the many other papers brought in to the Society.

Appendix

Surviving files in Classified Papers 3i and 24 referred to in this transcription are footnoted when possible. Sometimes the letters and/or numbers assigned by the committee remain visible on the document. In other cases, the identifications are not as certain, nor do they exhaust the surviving papers. Thus, the lack of a footnote does not indicate that a file has not survived.

184 On the system of registration, see Adrian Johns, *The Nature of the Book: Print and Knowledge in the Making* (Chicago: University of Chicago Press, 1998), 486.

185 Cf. Jason Scott-Warren, "Early Modern Bookkeeping," 155.

Meeting Minutes for the Committee for Dyeing, Royal Society, ClP/3i/27

May 28
Committee for dying met at Sir G. Ents

Sir G. Ent
Sr. Th. de Vaux
Mr. Hooke
Oldenburg
mr. Coxe
Mr. Colwall
Dr. Quatremain

a paper containing ~~all~~ the names of colors A. 1. sheet. M. Hook[186]
A receipt for dying of all sorts of colors. B. 1. sheet. M. Hook[187]
Questions for Mr. Fletcher concerning Dying C. 1 1/2 sheet. Sr G. Ent[188]
~~Mr. Fletcher's experiments: 2 sheets~~
Materials and Vessels used for dyeing. D. 1 1/2 sheet written. R. Hook
Sir Th. Vaux—Fletcher Tincture upon silk and wool with supel. n.1 two sheets and a loose paper[189]
Fletcher Stratford Bow n. 2. in 1. sheet Sr G. Ent.[190]
Fletcher of indigo. in. 2. sheets n. 3. Sr. G. Ent.[191]
Fletcher dictavit, 4 in 2 sheets. Dr. Quatremain[192]
Fletcher no. 5 in 2 sheets. Sr. G. Ent.[193]
Experiments made by Fletcher n. 6 in 2 sheets. Mr. Hook[194]

---

186 ClP/3i/29, "Colors."

187 ClP/3i/27 includes several recipes by Le Myre, Mr. Taylor, and Lannoy.

188 ClP/3i/30, "Questions for Mr. Fletcher." ClP/24/80/165, "Questions propounded unto Mr. Fletcher."

189 ClP/24/80/ unpaginated, "1. Fletcher 28 Jan. 1640 teincture sur soye & sur laine avec fusell.... 2. Extension de misteca. exaltation du violet. du long cochinelle. speculation novelle de [Mayerne] sur la fixation & exaltation du long cochenille &c."

190 ClP/ 24/80/172. "Dictavit Fletcher. From Stratford Bow.... All above written was told me by fletcher Ap. 20 1639."

191 ClP/24/80/178. "For Indico Tincture. Fletcher."

192 ClP/24/80/169. "17 May Mr. Fletcher Dictated. Indico."

193 [?] ClP/24/80/166–168, "Mr. Fletcher Aug 1 1639. Chelsay. Logwood."

194 ClP/24/80/unpaginated. "Articles that Mr. Fletcher gave me in writing April 12. 1639," containing six items.

Dr. Quatremain. Notes upon Keflers operacion nr. 7 in 1 sheet.
Dr. Quatremain. Peter van den Couter. Callico and other cotton stuffes, blews, n. 8 sheet 1.
Fletcher of Cochenille and safflor n. 9. sheet 1. Sir Th. de Vaux[195]
Fletcher of Cochenille and Turmerick Sir. G. Ent[196]
Fletcher of Scarlate n. 11. sheet 1[197]
Jean Davids of Scarlate n. 12. sheet 1. written.[198]
Grappenroot. n. 13. sheet 1.[199]
Black foxes. n. 14. Sheet 1/2.[200]
Memento of Fletcher ..... E. sheets 2. Sr The Vaux[201]
Black of Flanders by Lanoy. F. sheet 1.[202]
Flecher de Tincturis. G. sheet 1.
Lanoy, Waloon Dyer. H. sheet 2. Th. de Vaux[203]
Two Experiments. I. sheet 1. Sr Th. de Vaux.
Indico Fletcher K. 2 quarto pages. Sr Th. de Vaux[204]
What he hath seen and heard of M. K. (L. sheet 1/2)[205]

---

195 ClP/3i/31, "Escarlatte avec Safflor." "Fletcher, 13 September, 1639." ClP/24/80/176. "The tincture of Safflowre Tho. Fletcher." "To make the fair scarlet with safflower." ClP/24/80/unpaginated. "9. Fletcher dictavet et dixit 24 Aug. 1639. St. Martin Long Cochineel translation de safflor."

196 ClP/24/80/171, "ex ipsius ore. 14 August 1639. Fletcher. To dye wool with cochineel ....Turmericke."

197 ClP/3i/30, "Fletcher, 13 Sept. 1639. Escarlatte sur soye avec Safflor & annatto. Pourpre avec Long Cochenill & cognee d'huistres. Tanné. Aurange. Speculations sur la teincture de la soy en Escarlatte avec Cochenille. 11." The English translation is atClP/3i/38, "Fletcher 13 Sept. 1639. Scarlat upon silk with safflor and Annatto. Pourple with long Cochineel, and oystershels. Tawne. Orenge. Speculation upon the dying of silk and scarlat, with Cocheneel."

198 ClP/3i/34, "Jan Davids tencturier de Harlem, Teincture d'escarlate aux Cochenille, Avec Graine en poudre. Gutta gumii. Calico Imprimé. fiol de boeuf. Essay des choses teignantes. 15 September 1639.12." The English translation is at ClP/3i/37.

199 ClP/3i/34, "Grapproot singulieur, 20 Sepember 1639. 13."

200 ClP/3i/39, "Renards noirs."

201 ClP/24/81, "memento comme fletcher rougit."

202 ClP/3i/34, "Noir de Flandres" by "Mr. de Lanoy son teincturier à Londre. 4 May 1646. F." Lanoy's address is given inside as "southwarke, en Montague close."

203 ClP/3i/27, "Verdbrun avec Logwood de Mr. Lannoy. 1650." ClP/24/81/188v, "Mr. Lanoy Tincturier wallon en South wark. 17 1647." According to ClP/24/81, this was a Pierre de Lannoy.

204 ClP/24/80/169. "17 May Mr. Fletcher Dictated. Indico."

205 ClP/3i/35, "Ce que jay veu & appris de la bouche de M. K."

Four papers of ink. M.[206]
A Receipt to fix a cuve of Indigo N. sheet 1/2. M. Hook
Briot of metalline Colours. O. sheet 1/2 Sr G. Ent[207]
Articles of inventions not yet published. P. sheet 1/2.[208]
Le pape his paper. Q. sheet 1. Dr. quatremain
To take away all kinds of spots and to prevent moths. R. sheet 1/2 Sr The Vaux[209]
Opus of Fletcher 9 May 1639 S. sheets 2. Sr. G Ent[210]
Quaerenda. T. pag. 1.[211]
Experiments made upon ye tincture of Scarlate after Keflers way. Sheet. 1. U[212]
Dr. Cherevix propriò ore. sheets 3. all in one book. V. Mr. Coxe

Manuscript References

British Library. Additional 72897. Additional MS 78339. Additional MS 78340. Sloane 1139. Sloane 1512. Sloane 2001. Sloane 2079. Sloane 2052. Sloane 2057. Sloane 2222. Sloane 3292. Sloane 3423. Sloane 3426.
Bodleian Library. MS Aubrey 26.
Royal Society Archives. Classified Papers: ClP/2/24, ClP/2/25, ClP/3i/1, ClP/3i/6, ClP/3i/18 ClP/3i/24, ClP/3i/25, ClP/31/26, ClP/3i/27, ClP/3i/28, ClP/3i/29, ClP/3i/30, ClP/3i/31, ClP/3i/33, ClP/3i/34, ClP/3i/35,ClP/3i/36, ClP/3i/37, ClP/3i/38, ClP/3i/39, ClP/3i/40, ClP/3i/41, ClP/11i/9, ClP/14i/38, ClP/15i/50. ClP/15i/45, ClP/15i/46, ClP/15i/49, ClP/17/12, ClP/19/29, ClP/20/50, ClP/20/79, ClP/24/80, ClP/24/81, ClP/24/82. Domestic Manuscripts: DM/ 5/74a-b. Register Book Original: RBO/1/50, RBO/9/10, RBO/9/9, RBO/9/11. Letter Book Original: LBO/12/111.
Wellcome Library. MS 716.

206 ClP/3i/27, in German and French, 1630.

207 ClP/3i/30, an English translation of "Mr. Briot of metallic colors." See ClP/24/80/172, "Mr. Briot of divers tinctures of Acids and of mettalls."

208 ClP/3i/27, "Articles des Inventions non encor practiquee dans Albion publiquement 1639. P." ClP/24/80/179."Articles that Mr. Fletcher gave me in writing. April 12. 1629."

209 ClP/24/80/unpaginated. "Oster taches de draps de soye."

210 ClP/3i/30, "Opus Mr. Fletcher," and in English translation, ClP/24/80/174, "Opus Tho. Fletcher May 9 1639."

211 ClP/3i/38.

212 ClP/ 24/80/unpaginated. "1640. Deux Experiments … Dr Colladon, februar; Experiment sur la teincture d'escarlatte faict le 30 Janvier 1640."

CHAPTER 3

Accidental Archive: Samuel Hartlib and the Afterlife of Female Scholars

Carol Pal

In 1752, George Ballard was compiling an encyclopedic volume on learned Englishwomen. He had a wealth of material for the eighteenth century; but when it came to female scholars of the seventeenth century, he confessed he had been forced to give up:

> Those whose memoirs are here offered to the public I have placed in the order of time in which they lived, omitting none of whom I could collect sufficient materials. For as there may yet be some learned women of those times whose character I am an entire stranger to, so there are others whom I well know to have been persons of distinguished parts and learning, but have been able to collect very little else relating to them. Such are ... Lady Ranelagh ... Mrs. Makins (who corresponded in the learned languages with Mrs Anna Maria à Schurman) ... together with very many other learned and ingenious women since the year 1700.[1]

This documentary oblivion was a fate that had already been anticipated by some of Ballard's learned women themselves. At the age of 30, the celebrated Dutch scholar Anna Maria van Schurman (1607–1678) had thought about her role in the Republic of Letters and had been acutely aware of two things: first, that she was a full and active participant; and second, that all memory of her participation was destined for obscurity because she was a woman. The problem did not lie with her contemporary colleagues, the content of her correspondence, or with the seventeenth-century enterprise for the advancement of learning. The problem, she knew, was History.

In a letter to her mentor André Rivet, she had argued for the right of women to achieve, to become scholars, artists, and theologians. In effect, it had been an argument for her own identity, and a place for that identity in intellectual

1 George Ballard, *Memoirs of Several Ladies of Great Britain, Who Have Been Celebrated for their Writings of Skill in the Learned Languages, Arts and Sciences* (1752), ed. Ruth Perry (Detroit: Wayne State University Press, 1985), 54.

 | DOI 10.1163/9789004324305_005

history—because without that place, it would be as though she had never existed at all:

> In vain do we women boast of the nobility we have inherited from our ancestors, since it is soon enveloped by a useless obscurity. Thus we find that in reading the history of great expanses of time, the memorials to women's names are no more in evidence than the traces left by a ship crossing the ocean.[2]

Van Schurman had known from childhood that she would not marry and would not have children. Who, then, would remember her? And for what achievements would she be remembered?

The answers to van Schurman's implicit questions require a reformulation of the premise. The real question here is not, "Who is remembered?" but, "Who is doing the remembering?" and "Why?" Archivally, of course, this also leads to the question of: "Where?"

These answers have been difficult to find. Even three centuries later, a frustrated Virginia Woolf would find herself no better off than George Ballard, stymied by her inability to find any documentation of scholarly women prior to the eighteenth century:

> ... what I find deplorable, I continued, looking about the bookshelves again, is that nothing is known about women before the eighteenth century. I have no model in my mind to turn about this way and that.[3]

Thus it appeared that van Schurman's fears had been borne out—all memory of these women as active scholars had indeed sunk without a trace beneath the ocean of History.

However, as this essay will show, those documentary traces had not really vanished. These female scholars would indeed have an archival afterlife, but it would be in an unintentional archive—a place where they would be both unexpected and unlooked-for. Seventeenth-century female scholars were highly

2 "Frustra iactamus tanquam nostram quam a maioribus accepimus nobilitatem: siquidem mox eandem ignava involvit obscuritas. Hinc est quod in historia legenti longissimo saepe temporis tractu nostri nominis monumenta non magis comparent, quam navis per mare transeuntis vestigium." Anna Maria van Schurman to André Rivet, 6 November 1637. Koninklijke Bibliotheek MS. 133 B 8, no. 14. Later printed in *Nobiliss. Virginis Annae Mariae à Schurman, Opuscula Hebraea, Graeca, Latina, Gallica: Prosaica & Metrica* (Leiden, 1648), 69–70.

3 Virginia Woolf, *A Room of One's Own* (New York: Harcourt, Brace & World, 1929), 46–7.

active in their intellectual worlds, and well-known in the Republic of Letters; at the same time, however, they could not belong to any of the scientific, educational, or religious institutions that would eventually curate the papers of their male correspondents, and house their archives. Where, then, would their papers find their archival home? For a number of female scholars—including Bathsua Makin (c.1600–1680), Dorothy Moore (c.1612–1664), and Katherine Jones, Lady Ranelagh (1615–1691)—that home would eventually be found in the papers of the intelligencer Samuel Hartlib (c.1600–1662).

Hartlib's papers document his work in the service of a combination of seventeenth-century projects—the advancement of learning, the pan-Protestant cause, and a general program of improvement—and are a rich repository of letters, copies, excerpts, pamphlets, and notes. However, they were never intended to function as an archive. They were intended for himself alone; and unlike collections assembled with a memorial or historical sensibility, their animating logic was functionality in the present. Yet for the female scholars in his network, whose intellectual work was not part of the archival logic of other collections, Hartlib's papers became their de facto archival home. And this brings up archival questions—questions regarding the categories defined and created by an archive, and the challenges presented by the categories that otherwise defined seventeenth-century women's lives. This essay examines the presence of female scholars in the Hartlib Papers, and the ways in which that presence constructs an unintentional archive of their intellectual identities.

1 Hartlib's Papers

The papers of the intelligencer Samuel Hartlib constitute a significant collection. They document his astounding industry in gathering knowledge of all kinds—especially in philosophy, religion, science, and education—and then disseminating that knowledge in various forms in order to bring about a worldwide reformation of faith and learning.[4] Most of Hartlib's papers are at the University of Sheffield, and together with holdings elsewhere, these total

4 On Hartlib, see especially G. H. Turnbull, *Hartlib, Dury and Comenius: Gleanings from Hartlib's Papers* (London: Hodder & Stoughton, 1947), hereinafter "*HDC*;" Charles Webster, *The Great Instauration: Science, Medicine, and Reform 1626–1660* (London: Duckworth, 1975), and Mark Greengrass, Michael Leslie, and Timothy Raylor, eds., *Samuel Hartlib and Universal Reformation: Studies in Intellectual Communication* (Cambridge: Cambridge University Press, 1994).

over 5,500 items.[5] To this total, one must add the "scribal shadow" of Hartlib's presence in the archives of his closest collaborators—in this case, the Scottish pastor and ecumenical advocate John Dury (1596–1680), and the Moravian educationalist Jan Amos Comenius (1592–1670).[6] But as recent work has shown, this impressive remnant probably represents less than half of Hartlib's original correspondence, which would thus have totaled over 11,500 items—putting him on a level with intelligencers like Nicolas-Claude Fabri de Peiresc (1580–1637).[7]

Yet to categorize this significant collection as Hartlib's archive is problematic on at least two levels. First, because the collection was not created for the reasons we usually associate with archives. Hartlib's network functioned quite literally as an information factory, creating what he deemed useful knowledge—not just knowledge for its own sake, but knowledge that might be useful in constructing a new world. Some categories of knowledge were dependent on a genealogy of secrets, and this was especially true of alchemical practices and certain medical recipes.[8] However, Hartlib's own interests lay in sharing and expanding knowledge, rather than keeping it intact. Knowledge,

5 There are 3,484 of Hartlib's letters or letter extracts at the University of Sheffield, and 1,411 more in the British Library, Yale University, and elsewhere, for a total of 4,895—to which one can add various memoranda, enclosures, etc. See Mark Greengrass and Leigh Penman, "L'ombre des archives dans le cultures du savoir du XVIIe siècle: Le cas des papiers de Samuel Hartlib (c.1600–1662)," *Bibliothèque de l'École des Chartes* 171 (2013): 51–64; and Leigh Penman, "Omnium Exposita Rapinae: The Afterlives of the Papers of Samuel Hartlib," *Book History* 19 (2016): 1–65.

6 Mark Greengrass, "Archive Refractions: Hartlib's Papers and the workings of an Intelligencer," in *Archives of the Scientific Revolution: The Formation and Exchange of Ideas in Seventeenth-century Europe*, ed. Michael Hunter (Woodbridge: Boydell Press, 1998), 35–47, here 36–8.

7 Leigh Penman made this calculation by working with a small representative sample from Hartlib's papers. He mined that sample for all references to other letters, compared that total with what was extant, and developed an algorithm to express the result: L = S(x/y). "L" represents all letters received by Hartlib, while "S" is the sample being used; "x" represents the number of references in the sample to letters no longer extant, and "y" the references to existing letters. The extant papers today would thus represent 42% of the original total. Greengrass and Penman, "L'ombre des archives." On Peiresc, see especially Peter Miller, *Peiresc's Europe: Learning and Virtue in the Seventeenth Century* (New Haven: Yale University Press, 2000).

8 On the specifics of dealing with secret knowledges, see Michelle DiMeo, "Openness vs. Secrecy in the Hartlib Circle: Revisiting 'Democratic Baconianism' in Interregnum England," in *Secrets and Knowledge in Medicine and Science, 1500–1800*, eds. Elaine Leong and Alisha Rankin (Farnham, UK: Ashgate, 2011), 105–124.

in his view, was a living thing, and if it could not spread and grow, it might literally perish. As he wrote in his work diary, *Ephemerides*:

> A great injury that stationers have copies for ever. It should suffice they should enioy them for 5. 10. 15 *years*. Otherwise they never reprint them and by this meanes many good books are suppressed or perish altogether.[9]

Clearly, Hartlib believed strongly that texts must either circulate or die. This applied to publications, as well as manuscripts and papers—thus the responsibility for preserving textual knowledge lay not only with scholars, authors, and collectors, but with printers, booksellers, and stationers as well.

Given this circulatory imperative, Hartlib's papers could plausibly be seen as having functioned during his lifetime as an anti-archive. As Elizabeth Yale has argued, scientific archives in Stuart England reveal the development of a particular "historical consciousness" at the end of the seventeenth century, and the desire of naturalists and antiquaries to preserve their work for the use of future generations in the field. Hartlib, however, died in 1662; he was thus part of an earlier generation, with a very different focus. He did not gather his papers for the purposes of memory, preservation, or history; instead, he gathered them together in order to redistribute them again, putting them to immediate use for the public good.[10] The focus was not on informing the future, but rather on building it.

Secondly, as Leigh Penman has shown, the collection itself is at best a "rump."[11] In the last year of Hartlib's life, all his "best papers"—including, most importantly, the catalogue he had created—somehow disappeared from the home of the person he had entrusted with their safety. This was followed by a fire several months later, thus even Hartlib's less-than-best papers were damaged and reduced. Then, after Hartlib's death, the surviving remnant of his

9 From *The Hartlib Papers. A complete text and image database of the papers of Samuel Hartlib (c.1600–1662)*. Second edition. University of Sheffield, 2002 (hereinafter "HP"). Ephemerides for 1641, HP 30/4/76B.

10 As Yale points out, Hartlib's practice stood "in strong contrast to that of the next generation of English naturalists, who tended to deposit papers in public repositories rather than disseminate them." Elizabeth Yale, "With Slips and Scraps: How Early Modern Naturalists Invented the Archive," *Book History* 12 (2009): 1–36; here 12–13. See also Stephen Clucas, "Samuel Hartlib's *Ephemerides,* 1635–59, and the Pursuit of Scientific and Philosophical Manuscripts: the Religious Ethos of an Intelligencer," *The Seventeenth Century* 6,1 (1991): 33–55.

11 This discussion is taken from Penman, "Omnium Exposita Rapinae," esp. 4–9, 48–52.

papers suffered a variety of fates, being reorganized, pilfered, extracted, and sold off by an assortment of relatives, collectors, and erstwhile collaborators. Thus the Hartlib papers as we know them now, no matter how impressive a collection, are at best a vestige and a shadow of what was once there and can in no way represent what Hartlib might have thought to designate as his archive.

Nevertheless, for the female scholars who entered Hartlib's orbit, this remnant is perhaps their only archive, the only location in which they are remembered for their work alone. Their names might be recorded in family or church archives for their births, their marriages, or their descendants, and their letters and artistic creations might have entered other collections. But only in the Hartlib papers are they contained in a single place as part of an intellectual collectivity. The following sections examine this presence.

2 Princess Elisabeth of Bohemia (1618–1680)

We begin with Princess Elisabeth of Bohemia. Elisabeth's presence in the Hartlib papers is not extensive; however, that small presence provides an indicator of how a female scholar might have given consideration to her own archival afterlife and attempted to control it. Among the female scholars whose intellectual careers are documented in the Hartlib papers, she is the only one to evince any attentiveness to her scholarly legacy; that attentiveness, coupled with her awareness of the value of good editorial practices, makes her a counter-example that serves to demonstrate the range of female scholarship represented in this collection. The following exchanges are taken from what is now referred to as the correspondence of Hartlib's friend and collaborator John Worthington (1618–1671); however, it is quite probable that these letters originated in Hartlib's papers, since after Hartlib's death, Worthington organized and sorted much of the collection, removing some of his own material in the process.[12]

---

12 Interestingly, it appears that Worthington's correspondence might have met the same fate as Hartlib's papers. According to Worthington's executor Edward Fowler (later Bishop of Gloucester), some of Worthington's papers also seem to have disappeared for a while, only to reappear later: "As to the rest of the Doctor's (truly valuable) Papers; Such as I found in his Study, were carefully preserved, and faithfully disposed of according to the express Intent of his Will. But some were never in my Custody, nor heard I of them for many Years; they lying concealed in unknown Hands, out of which they were not retrieved till long after the Expiration of my Trust: And such are those contained in this Volume." One wonders whether during that time of abstraction, some items might have disappeared

Elisabeth, the granddaughter of King James I, has entered intellectual history primarily through her correspondence with René Descartes.[13] Their exchanges embraced both mathematics and philosophy, giving clear evidence of the depth and breadth of her intellect. Her scholarly life, however, was always secondary to her life as the member of a royal house in exile.

It had not been easy to find a permanent home for the resolutely celibate and scholarly Princess Elisabeth. Based on the reports of family members, she was generally a difficult guest, who could not refrain from trying to manage affairs wherever she happened to be. After the Restoration in 1660, when her mother, the Queen of Bohemia, returned to live in England, there had been speculation that Elisabeth might join her. Hartlib and Henry More were especially excited that the "Cartesian princess" might soon be on their shores.[14] Worthington wrote to Hartlib in October, 1660, asking when Elisabeth was due to arrive, since More was anxious to see her, and having her there "would make ye world a more desirable place."[15] Hartlib was equally

---

forever, or found themselves in new homes. From Fowler's Preface to John Worthington, *Miscellanies ... Also, A Collection of Epistles, Written to Mr. Hartlib of Pious Memory ... With the Author's Character, by Arch-Bp. Tillotson* (London: Printed for John Wyat, 1704), 291–5. The *Miscellanies* were edited by Worthington's son. See Richard Copley Christie, ed., *A Bibliography of the Works Written and Edited by Dr. John Worthington. Remains Historical & Literary Connected with the Palatine Counties of Lancaster and Chester*, Vol. 13, New Series (Manchester: Printed for the Chetham Society, 1888),63. On Worthington's involvement with Hartlib's papers, see Penman, "Omnium Exposita Rapinae,"17–25.

13 Recent work has begun to assess Elisabeth as a philosopher, as well as a royal. See especially *The Correspondence between Princess Elisabeth of Bohemia and René Descartes*, trans. Lisa Shapiro (Chicago and London: University of Chicago Press, 2007). See also: Beatrice H. Zedler, "The Three Princesses," *Hypatia* 4,1 (1989): 28–63; Erica Harth, *Cartesian Women: Versions and Subversions of Rational Discourse in the Old Regime* (Ithaca: Cornell University Press, 1992); Anna Creese, *The Letters of Elisabeth, Princess Palatine: A Seventeenth-Century Correspondence* (Ph.D. Dissertation: Princeton University, 1993); Deborah Tollefsen, "Princess Elisabeth and the Problem of Mind-Body Interaction," *Hypatia* 14,3 (1999): 59–77; and a somewhat romanticized treatment by Andrea Nye, *The Princess and the Philosopher: Letters of Elisabeth of the Palatine to René Descartes* (Lanham: Rowman & Littlefield, 1999). See also Carol Pal, *Republic of Women: Rethinking the Republic of Letters in the Seventeenth Century* (Cambridge: Cambridge University Press, 2012), esp. Chapter One.

14 Hartlib to Worthington, 15 October, 1660. Cambridge University Library, Baker MSS 29, ff. 209–10. Also printed in James Crossley, ed., *The Diary and Correspondence of Dr. John Worthington. Remains Historical & Literary Connected with the Palatine Counties of Lancaster and Chester*, Vol. 13 (Manchester: Printed for the Chetham Society, 1847–1886), 210–5.

15 Worthington to Hartlib, 25 October, 1660. In Crossley, vol. 13, 225–6.

enthusiastic, and in a rather whimsical moment wished that Elisabeth could marry More's ideas, and that the offspring of this marriage would be a wonderful Cartesian book:

> I wish she were in England, that she might marrie Dr. More's Cartesian Notions, which would beget a noble offspring of many excellent and fruitfull truths.[16]

In the end, Elisabeth did not join her mother in England, and Hartlib's hopes for this marriage of true Cartesian minds would come to naught.

However, in voicing their mutual disappointment, Worthington and Hartlib also provide us with further information regarding Elisabeth's archival awareness, and her ongoing control over the publication of her correspondence. In October, 1661, a new volume of Descartes' letters was published, leading to further disappointment on Worthington's part. He wrote to Hartlib:

> I suppose you have seen or heard of Des-Cartes his Second Volume of Letters; wherein many, or most of them, are about Matters betwixt him and Mersennus. They are all in French that are in this 2nd volume: No Letters to the Princess Elizabeth ...[17]

Then we find a passage that reveals much more than Worthington's wishes concerning the publication of Descartes' letters. Worthington's letter continued:

> I have read in some of your Papers an Extract of a Letter of hers, wherein she mentions some Letters of Des-Cartes to herself, which are not in the First Volume of his Letters, and are more worthy to be printed than several others in that Volume. She also thought, that the Methodizing and placing of the Letters might have been to better Advantage. If those Letters unprinted might be imparted to the Publick, they would be a great Ornament to the Second Edition of those Epistles.[18]

---

16 Hartlib to Worthington, 28 May, 1661. Cambridge University Library, Baker MSS 29, ff. 218–9. Also printed in Crossley, v. 13, 314–8.

17 Worthington to Hartlib, 7 October, 1661. Epistle XVIII in John Worthington, *Miscellanies*, 291–5. Also printed in Crossley, v. 36, 47–54.

18 Worthington to Hartlib, 7 October, 1661. Epistle XVIII in John Worthington, *Miscellanies*, 291–5. Also printed in Crossley, v. 36, 47–54.

And Hartlib agreed: "I wish heartily that both these volumes of letters were put into one book and published."[19]

Two points stand out in this passage. First, we see how Princess Elisabeth is thinking about her intellectual legacy; she is both withholding her own Cartesian letters, and voicing her criticism of the way in which the newly-published letters have been handled.[20] Clearly, Clerselier's editorial practices would not do for her own scholarly archive—and in the end, her correspondence with Descartes was not published until the end of the nineteenth century.[21] And secondly, we see the circulatory function of Hartlib's papers at work. Worthington's source for his knowledge of the missing letters is an extract from one of Elisabeth's letters—and he reminds Hartlib that this was an extract "I have read in some of your Papers."

Elisabeth's attentiveness to her intellectual legacy stands in stark contrast to the lament of her friend Anna Maria van Schurman, and is perhaps a response to the Dutch scholar's observation that the historical record of female scholarship amounted to no more than "the traces of a ship crossing the ocean." Elisabeth's presence in Hartlib's papers is therefore both suggestive and anomalous, because it documents a learned woman who is fully conscious of her intellectual legacy, and actively attempting to construct her own archival afterlife.

3 Bathsua Makin (c.1600–1680)

Bathsua Makin was an educator—a teacher, scholar, and writer, who made a forceful argument for the education of women.[22] She was the daughter of a

---

19 Hartlib to Worthington, October, 1661. In Crossley, v. 36, p. 57.

20 Descartes died in 1650. The reference here is to the first published edition of his correspondence: Claude Clerselier, ed. *Lettres de M. Descartes*, 3 vols. (Paris, 1657–1667).

21 Alexandre Foucher de Careil, *Descartes, la Princesse Elisabeth et la reine Christine, d'apres des lettres inedites* (Paris, 1879).

22 For a contemporary's view of Makin, see Sir Simonds D'Ewes, *The Autobiography and Correspondence of Sir Simonds D'Ewes, Bart., during the Reigns of James I. and Charles I*, ed. James O. Halliwell, 2 vols. (London: R. Bentley, 1845). Among the more important studies on Makin are: Jean R. Brink, "Bathsua Makin: Educator and Linguist," in *Female Scholars: A Tradition of Learned Women Before 1800*, ed. J. Brink (Montreal: Eden Press, 1980), 86–100; Vivian Salmon, "Bathsua Makin (1600-c.1673): A Pioneer Linguist and Feminist," in *Language and Society in Early Modern England: Selected Essays 1981–1994*, ed. V. Salmon (Amsterdam: John Benjamins Publishing, 1996), 239–60; Frances Teague, *Bathsua Makin, Woman of Learning* (Lewisburg: Bucknell University Press, 1998); and Anne Leslie Saunders, "Bathsua Reginald Makin (1600–1675?)," in *Women Writing Latin: From Roman Antiquity to Early Modern Europe*, ed.

schoolmaster, Henry Reynolds, about whom little is known. However, he was probably responsible for educating his remarkable daughter, who was recognized while still in her teens as the true scholar of the family. Henry Reynolds was an expert in cipher, and his daughter described him as "a schoolmaster and language lover."[23] He was not, however, a notable scholar; that honor, according to the parliamentarian Sir Simonds D'Ewes (1602–1650), truly belonged to his daughter Bathsua. D'Ewes had been a pupil at Reynolds' school in St. Mary Axe parish in London, where he observed that Reynolds had "a pleasing way of teaching," but that his real claim to fame was the remarkable Bathsua.[24] In fact, according to D'Ewes, she was "the greatest scholler, I thinke, of a woman in England."[25]

Makin worked primarily as a tutor, with a focus on languages; she tutored Princess Elizabeth Stuart (1635–1650), Lucy Davies Hastings, the Countess of Huntingdon (1613–1679), and Lucy's daughter Elizabeth Langham (1635–1664). Makin also appears to have published a pamphlet outlining a new shorthand system. Nothing survives of her 1619 *Index Radiographia* except for the title page, which gives a small sample of her system of strokes, dots, and semi-circles.[26]

Laurie J. Churchill et al., 3 vols. (New York and London: Routledge, 2002), 3:247–270. See also Pal, *Republic of Women*, esp. Chapter Six.

23 On the title-page of her first publication, Makin described herself as "Bathsua R. (daughter of Henry Reynolds, Schoolmaster and Language Lover in London)." Bathsua Reynolds (Makin), *Musa Virginea Graeco-Latino-Gallica, Bathsuae R. [filiae Henrici reginaldi Gymnasiarchae et philoglotti apud Londonienses]. Anno aetatis suae decimo sexto edita* (London: Edward Griffin, 1616). Her marriage to Richard Makin took place on 5 March, 1622.

24 "He had a daughter named Bathsua, being his eldest, that had an exact knowledge in the Greek, Latin, and French tongues, with some insight also into the Hebrew and Syriac; much more learning she had doubtless than her father, who was a mere pretender to it; and by the fame of her abilities, which she had acquired from others, he got many scholars which else would neither have repaired to him nor have long staid with him." D'Ewes, *Autobiography*, Vol. I, 63. On D'Ewes, see J. Sears McGee, *An Industrious Mind: The Worlds of Sir Simonds D'Ewes* (Stanford: Stanford University Press, 2015).

25 *The Diary of Sir Simonds D'Ewes, 1622–1624: Journal d'un étudiant londonien sous le règne de Jacques 1er*, ed. and trans. Elizabeth Bourcier (Paris: Didier, 1974), 68–9.

26 *Index Radiographia*'s dedication to Queen Anne places its publication before 1619. Although John Westby-Gibson claimed to be in possession of a 1617 manuscript version of the entire text, it disappeared when his collection was dispersed after his death. John Westby-Gibson, *The Bibliography of Shorthand* (London: Isaac Pitman and Sons, 1887), 188. The title page of *Index Radiographia* survives as Plate IV in R. C. Alston, *Treatises on Short-hand* (London: E.J. Arnold & Son, c.1866). E.H. Butler called Makin "the first lady shorthand inventor." See E. H. Butler, *The Story of British Shorthand* (London: Isaac Pitman and Sons, 1951), 212–14.

Her most well-known work, however, was published anonymously. Her *Essay to Revive the Antient Education of Gentlewomen* (1673) was a hybrid publication. In fact, it was an advertising brochure, intended to recruit students for a proposed double school for young gentlemen and gentlewomen. It featured a 34-page feminist argument—using historical examples, current scholarship, and a liberal dash of sarcasm—for giving women full access to the "whole Encyclopaedia of Learning," while also reassuring men that this would not threaten their social superiority.[27] The argument was followed by nine pages of pedagogical descriptions written by the educationalist Mark Lewis (1622–1681), Makin's proposed partner in this double school.[28]

Makin's connections with the Hartlib network do not appear to have been extensive. Some of Makin's intellectual contacts may have been made through her brother-in-law, the mathematician John Pell (1611–1685), who was a long-time Hartlib friend and collaborator; however, it is highly unlikely that he actually helped to facilitate her career. The distance was in part personal—Pell felt she was a bad influence on his wife Ithumaria—and in part political.[29] Makin shared the Hartlib network's enthusiasm for the pedagogical methods of Jan Amos Comenius, which she would vigorously advocate and promote in her printed work; however, she gave no evidence of a similar affinity for his world-changing agenda of reform. Instead, during these revolutionary years, Makin pursued a radical reformation of learning within a traditional

---

27 [Bathsua Makin], *An Essay to Revive the Antient Education of Gentlewomen, in Religion, Manners, Arts & Tongues. With an Answer to the Objections against this Way of Education* (London, 1673), 24.

28 There are entire passages in the pedagogical portion of Makin's *Essay* which are identical to passages in the work of Mark Lewis. This is because the *Essay* was a co-authored, hybrid text. It does not mean that we are looking at a case of early modern plagiarism; nor, as has recently been alleged, does this indicate that Mark Lewis was probably the true author of the entire *Essay*. Makin wrote the feminist argument, which comprised approximately 80% of the text; Mark Lewis was responsible for the remaining 20%, which presented the pedagogical program for their proposed school. For a different view, see Noel Malcolm and Jacqueline Stedall, *John Pell (1611–1685) and his correspondence with Sir Charles Cavendish: The mental world of an early modern mathematician* (Oxford, 2005), 232.

29 Pell was particularly worried about Makin's bad influence on Ithumaria while he was away from home as an agent for Cromwell on the Continent. In a letter criticizing his wife for her money management, he wrote: "I perceive Mrs. Mak. hath found the way to you againe. Her counsel, added to your owne inclinations, will make you altogether unfit to meddle with any mony of mine. So that I shall be forced to seeke some other, who will dispose of it punctually, according to my order." 25 March [1657]. BL, Add. MS. 4280, f. 214.

context—that of tutor to the children of the aristocracy. However, Makin does appear a number of times in Hartlib's *Ephemerides,* documenting her active presence in several learned networks that were of interest to Hartlib. Hartlib first mentions Makin in 1637, in an excerpt transcribed from a report in Pell's *Repraesentationes Rerum*:

> In Mr. Gelebrand's study amongst other things there was a description of Mr. Wright's Uranicum, with the construction of it, and the measures of every wheel as he told me. Another description of it and the use of it fairly written with a picture of it is I believe in my Sister Makin's library, for she had all her father's books ...[30]

Given Henry Reynolds' long career as an educator, the fact that Makin had inherited "all her father's books" makes it likely that she was the possessor of a fairly impressive library.

We also learn from Hartlib of Makin's participation in medical networks. Notations in Hartlib's *Ephemerides* for 1650 record the following for "Mris. Makin":

> Very well acquainted with Sir W. Raleigh's son a Parl*iament* man now. From the old Lady shee got all Sir W. R*aleigh's* Rece*ipts*.
>
> Mris Makin cured Mr Holsworth of the Palpitation of the heart. So did Dr Gurdain with his spirit of Amb*er Greece*.[31]

It is not clear how Makin became "very well acquainted" with Sir Carew Raleigh and "the old Lady," his mother.[32] It is also unclear whether the medical recipes had been a gift or a purchase. However, two years after Hartlib recorded this note, these medical recipes were back with the Raleigh family. In 1652, Carew Raleigh wrote in a letter to Sir Edward Conway, the husband of the philosopher Anne Conway, that he had just sent Conway a "parcell of papers," primarily medical receipts written in his father's own hand. Thus if Makin had been in possession of "all Sir W. Raleigh's Receipts" in 1650, it would seem that her

30 16 September, 1637. HP 22/15/1A.

31 *Ephemerides* 1650, Part I (Jan-Feb). HP 28/1/48A-B

32 Sir Walter Raleigh (1554–1618) had a reputation for his skill in chemistry and medicine. Aubrey tells us that "by his greate Skill in Chimistry, he made himself like a Leper: by which meanes he thought he might retard his journey to a Prison." The tactic was clever, but ultimately unsuccessful. See *Aubrey's Brief Lives*, edited from the Original Manuscripts by Oliver Lawson Dick (Reprint, Ann Arbor: University of Michigan Press, 1957), 253–60.

ever-present financial woes had finally resulted in her having to return the collection to the family.[33]

Because Makin published anonymously, the record of her intellectual activity has been hard to trace; it survives primarily in her correspondence with another female scholar, the learned Anna Maria van Schurman of Utrecht. However, her presence in Hartlib's papers documents an active scholar, the possessor of a good library, and someone participating in the circulation of medical knowledge. Thus it is in the Hartlib papers that we find Makin's intellectual afterlife.

4 Dorothy Moore (c.1612–1664)

The Anglo-Irish reformer Dorothy Moore is very visible in the Hartlib papers, primarily for her activities in Protestant reform, her scholarship in Hebrew, and her unwavering support of Hartlib himself.[34] In the papers, we see her arguing with the theologian André Rivet, hoping to carve out for herself a possible career as a preacher. The issue at stake was the role that women could play in serving God.[35]

---

33 Raleigh wrote: "My Lord, I send you a parcell of papers which perchance your selfe will scorne, and I am ashamed off, old, rotten, durtie, and torne things and such as a person less intelligent then your selfe would hardly understand, but they are all of this kynd (recepts) which I have, that you have not alreddy seene: they are most of them in my Fathers owne hand, and therefore I thinke approved ones." Carew Raleigh to Edward Conway, 28 June, 1652. In *The Conway Letters: The Correspondence of Anne, Viscountess Conway, Henry More, and their Friends 1642–1684*, ed. Marjorie Hope Nicolson, revised edition by Sarah Hutton (Oxford: Clarendon Press, 1992), 19–20.

34 On Moore see: Turnbull, *HDC*, 219–99 passim; James G. Taafe, "Mrs. John Dury: A Sister of Lycidas," *Notes and Queries* 207 (1962): 60–1; Lynette Hunter, "Sisters of the Royal Society: The Circle of Katherine Jones, Lady Ranelagh," in *Women, Science and Medicine 1500–1700: Mothers and Sisters of the Royal Society*, ed. Lynette Hunter and Sarah Hutton (Phoenix Mill: Sutton Publishing, 1997), 178–197; and Lynette Hunter, ed., *The Letters of Dorothy Moore, 1612–64: The Friendships, Marriage, and Intellectual Life of a Seventeenth-Century Woman* (Aldershot, UK: Ashgate, 2004). See also Pal, *Republic of Women*, esp. Chapter Four.

35 The extant correspondence consists of three letters from Moore, and two replies from Rivet. Copies of these letters, with notations in what appears to be Dury's hand, are preserved in the Hartlib Papers, 21/3/1A-13B. The letters were calendared by Paul Dibon in *Inventaire de la correspondance d'André Rivet (1590–1650)* (The Hague: Martinus Nijhoff, 1971). Also translated in Lynette Hunter, *Dorothy Moore*, 21–35.

We also see her exploring the possibility of a chemical career, and acting as a political go-between.[36]

Moore was born in Dublin around 1613, one of the nine children of Sir John King, a wealthy Irish administrator, and Catherine Drury. Moore and her sisters Margaret and Mary were all spoken of as being learned, and their brother Edward King was also a scholar, a schoolmate of John Milton and Henry More at Cambridge. When Edward was tragically drowned in a shipwreck at the age of twenty-five, a book of elegies appeared in tribute, and Milton immortalized the young man as his "Lycidas."[37]

The Kings, like their neighbors the Boyles, belonged to a social rank that educated sons and daughters very differently. In general, the girls were raised to marry early, marry well, and have the necessary skills to run a large household, while their brothers were educated at Cambridge.[38] Certainly, the King sisters did not receive any schooling of their own beyond the basic noblewomen's curriculum; in fact, Moore later wrote a short treatise on the education of girls, in which she lamented this regrettable education in what she considered completely useless and frivolous skills.[39] Yet apart from the unfortunate instruction she lamented, Moore had somehow acquired proficiency in multiple languages, and was particularly adept as a Hebrew scholar; both John Dury and Samuel Hartlib would refer to her in correspondence as "Mrs. Aethiop."

Dorothy King's formal education and proxy lessons would both have ended in the late 1620s, when she was still in her teens. It was then that she married Arthur Moore, the younger son of Garrett Moore, Earl of Drogheda. This marriage also made her the aunt of Katherine Jones (née Boyle), Lady Ranelagh, although the two women were contemporaries who had probably been acquainted since their childhood in the Anglo-Irish community of Dublin.[40]

---

36 In a letter to Hartlib, the chemist Benjamin Worsely discussed the range of appropriate options for Moore in terms of a possible career in chemistry. Worsely to Hartlib, 22 June, 1649. HP 26/33/1A-3B. The key objection to Moore's having a career of this sort would have originated with her social rank, rather than her gender.

37 The opening address of this volume speaks of Edward's sisters as being learned: "*sorores, foeminas lectissimas.*" On Milton and Edward King, see David Masson, *The Life of John Milton: Narrated in Connexion with the Political, Ecclesiastical, and Literary History of His Time* (London: Macmillan and Co., 1873–1894): v.1, 647–59.

38 Nicholas Canny, *The Upstart Earl: A Study of the Social and Mental World of Richard Boyle, first Earl of Cork 1566–1643* (Cambridge: Cambridge University Press, 1982), 88–9, 103.

39 BL, Sloane MS. 649, ff. 203–5.

40 Arthur Moore's sister Frances and her husband, Roger Jones, had two children—their son, Arthur Jones, married Katherine Boyle.

With her husband Arthur Moore she moved to Utrecht in the late 1630s, and began to associate with a network of Protestant reformers in the Netherlands. As a subset within the Republic of Letters, this group included scholars who were intricately integrated with several other networks—networks whose aims were very different, but whose memberships overlapped for purposes of knowledge and learning. These included Royalist exiles living on the Continent, reformers associated with Parliament during the Interregnum, Anglo-Irish Protestants who fled Ireland during the Rebellion of 1641 to sit out their exile in London, young men in Europe for their Grand Tour, and an ever-circulating cadre of Huguenot ministers. The members of these groups frequently travelled back and forth across the Channel, carrying news reports, correspondence, and newly published books and treatises. They included John Dury, Lady Ranelagh, Robert Boyle, Benjamin Worsley, Samuel Hartlib, Pierre, Peter, Louis, Cyrus, and Marie du Moulin, and the brothers Gerard and Arnold Boate. It was also during this period that Moore must have been introduced to the Hartlib network.

Moore spent a great deal of time and energy trying to discern the precise nature of what she referred to as her "calling": the role God truly wanted her to play in serving "the advance of Christs kingdome."[41] More than anything else, Moore wanted to preach. She wrote to Ranelagh, her kinswoman and colleague, about the anguish of trying to determine how this could be done. Apparently, there had been setbacks, but Moore was undaunted:

> ... yett I repent not my Omissions; my preachings being governed then according to the light I had to cleere my way and such it represents my Actions still ...[42]

Moore's model was the primitive church, the original apostolic community of St. Paul. Women had fulfilled prominent roles in early Christianity, and had been able to have ministries not constrained by her gender—which was precisely what Moore wanted for herself.[43]

---

41 Moore to Ranelagh, 5 May 1645. HP 3/2/119A. Also in Hunter, *Dorothy Moore,* 73–4.

42 Moore to Ranelagh, 8 July 1643. HP 21/7/1A-2B. Also in Hunter, *Dorothy Moore,* 17–20.

43 According to Elisabeth Fiorenza, "The prominence of women in the early Christian movement is confirmed by examining the list of names in Romans 16. Of the thirty-six persons mentioned, sixteen are women and eighteen men." See Elisabeth Schüssler Fiorenza, "Word, Spirit and Power: Women in Early Christian Communities" in *Women of Spirit: Female Leadership in the Jewish and Christian Traditions,* ed. Rosemary Ruether and Eleanor McLaughlin (New York: Simon and Schuster, 1979), 33–4, 36.

However, she was still unsure whether she was truly attending to God's call, or only listening to the sound of her own willfulness and desire. She awaited a sign. In the meantime, as she told Ranelagh, she was willing to wait. And while she was waiting, she was willing to entertain several other options; she might seek a position at the exile court in The Hague, or "take up a course of Instructing youth, of my owne Sex."[44]

Moore's commitment to the Hartlib circle's reformationist agenda, coupled with her own refusal to accept the limitations imposed on her sex, had led her to develop some ideas for changes in women's education. She jotted these down and sent this short treatise on the education of girls to Ranelagh, who had requested it "as the instrument of others"—presumably Hartlib.[45] In the accompanying letter, she lamented her own regrettable education in the traditional skills of young noblewomen:

> Upon this ground I have in this litle draught, left out the teaching of youth dauncing and curious workes; both which serve onely, to fill the fancy with unnecessary, unprofitable and proud imaginations. My experience as well as my reason tells me this; for my owne education was to learne both, and all I got by them was a great trouble to forget both, that so I might stand lesse in opposition to the simplicity of the Gospell in my affections and practice.[46]

Thus for Moore, the goal was not educational reform per se. Her goal was rather to make women's education conformable to a godly reformation—a goal toward which she and her colleagues in the Hartlib network were devoting so much of their energies. In their Comenian approach to educational reform, boys and girls would both be trained toward the same end. Virtue and practical skills came first, while those of either sex who had intellectual abilities for more learned pursuits would be encouraged to pursue them.

When Moore turned her attention from preaching to politics, her aim was to heal the many breaches in the Protestant world. In particular, her focus was on the end of the civil wars in England, and the restoration of the exiled Queen of Bohemia, daughter of King James I. Moore deplored the Queen's poverty, and the fact that her birth nation had done so little to help her; and rather than sit on the sidelines, she decided to contact everyone in her network who might be

---

44 Moore to Ranelagh, 8 July 1643. HP 21/7/1A-2B. Also in Hunter, *Dorothy Moore,* 17–20.

45 BL, Sloane MS. 649, ff. 203–5. This is a cover letter; the treatise itself does not appear to be extant.

46 BL, Sloane MS. 649, ff. 203–5.

able to help. Writing to Hartlib, she bemoaned the fact that the exiled queen had been all but abandoned by her countrymen:

> Mr Hartleb. I have heertofore intreated mr Durye, to acquaint you with the present condition of the Queen of Bohemiah, which is soe sadd in all respects, that it cannot be related or heard, without great sence of her miserye, by any who can trewly compationat the afflictions of others but alas! most men are such lovers of themselves, and the rest are now so involved in publicke and privet distresses, as 'tis hard to finde any to whom probably an effectuall application can be made for her advantage.[47]

The Queen of Bohemia was in dire financial straits. Moore blamed Parliament, and she especially blamed the inadequacies of the Queen's chaplain Samsson Johnson.[48] So she decided to begin these negotiations with Parliament herself, meeting with little success:

> I must tell you I have negotiated in her Majestys behalf, with sume of members of the howses, who say there is little hope of … receiving any help from them.[49]

Yet Moore's political negotiations did not end there. A year later, she was still working to effect change through a closer relationship between Britain and the Palatine house, but she was now approaching from another direction; and this time, she was addressing herself directly to the Elector Palatine—the Queen of Bohemia's son, Karl Ludwig—urging him to present the "enclosed paper" to the appropriate body.

> Your Highness being upon the place will easily discerne how convenient it may bee to make the subject of this Discourse a request as also where to make it, whither to the Committee of both kingdoms or Parlament and Assembly.[50]

47 Moore to Hartlib, 17 March 1643. HP 21/5/3A-4B. Also in Hunter, *Dorothy Moore,* 14–17.

48 Moore deemed Johnson far too morally lax, and held him responsible for the fact that the Lord was still afflicting the luckless Palatine family. She had in fact gone so far as to call him a "dumbe dogg" and "a barking kurr." Moore to Hartlib, 24 August 1642. HP 21/5/1A-2B. Also in Hunter, *Dorothy Moore,* 9–11.

49 Moore to Hartlib, 17 March 1643. HP 21/5/3A-4B. Also in Hunter, *Dorothy Moore,* 14–17.

50 Moore to Karl Ludwig, Prince Elector, 29 September 1644. HP 21/6A-B. Also in Hunter, *Dorothy Moore,* 49.

However, there is yet another facet to Moore's political networking, because on the same day that she wrote to Karl Ludwig, Moore wrote the following letter to Hartlib:

> I have a great desire to bringe an acquaintance to pass between the Prince Elector and my Lady Ranalaugh; and therfore I thincke it very expedient that you give her my Leter to his Highness, that shee may give him oppertunity to make his acquaintance.

The point of bringing Ranelagh and the Elector together was not social; it was, in fact, political:

> Which being done you will finde her very capable to be instrumentall, for the effecting faithfully and prudently, whatsoever shall be by you or others committed to her trust; eyther concerning the busines therin inclosed or any other, that shall heer after bee fitt to communicate to his Highnes.[51]

Thus it is not the Elector who can be of use to Ranelagh, but rather Ranelagh who can be of use to the Elector.

Moore obviously deemed Ranelagh to be an effective "instrument" for political negotiation, and felt certain that it would be of benefit to Karl Ludwig to get to know her—so certain, in fact, that she was making Ranelagh the intermediary for the letter itself. And Moore was not the only one to have this opinion. In fact, Ranelagh's effectiveness in the realms of politics, religion, and revolution was well known and highly respected.

The documentary record of Moore's ambitions—religious, political, educational, and chemical—is the record of a remarkable woman. She learned Hebrew, wrote a treatise on educational reform, and made an attempt at a chemical career; using her connections in London, Dublin and The Hague, she initiated political negotiations with both members of Parliament and the Elector Palatine; and throughout it all, she continued to dream of one day being a preacher. A remarkable record indeed—but one that would never have been made visible if her letters had not been gathered together into one place in the Hartlib papers.

---

51 Moore to Hartlib, 29 September [1644]. HP 21/5/16A-B. Also in Hunter, *Dorothy Moore*, 50. A misprint in Hunter erroneously references the letter as HP 1/5/16A-B.

5 Katherine Jones, Lady Ranelagh (1615–1691)

The most visible female scholar in the Hartlib papers is Katherine Jones, Lady Ranelagh. Hartlib also corresponded frequently with her younger brother, the scientist Robert Boyle; but it was with Ranelagh that he had the strongest connection. In fact, she is often referred to without a name, existing in the papers as "the incomparable Lady," "the excellent Lady," or simply "my lady."

Ranelagh lived a long and productive life.[52] For fifty years, she was involved in politics, ethics, education, medicine, and godly reform. She associated with preachers, poets, Parliamentarians, royalty, and practitioners of experimental science. She was known as a passionate partisan in terms of faith, an indispensable ally and patron to the members of her circle, and a force to be dealt with in the business of politics. And among all of these, the epithet used to refer to her was always "the Incomparable." It was a term that made her very much her father's daughter, and the ubiquity of "the Incomparable Lady Ranelagh" resonates with that of "the great Earl of Cork"—"great" being a seemingly unavoidable adjective when referring to the patriarch of the Boyle clan.[53] As Sir John Leeke remarked, in speaking of young Katherine: "Believe it ould Corke could not begett nothing foolish."[54]

Her relationship with her brother Robert, often called "Robin" in her letters, had always been close. Boyle finally moved in with his sister in 1668, and they spent the rest of their lives together. When she died in 1691, he only survived her by a week, thus they would be eulogized

52 Ranelagh was born Katherine Boyle, and in 1631 she married Arthur Jones, thus becoming Katherine Jones. Her husband succeeded to the title of Lord Ranelagh in 1643. On Ranelagh, see: Kathleen M. Lynch, "The Incomparable Lady Ranelagh," in *Of Books and Humankind*, ed. John Butt (London: Routledge, 1964), 25–35; R. E. W. Maddison, *The Life of the Honourable Robert Boyle* (London: Taylor & Francis, 1969), passim; Lynette Hunter, "Sisters of the Royal Society," in *Women, Science and Medicine 1500–1700*, 178–197; Alan Cook, "Ladies in the Scientific Revolution," *Notes and Records of the Royal Society of London* 51,1 (1997): 1–12; Ruth Connolly, "'A Wise and Godly Sybilla': Viscountess Ranelagh and the Politics of International Protestantism," in *Women, Gender, and Radical Religion in Early Modern Europe*, ed. Sylvia Brown (Leiden: Brill, 2007), 285–306; and Michelle DiMeo, "'Such a sister became such a brother': Lady Ranelagh's Influence on Robert Boyle," *Intellectual History Review* 25,1 (2015): 21–36. See also Pal, *Republic of Women*, esp. Chapter Five.

53 Richard Boyle, First Earl of Cork (1566–1643).

54 Sir John Leeke to Sir Edmund Verney, 13 September, 1638, in Frances P. and Margaret M. Verney, *Memoirs of the Verney Family During the Seventeenth Century*, 3 vols. (London: Longmans, Green and Co., 1925), 1:124.

together by Bishop Gilbert Burnet (1643–1715). Burnet began by recounting the touching circumstances of their lifelong devotion to each other.[55] He especially noted how Ranelagh had been a force in seventeenth-century political life:

> She lived the longest on the publickest Scene, she made the greatest Figure in all the Revolutions of these Kingdoms for above fifty Years, of any Woman of our Age.[56]

He talked about her political acumen, and her ability to make use of her vast web of connections in helping those whose causes she supported. Burnet then concluded by praising the other side of Ranelagh—her intelligence, spirituality, and kindness.

This eulogy, rather than being an excuse for undeserved praise, seems to have been a fitting and accurate description of Ranelagh's life and career. However, without her presence in the Hartlib papers, her documentary life would still be telling a somewhat familiar story—that of a remarkable daughter and sister who had been born into a remarkable family. The Hartlib papers tell us something else, and it is not familiar at all.

Ranelagh was an active participant in Hartlib's dreams and schemes, and his papers contain nearly two hundred items related to her. Apart from dozens of letters to or from Ranelagh herself, the collection includes scores of letters in which she is excerpted, quoted, or referred to. This correspondence was copied out, translated, and distributed throughout Europe, and there are also numerous references to her in Hartlib's *Ephemerides*.[57] These were not social

55 "His Sister and he were pleasant in their Lives, and in their Death they were not divided; for as he lived with her above Fourty years, so he did not outlive her above a Week ... Such a *Sister* became such a *Brother*; and it was but suitable to both their Characters, that they should have improved the Relation under which they were born, to the more exalted and endearing one of *Friend*." *A Sermon Preached at the Funeral of the Honourable Robert Boyle; at St. Martins in the Fields, January 7, 1691/2. By the Right Reverend Father in God, Gilbert Lord Bishop of Sarum* (London: Richard Chiswell and John Taylor, 1692), 32–4.

56 Burnet, *A Sermon Preached*, 32–4.

57 The extant diary covers the years 1634 to 1660, with gaps from 1636–1638, and 1644–1648. On Hartlib's *Ephemerides*, see especially Chapter 4, "Taking Notes in Samuel Hartlib's Circle" in Richard Yeo, *Notebooks, English Virtuosi, and Early Modern Science* (Chicago: University of Chicago Press, 2014). See also Clucas, "Samuel Hartlib's *Ephemerides*, 1635–59," and Ann Blair, *Too Much to Know: Managing Scholarly Information before the Modern Age* (New Haven: Yale University Press, 2010). Chapter Two, "Note-Taking as Information Management," is especially applicable to Hartlib's *Ephemerides*.

exchanges. The items fall rather into three categories: political negotiations, medical information, and religious correspondence.

Ranelagh was well known for her expertise in medicine, and for over fifty years she was an active participant in medical manuscript networks. Throughout northern Europe and Britain, the circulation of medical receipts was a widespread and long-lived phenomenon.[58] It involved men as well as women from many walks of life, from the nobility to the middling classes. In Hartlib's correspondence and in his *Ephemerides,* Ranelagh is asked to provide her medical recipes, and to compare and assess the remedies of other practitioners; she is also involved in discussions of how to obtain the necessary ingredients, asking her brother Robert for certain chemicals, and noting where to find healing herbs. In addition, when medicine runs out, she has a new batch concocted herself.

Moreover, since the Hartlibean agenda for the advancement of learning included medical knowledge, we should not be surprised to see that medical and political networks crossed paths at Ranelagh's home. This is evidenced by a letter from the physician William Rand. Rand was a member of the Hartlib network and a Paracelsian chemist, part of a movement trying to introduce chemical knowledge and experiment into the practice of medicine. However, the medical profession was still very tightly in the grip of the College of Physicians—a resolutely resistant, conservative, and highly Galenic group that would not license those who practiced chemical medicine.[59] Thus a new idea was proposed. The idea was to set up an alternative medical society; Rand wrote up a prototype constitution, and then sent his *Proposals relateing to those Graduate Physitians of any Universitie, that have bin there licentiated, & are*

58 Recent scholarship has established women's extensive involvement in medical manuscript networks at every level, and demonstrated the profound blurring of the boundaries between printed and scribal publication. See, for example: Jennifer K. Stine, "Opening Closets: The Discovery of Household Medicine in Early Modern England," (Ph. D. Dissertation, Stanford University, 1996), esp. Chapter 4; Alisha Rankin, *Panaceia's Daughters: Noblewomen as Healers in Early Modern Germany* (Chicago: University of Chicago Press, 2013); Mary Fissell, "Introduction: Women, Health, and Healing in Early Modern Europe," *Bulletin of the History of Medicine* 82,1 (2008): 1–17; and Elaine Leong, "Making Medicines in the Early Modern Household," *Bulletin of the History of Medicine* 82,1 (2008): 145–168.

59 William Rand (1617–1663) was the son of the apothecary James Rand. He was educated at Cambridge and in the Netherlands, where he worked with Hartlib's correspondent Johann Moriaen, and became familiar with Helmontian chemistry. See Charles Webster, "English Medical Reformers of the Puritan Revolution: A Background to the 'Society of Chymical Physitians,'" *Ambix* 13 (1965–6): 16–41.

now resident in London & not incorporated nor desirous to incorporate with the present College to Hartlib. He reminded his friend of the conversation in which this idea was first discussed:

> I have sent yow the Propositions I told you of, when we were together at my Ladie Ranulaghs. Be pleased to com*m*unicate them only to such as yow know are averse to the Colledge, & desire that the matter may be kept secret, least we be countermined.[60]

Given Ranelagh's extensive participation in both medical and political networks, one must presume that she would also have been involved in the discussions for the promotion and development of this particular project.

Because her home was the physical setting for so much activity—such as meetings of the Hartlib network, book exchanges, and her brother Robert's experiments—scholars have generally characterized Ranelagh's role in these projects as that of a "hostess" or "patroness." However, performance in the role of "patroness" would require the ability to provide some form of material support. Ranelagh had some influence, but no funds. In fact, her actual patronage probably extended no further than her hiring of John Milton and Henry Oldenburg as tutors for her son Richard Jones. Moreover, it is clear from the documentation in the Hartlib papers that these labels are both inadequate and inaccurate. When Ranelagh is viewed through lenses derived from other models of female intellectual activity—such as aristocratic patroness, mother hen, or *salonnière*—the result is distortion. Ranelagh was a female scholar and activist who was quite emphatically the product of her revolutionary times: thus she followed no real model whatsoever.

In fact, just as "everybody knew Hartlib," it would probably be accurate to say that everybody knew Ranelagh.[61] And I would like to suggest that this statement might be true for many of the same reasons. There is a logical explanation for why Ranelagh seemed to bridge so many disparate networks—agriculturalists and aristocrats; Irish and English; millenarians and archbishops; physicians and philosophers; chemists and poets; royals and reformers. If we frame her activities as a corollary of her family connections and patronage, we encounter immediate difficulties. Ranelagh occupied a position considerably weakened by war. She was a political refugee and an abandoned wife, for

60 Rand to Hartlib, 15 August, 1656. HP 42/10/1A-4B.

61 The phrase "everybody knew Hartlib" is from David Masson, *The Life of John Milton: Narrated in Connexion with the Political, Ecclesiastical, and Literary History of His Time*, 7 vols. (London: Macmillan and Co., 1873–1894), 3:193.

whom massive fluctuations in the political situation had problematized her social status, and constrained her access to the processes of largesse and preferment. And if we frame Ranelagh's activities in terms of one network alone, we are privileging one group of connections at the expense of all the others. Neither of these approaches can help us to understand Ranelagh's activities as a coherent career of intellectual, political, and spiritual practice. Instead, I would suggest that we consider Ranelagh's activities to be those of an early modern intelligencer—not just a family informant, but a partner in Hartlib's far-flung network of activists and correspondents.[62]

Ranelagh achieved a high level of activity, visibility, and respect in a disparate array of networks—the Hartlib network, the Great Tew circle, medical manuscript networks, scientific groups, and Anglo-Irish reformers. But it also appears that she was known as an intelligencer among the royal families of Europe, specifically the Stuarts and the exile court in The Hague. This is evidenced by a remarkable letter she wrote to the Queen of Bohemia in 1646. In that letter—clearly not the first one she had exchanged with the exile queen—she provides war-related intelligence, castigates the Queen's brother King Charles I for his intransigence, and pardons herself for so "saucylie" writing to the Queen again after such a brief interval:

> I beleeve that your Majesties former inteligences have soe informed you that the affayres of this kingdome are now at a Determining point ... And therefore I very saucylie venture to present your Majesty with as much satisfaction in that poynt as my information of matters here makes me able to doe though I have soe lately adressed my selfe to your Majesty in one of those troubles, that it were fitter for me to beg a pardon, than fall to a repetition so sudenly.[63]

Ranelagh's reporting was not neutral, nor did she hold back on her judgments. The letter then ended with a stern and pious reflection; she hoped the Queen might "be secured from a share of our ruine" (perhaps

62 Recent scholarship has begun to interrogate the gendered assumptions that underlie the arbitrary categorization of news as either "intelligence" or "gossip." See, for instance, James Daybell, "'Suche newes as on the Quenes hye wayes we have mett': the News and Intelligence Networks of Elizabeth Talbot, countess of Shrewsbury (c. 1527–1608)," in *Women and Politics in Early Modern England, 1450–1700*, ed. James Daybell (Aldershot: Ashgate, 2004), 114–31.

63 Ranelagh to Queen of Bohemia, 7 August [1646]. National Art Library, Victoria and Albert Museum. Forster Collection, MS. 454, Pressmark 48.g.25, f. 74.

the consequence of her brother Charles I's unfortunate behavior), while she also pointed out that this same ruin might actually benefit her in the end "if God be pleased to make up to your soule what he takes from your Estate."[64]

Taken together with Moore's earlier suggestion that the Prince Elector (the Queen of Bohemia's son) would do well to contact Ranelagh for help in negotiating with Parliament, this letter suggests that Ranelagh's intelligencing activities constituted a well-known and frequently utilized link between royal circles, the exile court, and the Hartlib network. Ranelagh appears to have arrived in London in 1643 with her noble networks, medical contacts, and Anglo-Irish associations already fully functional. Furthermore, while she never appeared in print, her work was identified, translated, excerpted, copied, and distributed widely throughout the international community served by Hartlib's network of scribal publication.[65]

A large part of Ranelagh's work with Hartlib involved linking their reformist associates with her network of royals, nobles, and Anglo-Irish scholarship. As Hartlib put it in a letter to Worthington, Ranelagh was a woman of "piety and parts, wth power and Interest, wch she hath and will always have with all the Great ones, let them change never so often."[66] This, then, becomes an intelligencing strategy. Hartlib and Ranelagh, as two intelligencers with shared values and yet a vastly different network of contacts, would have made an exceptionally powerful team, and Ranelagh's extensive presence throughout the Hartlib Papers is the presence of a co-intelligencer.

An example here might be made using some of Hartlib's correspondence with Robert Wood (1621–1685). Wood was a mathematician, astronomer, and

64 Ranelagh to Queen of Bohemia, 7 August [1646]. National Art Library, Victoria and Albert Museum. Forster Collection, MS. 454, Pressmark 48.g.25, f. 74.

65 Harold Love articulated scribal publication as a function of English networks in the seventeenth century; the intent was for manuscripts to be copied and circulated among the members of a "scribal community," in order to "nourish and articulate a corporate ideology." Love was among the first to dismantle the assumption that by the beginning of the seventeenth century, the printing press had completely done away with manuscript culture. Harold Love, *Scribal Publication in Seventeenth-Century England* (Oxford: Clarendon Press, 1993), reprinted as *The Culture and Commerce of Texts* (Amherst: University of Massachusetts Press, 1998), 180–181.

66 Hartlib clarified further: "I mean the Earl of Corkes, and Mr Boyles incomparable sister the Lady Viscountess Ranalagh." Hartlib to Worthington, 30 January, 1659. Cambridge University Library. Baker MSS 29, ff. 203–5; see also Crossley's transcription in BL Add. MS. 32498, ff. 30r-35r, here 30r. Also printed in Crossley, *The Diary and Correspondence of Dr. John Worthington*, v. 13, 162–177, here 164.

author of "Ten to One," a pamphlet on decimalizing British currency.[67] In these exchanges, we see Ranelagh performing a number of intelligencing activities, promoting the circulation of knowledge that could advance the fields of science, agricultural reform, economic reform, and politics.

In terms of science, one of Wood's letters to Hartlib provides us with a glimpse of Ranelagh functioning as a conduit to help her brother Robert with his researches. She had been asked to provide Robert with some information from Ireland on the nature of barnacles; however, as Wood reports, she was clearly frustrated with the insufficient response of her countrymen, and their lack of interest in the Book of Nature:

> My Lady Ranalaugh writes that her brother our noble friend at Oxford having with some earnestnes desired what information she could get in this country about the generation of Barnicles, & her selfe having tried to get some here with little successe by reason of the generall unobservance of the workes of God, desires me to speake to Major Symner for what either of us know therein.[68]

The eventual results of Ranelagh's efforts are perhaps represented in Boyle's "About Barnacles," where he writes that he had tried without success to ascertain whether certain reports he had heard in Munster—for instance that barnacles had been seen swimming about like "wild fowle"—were at all true.[69]

In terms of agricultural reform, we learn from a 1656 letter to Hartlib that Wood had been concerned about his experiments on the cultivation of clover grass. In order to get feedback, he needed to get his work into circulation, and he did so via a combination of Hartlib and Ranelagh:

> The last weeke I sent, according to your desire, An Answer to that Objection against Clauvergrass that it eats out the heart of the land, which I hope you will not altogether dislike, tho it were writ somewhat loosely as to the stile; yet was there nothing in it but Truth as to matter of fact, & for the rational part I must submitt it to the judgement of others,

---

67 Wood was elected a Fellow of the Royal Society in 1681. On Wood and decimal coinage, see Webster, *The Great Instauration*, esp. 416–420.

68 Wood to Hartlib, February 1659. HP 33/1/44A-B.

69 Boyle concluded by noting: "I could not find that any serious and credible person would positively affirme that he (or she) had ever seen any Barnacles come out of the shell alive, and either swim, or fly, away." See "About Barnacles," Royal Society, Boyle Papers, vol. 17, f. 145.

> I sent it in the incomparable Lady Ranelagh packet (for my acquaintance with whom I must again thanke you, & ever remain your debtor).[70]

Then a year later, we find Wood discussing the vicissitudes of the post, looking for letters and books from Hartlib that might have been delayed or lost. The intermediary, again, was Ranelagh:

> I received both your welcome letters of Dec 22th & Jan 22th. the latter came not to my hands till the middle of feb. by reason of a continued constant west wind which kept 3 packets togeather at Holy head, as it sometimes falls out ... Upon the receipt whereof, I had within a few dayes the given me the happinesse of an opportunitie to wayte upon the incomparable Lady Ranalaugh & to enioy above an hours discourse with her, which I must heartily thanke you for making me a way unto. The Bookes you did me the favour to send were not then heard of, but my Lady promised me I should have notice so soon as she received them, & accordingly on friday last she sent them unto by her servant.[71]

In terms of economic reform, it apparently was also the case that once some writing met with Ranelagh's "approbation," her efficiency in putting it into circulation might sometimes have outstripped the author's own wishes—a complaint frequently voiced with respect to Hartlib himself. In this case, the publication was Wood's work on decimal coinage:

> I thought I ought at least for yo*ur* sake so much to regard it, as to get a Copie thereof to have shewed to my Lord, but my Lady had bin so forward & punctuall in oberving your Desires, *tha*t she had transmitted it to her Brother my Lord Broghill, before I had an opportunity to request it from her; I write not this to desire of you the trouble of sending any other Copie, for though that excellent iudgement my Lady Ranulaugh was pleased to passe a sentence of approbation upon it, as to publick use, yet my thoughts were I confesse grown so coole towards it; *tha*t I can scarce yet conceive hopes enough to make me thinke it in any degree of likelihood to take effect.[72]

70 Wood to Hartlib, 13 May, 1656. HP 33/1/1A-2B.

71 Wood to Hartlib, 3 March 1657. HP 33/1/11A-12B.

72 Wood to Hartlib, 8 April 1657. HP 33/1/13A-14B. Patrick Little also notes this episode as a reflection on Ranelagh's intellectual influence, but misconstrues the letter as referring to a book by Hartlib, rather than Hartlib's copy of a book by Wood. And in Webster's reading

In fact, we might pause over this letter to note how it reveals the unique combination that characterized Ranelagh's work as an intelligencer. On the one hand, like Hartlib, she was an intellectual broker: she received books, pamphlets and letters; she read and evaluated them; and then she sent the ones she deemed worthy on to those she thought might make use of them for the public good. And in this case, having read and approved of Wood's book on decimal coinage, she had wasted no time sending it on to her brother Roger, Lord Broghill. But on the other hand, unlike Hartlib, Ranelagh was additionally a political broker; and the timing of the letter suggests that this role was also very much in play.

Ranelagh's brother Broghill had a strong working relationship with Oliver Cromwell; he was appointed Lord President of the Council for Scotland in Edinburgh in 1655, and was a Member of Parliament for Cork in both the First and Second Protectorate Parliaments. But in April 1657, at the very time when Ranelagh was sending him Wood's book, Broghill was perhaps at the height of his political influence, attempting (without success) to convince Cromwell to accept the crown.

For Ranelagh, however, this was not only about helping family. She had her own relationship with Cromwell and had written to him on other occasions trying to influence his decisions on the progress of religious reform.[73] This raises the intriguing possibility that Ranelagh might have hurriedly sent Wood's book on economic reform to Broghill for political reasons. She might have hoped that her brother would use it to bolster his standing with Cromwell, thereby reinforcing the case for having the Protector accept the crown. And Ranelagh would have done this not only for the sake of Broghill's career, or for Wood's, but for the good of the nation.

In fact, while Ranelagh was known to be a powerful supporter of her family members, she was equally well known for her political work apart from them—and, in the case of her negotiations with Broghill's rival Lord Inchiquin, she was clearly respected as an individual and independent political actor. Broghill

---

of this same letter, he mistakes Ranelagh for her sister-in-law, stating that it was Margaret, Lady Broghill, who had passed the copy to her husband Roger Boyle. See Patrick Little, *Lord Broghill and the Cromwellian Union with Ireland and Scotland* (Woodbridge: The Boydell Press, 2004), 202; and Webster, *The Great Instauration*, 419. My thanks to Michael Hunter for this reference.

73 Ranelagh had written to Cromwell in 1656, urging him to let Broghill return to England from Scotland to secure religious reforms in some of the northern areas. Broghill reminded Cromwell of Ranelagh's efforts in this regard in a letter of 27 June, 1656. See Little, *Lord Broghill*, 204.

had asked Ranelagh to act as a go-between in initiating a reconciliation with Inchiquin. Ranelagh had done so, but then Broghill reneged. Inchiquin wrote to Ranelagh afterward to express his disappointment with her brother, while also making it clear that this in no way affected his respect for her as an individual:

> The reason why I make this discourse, unusual concerning a brother to a sister who I know loves him and has no reason to regard me, is because I know the distance between us will come to your ears by other means, and then perhaps you may imagine that the malice I bear him might be of force to lessen the honour I owe to your ladyship, which really, Madam, I shall ever nourish, being fixed upon no other object or ground but those gifts which God has given you in more than ordinary measure.[74]

Clearly, Inchiquin's concern here is not with Broghill. Inchiquin is concerned over his working relationship with Ranelagh herself; and no matter what transpires with her brother, he is anxious to remain in her good graces.

In addition to her ability to bridge both ends of the political spectrum—corresponding with Cromwell on the one hand, and the Queen of Bohemia on the other—Ranelagh had no hesitation about trying to influence politics through her decades-long friendship with Edward Hyde, the Earl of Clarendon and future Lord Chancellor. Her boldness did, however, have its downside. The fact that she had so much personal influence at the highest levels of government was completely incommensurate with the fact that—in addition to being a woman—she had no official position whatsoever. Even Clarendon would eventually become uncomfortable with the realities of this situation, as Lord Clifford reported in a letter to his father, Ranelagh's oldest brother Richard:

> I am told yt my Ld. Chancellor does not thinke it proper to have my Aunt Ranelaugh consulted with about what relates to ye businesse of ye country, but I would not upon any account yt yr. Lop. should take notice of this to any boddy; but I thought it fitt (yt haveing bin told such a thing) yr. Lop. may make so much use of this intimation, as to observe hereafter, whether it bee so or no.[75]

---

74 Inchiquin to Ranelagh, 16 March, 1656. Historical Manuscripts Commission, *Report on the manuscripts of the Earl of Egmont* (London, 1905), I: 374–5. Also cited in Little, *Lord Broghill*, 203.

75 Charles, Lord Clifford, to Richard Boyle, First Earl of Burlington, 1 June [1667]. BL, Add. MS. 75354. Althorp Papers, B5.

We can see in this letter the obvious discomfort arising from Ranelagh's ability to influence "ye businesse of ye country;" yet we can also see the obvious strategic power of Hartlib and Ranelagh working together as co-intelligencers. Taken together, these letters demonstrate Ranelagh's intelligencing activities and political brokering at a very high level—thus identifying her as Hartlib's co-intelligencer is both tantalizing and plausible. It is a powerful and suggestive example of how a female scholar's presence in the Hartlib papers can present us with a new and unlooked-for identity, one whose existence we would not even suspect without the archival afterlife they received in this collection.

6 Conclusion: Afterlives and Identities

Archives do not construct themselves. If they are not just ad hoc collections of papers, then they are consciously constructed sites of documentary memory, organized around what Andreas Petter has called "an intent to represent."[76] Thus while the papers might have originated as the correspondence of an individual, the term "archive" usually implies that the intent is now to represent some form of institution—anything from a family or a church, to officially constituted organizations such as scientific societies, universities, towns, or nations—and it is from the institution that these records will derive their perceived identity and historical significance. However, when we are confronted with evidence of intellectual activity and collegiality from a group not associated with any institution at all, it is incumbent upon us to start asking new questions.

In particular, we must begin asking ourselves how we understand the formation of archival memory. Institutions have no historical mandate other than to preserve institutional memory, through the formation of archives and the creation of associated documents. Moreover, since an institution's archive will be formed with an organizational sensibility at its core, that sensibility will become its archival logic. Not all records will be kept; some documents, which might seem irrelevant to the institutional mandate, could find themselves without a home.

76 Andreas Petter, "Schriftorganisation, Kulturtransfer und Überformung—drei Gesichtspunkte zur Enstehung, Funktion und Struktur städtischer Ambtsbuchüberlieferung aus dem Mittelalter," in J. Sarnowsky, ed., *Verwaltung und Schriftlichkeit in den Hansestädten* (Trier: Porta Alba Verlag, 2006), 17–63. Cited in Eric Ketelaar, "Records out and archives in: early modern cities as creators of records and as communities of archives," *Archival Science* 10 (2010): 201–210, here 205.

For the female scholars associated with Samuel Hartlib, this sort of institutional identity formation was the beginning of the end for their historiographical presence. The only intellectual organization with which they were associated was Hartlib's intelligencing network—an immaterial institution. It had no physical home, no official archive, and would not survive him.

We are left, then, with a case that problematizes the archive as a coherent category of analysis. Can Hartlib's surviving papers rightly be called an archive at all? For these female scholars, this essay has shown that the answer is both yes and no. Yes, because Hartlib's papers are the repository of their intellectual legacy. His papers are the only location in which these female scholars are identified primarily through their work, and it is therefore their archive. And no, because Hartlib's papers were gathered together in order to circulate and generate; they were meant to function as an anti-archive, and have become a static collection only in the light of actions and decisions made by others.[77] Thus it is clear that it is time to expand the analytical category of the archive, and perhaps develop a more complex and layered archival lexicon—one that would have a place for the yes-and-no status of Hartlib's papers. In the meantime, as this essay has shown, Hartlib's papers have indeed established an intellectual afterlife for the female scholars in his network, giving them an unexpected home in an unintentional archive.

77 Penman argues for analyzing Hartlib's papers using Burke's concept of "microsociologies"—an approach based on the users of the documents as much as the documents themselves. See Peter Burke, *A Social History of Knowledge* (Cambridge: Blackwell Publishing, 2000); and Penman, "Omnium Exposita Rapinae," 48–52.

CHAPTER 4

Fossilized Remains: The Martin Lister and Edward Lhuyd Ephemera

Anna Marie Roos

1 Introduction

While the current archival turn in the history of science emphasizes paper and ink, archives engaged in a culture of more diverse materiality.[1] The Lister ephemera provide a case in point. Martin Lister (1639–1712), the first scientific arachnologist and conchologist, entrusted his daughters Susanna and Anna to illustrate his landmark *Historiae sive synopsis methodica Conchyliorum*, assembled between 1685 and 1692. (2nd edition, 1692–1697) (Fig. 4.1). The Lister ephemera were analysed and repurposed by Ashmolean Museum Keeper William Huddesford to produce an updated second edition of the *Historiae,* which was in continual demand by naturalists, incorporating new Linnaean taxonomy. We argue that Huddesford's activities reflected increasing antiquarian interest in scientific archives, as well as his desire to raise the status of the Ashmolean Museum as an institution. The preservation and use of the ephemera to create a new edition thus had practical and tactical uses.

We will then argue that an analysis of the afterlives of the Lister ephemera used to create the *Historiae* elucidates the book's complex bibliographic history and construction, and the enterprise of early modern scientific illustration. An examination of the afterlives of the ephemera also elucidates the intellectual and artisanal challenges presented to a naturalist in the transformation of field specimens into aesthetically pleasing illustrations and scientific objects conveying new developments in taxonomic information. We then consider the surviving Lister shell specimens figured in the book, as taxonomic classification involved both refined powers of visual apprehension, but

---

BL = The British Library, London.

Bodl. = The Bodleian Library, University of Oxford.

NHM = Natural History Museum, London.

1 For example, the Max Planck Working Group, "Working with Paper: Gendered Practices in the History of Knowledge."

 | DOI 10.1163/9789004324305_006

also tactile ones. Not only does having the specimen with the drawing next to it reveal roots of visual techniques in natural history, but as Aby Warburg stated, between the "imagination's act of grasping and the conceptual act of observing, there is the tactile encounter with the object ... which we term the artistic act."[2] The visual and material qualities of the Lister ephemera and specimens reveal to what extent the tactile artistic act was used for natural history classification; we argue that an embodied empiricism was necessary to determine type characteristics of species.

2 The Survival of the Lister Ephemera and Their Afterlives

The survival of the Lister ephemera was due to a unique set of circumstances. On 20 July 1758, Emanuel Mendes Da Costa (1717–91), botanist, malacologist and antiquarian, wrote William Huddesford (1732–1772), the Keeper of the Ashmolean Museum, that "about a year ago at a sale of a gentleman's effects, who it seems was a distant relation of Dr. [Martin] Lister's, there were found put up in band-boxes, confused like waste paper, several bundles of Dr. Lister's papers, consisting ... of letters from several learned men to him.... All these papers were bought by, and are now in the custody of Dr. John Fothergill." [3] Fothergill (1712–1780) was an English physician, Quaker, and F.R.S. who, in his leisure time, studied conchology and botany, so had a logical interest in the manuscript collection. Fothergill subsequently informed Huddesford by letter that he had bought the papers which "had been thrown aside in the dirt as wast[e] paper" to save them from annihilation in the "pastrycooks oven."[4] The papers were not just only letters but also a "collection of many, (not all)

2 Aby Warburg, "The Absorption of the Expressive Values of the Past," trans. Matthew Rampley, *Art in Translation* 1,2 (2009): 277.

3 "Letter of Emanuel Mendez Da Costa to William Huddesford, 20 July 1758," in John Nichols, ed. *Illustrations of the literary history of the eighteenth century* (London: Nichols, 1822), 4: 458; Arthur MacGregor, "William Huddesford, (1732–1772): his role in reanimating the Ashmolean Museum, his collections, researches and support network," *Archives of Natural History* 34 (April 2007): 58.

4 Bodl. MS Ashmole 1822, ff. 225r. In the next line, Fothergill went on to say that he also purchased "Dr. Loyds [Lhuyd's] letters" from "Da Costa, at least I accepted them as payment for a large debt." Da Costa had substantive financial difficulties and was dismissed as Royal Society secretary for the supposed embezzlement of subscription funds. See Geoffrey Cantor, "The Rise and Fall of Emanuel Mendes da Costa: A Severe Case of the Philosophical Dropsy?" *English Historical Review* 116, 467 (2001): 584–603. See also Elizabeth Yale's chapter on Lhuyd in this volume.

of the drawings of the Land Snails. They are coarsely done, but extensive."[5] Fothergill wrote that he should not "detain a treasure in my hands, that I cannot enjoy, merely on speculation that I may enjoy it in some future period."[6] He also lamented that he should "never have the leisure to peruse them," and wondered "what to do with them?"[7] Fothergill then mused that he "had best give them to some public Body—either to the Universities or to the Royal Society," asking Huddesford "What dost Thee think?"[8] Huddesford quickly replied, "You ask Dr an interested Man. I say to the University of Oxford and to the Ashmolean therein. But I will give you a reason also—The Papers consist of letters to ... Lister ... a very great Benefactor."[9]

These were the papers of Martin Lister had assembled for the *Historiae*.

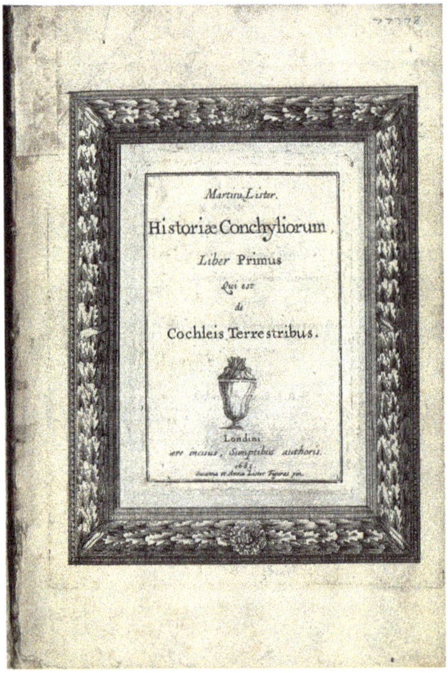

FIGURE 4.1
Title pages, Martin Lister, Historiae Conchyliorum, Liber Primus. *London: The author, 1685–92.*
WELLCOME LIBRARY, LONDON.

5 Bodl. MS Ashmole 1822, f. 226r.

6 Bodl. MS Ashmole 1822, f. 225r.

7 MacGregor, "Huddesford," 59 and 66; As MacGregor indicates, Fothergill's laments were related in an undated letter from Huddesford to the antiquarian John Loveday, received on 12 April 1769. The letter only exists in typescript in the collection of Robert William Theodore Gunther in the Museum of the History of Science, Oxford, MS Gunther 45/2, and reproduced in the appendix to B.F. Roberts, "A note on the Ashmolean collection of letters addressed to Edward Lhuyd," *Welsh History Review* 7 (1975): 183–185.

8 MacGregor, "Huddesford," 59.

9 MacGregor, "Huddesford," 59.

This first comprehensive study of conchology consisted of over 1000 copperplates portraying shells and molluscs that Lister had collected from around the world, as well as an appendix of molluscan dissections and comparative anatomy. As for Huddesford, he was described as the only "eighteenth-century Keeper known to be active in his office," began his post when the "museum was in decline, and immediately set to work to re-form the geological collections which were in disarray and to lay the foundations of new collections;" Huddesford also searched out new manuscript collections of past donors to the Ashmolean, such as those of Lister.[10] By doing so, Huddesford wished to repair the reputation of the Museum, so "it did not appear the nasty confused heap of trifles it has been invidiously represented to be."[11] Ordering such papers was also part of the impulse (discussed in the introduction) of the increasing antiquarian interest in the early modern period in preserving private papers for posterity by depositing them in public archives and, in some cases, publishing them.

As part of this restoration of the reputation of the Ashmolean, Fothergill and Huddesford hoped they could convey "the solid virtue of learned men of the past century in contrast to this frivolous age."[12] Fothergill recommended to Huddesford that he accession the material offered at the auction so he could create of a new edition of Lister's *Historiae*: "let the plates be retouched from as many originals" as can be procured. You have some shells no doubt in the Musaeum at Oxford. Procure an able engraver to compare the plates with these originals, and amend them when necessary.[13] Fothergill was prescient in his advice, as in 1715, Huddesford's predecessor, John Whiteside, (keeper of the Ashmolean from 1714 to 1729) had sought permission to reprint 30 copies of the *Historiae* due to the demand. [14]

Huddesford took Fothergill's advice, and subsequently gleefully recorded:

> ... came down one large Box, near a hundred weight. The contents as followeth

10 Bryn Roberts, "Memoirs of Edward Lhuyd, Antiquary and Nicholas Owen's British Remains," *National Library of Wales Journal* 19, 1 (1975): 74.

11 "Letter from Huddesford to Emanuel Mendez Da Costa, 30 November 1757," in John Nichols, ed. *Illustrations of the literary history of the eighteenth century* (London: Nichols, 1822), 4: 456.

12 Bodl. MS Ashmole 1822, ff. 225–226.

13 Bodl. MS Ashmole 1822, ff. 225v.

14 John Whiteside, "Account of Coins Missing before the year 1715," fol. 1., AMS 21, formerly MS Ashmole 1821.

1. 3 large Vols of letters to Lhuyd [Lister's colleague Edward Lhuyd] (Now Bodl. MSS Ashmole 1813, 1816, 1817b, 1820a, 1829, 1830. Letters of Martin Lister primarily to and from Edward Lhuyd, and Robert Plot; Bodl. MSS Lister 3, 35, and 36)
2. Several Bundles of Letters to Lister (Now Bodl. MSS Lister 2–4; 34–37)
3. Near 40 Books in 4to [quarto] of MSS annotations on, and extracts from various Authors—Lister's hand. (possibly Bodl. MSS Lister 5–8; 10–15; 23–26; 38–40)
4. Several Private Pocket Books in which Lister kept an account of the Fees He received in Practise. (Bodl. MSS Lister 27, 29, 30, 31, 32, 32*. Copies of printed almanacks, interleaved, with ms notes by Lister, forming a rough account book, chiefly of fees, for each year, with personal lists and notes.)[15]

In the course of the mid-nineteenth century, these collections passed from the Ashmolean to the Bodleian Library. Most of these papers were kept together, forming the bulk of what is known appropriately enough as MSS Lister. The collections of drawings of land snails that Huddesford mentioned, for instance, most likely comprise MS Lister 12, the first draft of Lister's *Exercitatio anatomica in qua de cochleis maxime terrestribus et limacibus agitur*.

However, four boxes of ephemera that had also come in the "large box," consisting primarily of loose papers and two stray copperplates, were separated from the bulk of MSS Lister. In March 2012, Diane Bergman, the Griffith Librarian at the Sackler Library at the University of Oxford, rediscovered them, archival remains not "lost" in a physical sense, but not catalogued.[16] Alexandra Franklin of the Bodleian Library subsequently found 20 more copperplates and drawings listed as "Lister uncatalogued."[17] Why were the ephemera in the Sackler? Huddesford set aside the ephemera because the ephemera directly relate to the *Historiae's* production and could inform his creation of another edition of Lister's *Historiae Conchyliorum*, which he accomplished in 1770.[18]

15 MacGregor, "Huddesford," 59. These documents comprise MSS Lister at the Bodleian Library, University of Oxford. I have added the numbers into the list for ease of reading.

16 Dianne Bergman. Email to Dunja Sharif, 19 March 2012, forwarded to author by Alexandra Franklin on 27 March 2012. A preliminary look at the ephemera can be found in Anna Marie Roos, "A Discovery of Martin Lister ephemera: the Construction of Early Modern Scientific Texts," *Bodleian Library Record* 26, 1 (2013): 125–36. This article will significantly add to my analysis.

17 Alexandra Franklin. Email to the author, 23 May 2013.

18 It is only recently that items that are not books or manuscripts have appeared in the public catalogue of the Bodleian Library.

The Sackler, which opened in 2001, incorporated the former holdings of the Ashmolean Library, which of course had been under Huddesford's control, and the ephemera thus became part of the Sackler's collection.

Huddesford had a long-held interest in Lister, seeing his research into early natural history as an escape from some of his endeavours in town-and-gown politics. He remarked when doing research about "Lister and the old Nat[ural] Historians," he became so engrossed in his research "conversing with Lister and the old Nat[ural] Historians I scarce know who is minister of state."[19] He was so obsessed with Listeriana that his friend James Granger (1723–76) wrote a fantasy in his honour occasioned by reading one of Huddesford's letters. Granger imagined Venus, "in all her Charms, just risen from the Sea, and seated in an ample shell. She was preceded by Tritons sounding their Buccinums (whelk shells) and attended by Nereids, adorned with Chains and Bracelets of Couris (cowries), intermixed with Pearls. ... Go, said she, with a significant, but ineffable Smile, to the nimblest Diver of all her train, and fetch me one of my own Shells. I intend it as a present to the Editor of the Synopsis Conchyliorum who is a great Lover of *natural Curiosities*."[20] The Nereid returned with a "most beautiful Concha Veneris," for presentation to "Huddesford for the Ashmolean Museum."[21] The goddess then accompanied Huddesford to the afterlife where he was able to meet and converse with the "respectable Shade of Dr Martin Lister, who expressed the greatest Joy upon the Discovery of a large Ventletrap." (a "staircase shell" in the family of gastropods, *Epitoniidae*)[22]

Huddesford also hoped to append Lister's biography to his new edition of the *Historiae*. In 1769, he wrote Granger, "In your work you mention a Mr Gregory [Edward Gregory, Fellow of Magdalene College, Cambridge and a nephew of Lister's] as your acquaintance, who has pictures, &c. of the Lister family. I could wish you would make known to him, that I am engaged in a work that will do honour to Dr Martin Lister ... having in my possession a considerable deal of his Philosophical correspondence, given us by Dr Fothergill."[23] In return, he gave Granger's wife "a little addition to

19 William Huddesford to William Borlase, 31 July 1770, Borlase papers in the Morrab Library, Penzance, MOR/BOR/3, f. 62.

20 James Granger to William Huddesford, 5 July 1770, Bodl. MS Ashmole 1822, f. 323r. James Granger was the author of *A Bibliographic History of England*, his name lent to the term "grangerizing" or adding additional prints, letters, engravings, etc to a book that were not included in the original volume.

21 Bodl. MS Ashmole 1822, ff. 323–324.

22 Bodl. MS Ashmole 1822, f. 323v.

23 Nichols, *Illustrations*, 4: 140. Gregory sent Huddesford accounts of the Lister family, something that Granger noted in his work *A Bibliographic History of England* (London: T.

her Collection of Fossils." Huddesford then sent a series of queries to the Gregory family of Harlaxton Hall, as well as Lister's descendents living in the family's manor house in Burwell Park, Lincolnshire, accomplishing "oral history interviews" by post. In a similar manner, Huddesford contacted the President of St John's College, Cambridge, Lister's *alma mater*.[24] Huddesford subsequently remarked to Granger, "I think from the Materials which I have at present, and from what he [Gregory] might be persuaded to add, I might be able to prefix some sort of Life to the *Conchyliorum*, and do his Worthy Ancestor some little Credit." [25]

Though Huddesford never finished the biography due to his premature death at the age of 40, he did successfully complete another limited 250-copy edition of Lister's *Historiae Conchyliorum* in 1770 using the ephemera, the original plates, as well as notes and observations written by Lister in a working copy of the first edition containing details of shells given by him to the Ashmolean.[26] In his new edition, Huddesford provided updated

Davies, J. Robson, 1775), 2: 289 and 388. The accounts in Granger's work that Gregory sent were of Sir Matthew Lister, Martin Lister's great uncle, and Susanna Temple, Lister's mother. In a letter of Grgory to Huddesford of 20 December 1769, Gregory noted, "I shall be extremely happy to give you all the information I can procure relative to Dr Lister; for which purpose, I will make a diligent enquiry among my Relations ..." Gregory also noted he had communicated information about the Lister family to "Mr Granger." Bodl. MS Ashmole 1822, f. 284r.

24 Lister held a fellowship by Royal Mandate at St John's College, granted in 1660. For Huddesford's letters concerning oral history interviews, see the following in BL Additional Manuscripts 22596:

1. Edward Gregory to William Huddesford; Magd. Coll. [Cambridge], 27 Feb. 1770, f. 86.
2. T[homas] Martyn [Professor of Botany at Cambridge 1761–1825], to Edward Gregory; Sidney Coll., 22 Dec. 1769, f. 88.
3. Susanna Gregory to her nephew, Edward Gregory; Short Hill, Nottingham, 31 Jan. 1770. With *seal*. Accompanied by a paper of answers to questions about Dr. Lister. ff. 90, 92.
4. Matthew Lister [of Burwell Park, near Louth] to Edward Gregory; Leadenham [near Sleaford], 22 Feb. 1770, f. 94.
5. Rev. George Ashby [President of St. John's College, Cambridge], to William Huddesford; "S[t] J[ohn's] C[ollege] C[ambridge]," 9 Mar. 1770, f. 95.

25 Bodl. Ms. Engl. Misc. d91, f. 146.

26 The biography's structure is only extant as an outline in manuscript. See "Dr Martin Lister: Life of, by Rev. W. Huddesford: 1769–1770 Imperfect," BL Additional Manuscripts 22596, f. 84. Lister's working copy in the Linnean Library, London is: *De Cochleis, tam*

indexes and transcriptions of Lister's relevant marginalia as well, along with two parallel tables and indices with Linnean classification for fellow natural philosophers and with common names for "English Naturalists" who "have, as yet, formed no system of Conchyiology in their own Language."[27] These tables were dedicated to Margaret Cavendish, the Duchess of Portland, as Huddesford used for taxonomic reference her renowned shell collections and the expertise of eminent conchologist John Lightfoot (1735–1788). In a 4 January 1770/1 letter to Huddesford, Lightfoot indicated he had spent three weeks comparing the Duchess' collection of shells to Huddesford's new index for the *Historiae* and Linnaeus's new taxonomic descriptions and found Huddesford's references "very correct and just."[28] In his new edition, Huddesford was able to draw attention to the legacy of Oxonian natural philosophers and the importance of the collections in the Ashmolean, and at the same time repurpose Lister's work as one still relevant and authoritative to eighteenth-century natural philosophers. Because of Huddesford's efforts, even when F. W. Martini and J.H. Cheanitz published their *Conchyliorum Cabinet* (1795), which became the standard text in the field, it was still seen as useful to publish a new index to Lister's *Historiae* in 1823. Dillwyn, the editor of the index, remarked that the *Historiae* was a work "which has been so long and universally referred to by every naturalist who has published on either recent or fossil shells."[29]

3 The Contents of the Boxes of Lister Ephemera and the Art of Science

When we look more closely at the contents of the boxes, we see that the papers within confirm and extend what we know about the *Historiae's* original construction, techniques of classifying the drawings of shells, and publication, all of which were of great interest to Huddesford as he prepared his new edition. The first three contain draft engravings of shells on white paper pinned

terrestribus, quam fluviatilibus exoticas, item de ijs quae etiam in Anglia inveniantur Libri II.(*Conchyliorum Marinorum Liber* III ... [&] *Buccinorum Marinorum Liber* IV); *De Cochleis* (London: the Author, 1685–92), MS 131, Case 2C.

27 Martin Lister, *Historiae Sive Synopsis Methodicae Conchyliorum et Tabularum Anatomicarum Editio Altera*, ed. William Huddesford (Oxford: Clarendon Press, 1770), Index II: 3.

28 Bodl. MS Ashmole 1822, ff. 291–292.

29 L. W. Dillwyn, Preface to *An Index to the Historiae Conchyliorum of Lister* (Oxford: Clarendon Press, 1823), 221.

to seventeenth-century blue and chocolate-brown paper for the *Historiae*. The blue paper is similar to that of MS Lister 9 in the Bodleian Library, the original sketchbook of shells that belonged to Martin Lister's second daughter, Anna; there are also drawings of shells on the same type of paper in another draft notebook for Lister's *Historiae* (Bodl. shelfmark RR.y.56).

The pins have verdigris on their shafts, indicating copper content; they presumably were straight pins. Such remains point to the domestic sphere in which the *Historiae* was produced by Lister's daughters; not only were pins used in sewing and mending, but they were also employed to fasten clothes together: a gentleman's neck cloth or the front bodice of a lady's gown; in elite dressing "pins were used in vast numbers, as is evidence by Elizabeth I ordering over 100,000 pins of different types in one six-month period in 1565, but they were also used as makeshift fastenings for those who could not afford buttons."[30] Here they are used for the purposes of arrangement of specimens and taxonomic classification. The draft engravings are pinned to different decorative borders; the press, possibly at his house or nearby, turned out images on Lister's correspondence paper.[31] The engravings in the boxes of ephemera also have the same watermarks as his stationery, dating them as draft images printed for the first or second seventeenth-century editions. The decorative baroque borders around the frontispieces and shell specimens that Susanna and Anna created added an extra step to the printing process. The Listeriana confirm Guy Wilkins' surmise that "borders of all sizes and designs were run off in quantity, to be over-printed with plates of figures, and varied at will."[32] In the final editions, as the sheets were run through the home press twice, the strain on the thin paper meant it was sometimes cut through the surface, and strips had to be pasted on the back of the sheets.

Lister left behind multiple copies of the *Historiae*, giving drafts to friends and colleagues as gifts and to invite their suggestions for improvement, and there were numerous printed variants. Lister's acquisition of new specimens

---

30 Abigail Shinn, "Cultures of Mending," in The *Ashgate Research Companion to Popular Culture in Early Modern England*, ed. Andrew Hadfield, Matthew Dimmock, Abigail Shinn (Aldershot: Ashgate, 2014), 246.

31 The watermark is extant on several pages of the 1685–1688 edition of the *Historiae Conchyliorum*, Shelfmark Gough Nat. Hist. 57, Bodleian Library, as well as in several pages of Lister's letters. The watermark's three circles, Griffins, Cross and Crown and Arms of Genoa is similar to ARMS.099.1 from the Thomas Gravell Watermark Archive. http://www.gravell.org. The source of the Gravell watermark was a 1666 London imprint from the Folger Shakespeare Library, Washington, D.C., shelfmark L.F. WM Coll 1.

32 Guy L. Wilkins, *A Catalogue and Historical Account of the Sloane Shell Collection* (London: British Museum, 1953), 39.

also necessitated rearrangement of the engravings for taxonomic purposes, with interleaving notes. A draft notebook in the Linnean Library of the first book of the *Historiae*, the *De Cochleis*, shows a similar approach that we see in the ephemera, using glue (probably wheat paste) sometimes instead of pins to rearrange the engravings of shells.[33] Another working copy of Lister's in the Bodleian Library also demonstrates his method.[34] The Listeriana in the box files show several numbering systems used to rearrange his shells, including pasted in numbers, annotations, and numbers that had been engraved on the copperplates themselves. We also find several prints of the same image, waiting to be cut, pasted and rearranged.

In addition to the visual evidence provided by his daughters, Lister's practices of processing and creating written information in the *Historiae Conchyliorum* shaped his conceptions of a natural order, blurring the boundaries between printed works and manuscripts. Lister' s system was not unique to him, but typical of many naturalists. The chapter by Arnold Hunt shows that the herbarium of James Petiver was a paper archive containing labels, and pasted loose papers on brown paper, his publication used as indexes and finding aids to his collection. Scholarship by Müller-Wille and Charmantier has demonstrated, from the early 1750s onward, Linneaus did his work similarly by interleaving "copies of his own publications, a method that had been endorsed by some Renaissance scholars, and was used by jurists and historians in the early eighteenth century to produce revised editions of their works ... Linnaeus perfected this method for natural history, designing his taxonomic publications accordingly."[35] Later, he used paper slips for botanical organisation, cross-referencing them with his herbarium sheets. It seems Lister used these techniques in interleaving, and in slip making, but in Lister's case the image of the shell rather than a Linnean *hortus siccus* was his reference point.[36]

On one of his grangerized *Historiae*, Lister scrawled Cicero's quotation about Pythagoras that the "imposition of names on things is the highest part

---

33 Working copy of Martin Lister's *De Cochleis*. MS 131, Case 2c, Linnean Society, London.

34 Lister, *Historiae Conchyliorum* Grangerised, Shelfmark RR. y. 56, Bodleian Library.

35 Isabelle Charmantier and Staffen Müller-Wille, "Carl Linnaeus's botanical paper slips," *Intellectual History Review* 24, 2 (2014): 219.

36 For other uses of paper slips in the work of classification in natural philosophy, particularly in the work of naturalist Conrad Gesner (1516–65), see Ann Blair, *Too Much to Know: Managing Scholarly Information before the Modern Age* (New Haven: Yale University Press, 2010); Lothar Müller, *White Magic: The Age of Paper*, trans. Jessica Spengler (Cambridge: Polity Press, 2014), 124.

of wisdom."[37] With the constant accumulation of new mollusc species from his international contacts, his improvised filing system allowed him to gain taxonomical sagacity. For instance, the engravings are supplemented on several pages by Lister's notes commenting on the shells' morphology and classification. He remarked for instance on the subtle depths of color of varieties of shells within a given species, or on its slightly different pattern.[38] Because shells were valued not only for their status as biological specimens, but as art objects, it is not entirely surprising that for Lister, the visual and the textual went hand in hand.

The fourth box of Listeriana not only contains draft engravings of shells pinned to album paper much in the same manner as in the other boxes, but two copperplates. The copperplates correspond to specimens 495 in his *Historiae* (now named *Pholadomya margaritacea*, a fossil bivalve specimen belonging to naturalist John Woodward) and 156 (the freshwater bivalve now named *Mytilus Cygneus*, the figure also taken from the Appendix to Lister's *Historia Animalium Angliae*, t.1. f. 3.).[39] The copperplates are wrapped in what seems to be their original papers, or papers later provided by Huddesford when he reprinted the *Historiae Conchyliorum*. Two copies of Huddesford's notice in French and English of his intent to re-publish Lister's *Historiae* are also in the box; Huddesford in particular indicates that he would be greatly "obliged to any Person who would favour him with the Use of any good Copy of his Work, which shall be carefully and punctually returned."

Huddesford's correspondence shows his reach was wide. Reverend George Ashby (1724–1808), a don at Lister's alma mater of St. John's College Cambridge, arranged for Huddesford to consult John Woodward's correspondence regarding Lister and fossilised shells.[40] James Granger mentioned Huddesford to the Duchess of Portland, Lady Margaret Cavendish, and on 12 December 1770, she wrote that she looked forward to the pleasure of

37 Linnean Society, London, Working copy of Martin Lister's *De Cochleis*. MS 131, Case 2c. Lister's inscription is in the front of the work, in the first endpaper. The quotation is from the first book of Cicero's *Tusculan Questions*. My thanks to Vera Keller for tracing the source of the quotation.

38 For instance, Lister wrote about depth of colour: *Huic color interdum fusius, quae altera tantum varietas esse videtur.* He then commented on pattern: *Rhombus C.P. lineis plurimus rufoscentibus et ex albu maculati distinctis clavicula leviter mericata, rostro purpurascente.* Bodl. Lister Ephemera, Box one, 17.

39 Martin Lister, *Historia Animalium Angliae* (London: John Martyn, 1678), table 1, figure 4.3.

40 Bodl. MS Ashmole 1822, ff. 278–279

showing Huddesford her shell collection.[41] As mentioned, John Lightfoot compared Lister's plates to specimens in the Duchess's collection to confirm Huddesford's updated taxonomic classification, a "long and arduous task."[42] Mr Evenus Hammer, a Danish colleague of Huddesford's studying in France, distributed Huddesford's advertisement to French intellectuals such as Joseph Mary Anne Gros de Besplas, Rector of the Sorbonne (1734–1783), Abbé Jean-Antoine Nollet (1700–1770), chair of physics at the Collège de Navarre and Director of the Académie des Sciences, and Jacques Laurent Berigny.[43] This assistance no doubt contributed to the great success of Huddesford's 1770 edition, which was hand-coloured.

The fourth box of ephemera also gives insights into other works in which Lister and Huddesford were involved. Interspersed in these papers are prints of engravings for Edward Lhuyd's *Lithophylacii Britannici Ichnographia* (1699), one of the first field guides to fossils (Table 1 and 3 fossilized corals; Table 5 *Lithophylaca*; table 9 and 10 *Ichthyodontes*). Although originally conceived as a guide to Oxford fossils, this study in Latin of "formed stones" was "arranged as a drawer-by-drawer guide to the cabinet of fossils which the author had collected and had deposited in the Ashmolean as deputy keeper and keeper."[44] Lhuyd and Lister were close friends, and about 200 pieces of correspondence between them are extant. Serving as a mentor, Lister helped Lhuyd bring his *Lithophylacii* to press, supervising the production of the illustrations and lent his friend copperplates that would be useful, telling him "Yor Booke is well printed for letter & paper, & cutts, as any ever was in England."[45] As Oxford University Press would not publish Lhuyd's work due to its expensive illustrations, Lister also encouraged his colleagues to subscribe, the cost of the 120 books produced being shared between ten virtuosi and noblemen, including Lister, Lord Somers, Lord Montagu, the Earl of Dorset, Dr Hans Sloane, Tancred

41 "James Granger to William Huddesford, 17 November 1769," in Nichols, *Illustrations*, 4: 139.

42 John Lightfoot to Huddesford, 3 December 1770, Bodl. MS Ashmole 1822, ff. 295–296.

43 Evenus Hammer to Huddesford 20 July 1769, Bodl. MS Ashmole 1822, f. 258r. Not a deal is known about Hammer. His signature appears in an *album amicorum* of Johann Thomas Ludwig Wehrs (1751–1811) of Göttingen dated 13 October 1770; Hammer was a student of maths. See Göttingen Stadtarchiv: Stabu Nr. 17, 55r. Shelfmark: WWS 11.6.2013. Also see W.W. Schaubel, Repertorium Alborum Amicorum, accessed October 25, 2014, http://www.raa.phil.uni-erlangen.de/index.shtml.

44 MacGregor, "Huddesford," 57. See Elizabeth Yale's chapter in this volume for more on Lhuyd's archives.

45 Bodl. MS Ashmole 1816, ff. 128–29. The letter from Lister to Lhuyd was dated 28 January 1698.

Robinson, chemist and professor Étienne François Geoffroy of Paris, Francis Aston and Isaac Newton. Because of Lister's efforts, Lhuyd wrote "you have been at a great deal of trouble and expence about the graving which is a kindnesse I am troubled I know not how to make any amends for, tho I know your goodness never expected any."[46] Subsequently, Lhuyd dedicated his book to Lister, describing him as the "Fundator munificus" of the Ashmolean Museum as well as a great encourager of the study of British fossils.

Huddesford was also interested in the *Lithophylacii*, and as with the *Historiae*, he decided to publish a new edition as the original text had contained newly coined terms for species which eighteenth-century readers found incomprehensible, and it was in some places in an unfinished state. The imperfect nature of the original edition was due to several problems with the printers and compositors. "During one of Lhuyd's extended visits to Wales, the text had in fact been seen through the press in London by Tancred Robinson, who clearly found it a frustrating exercise."[47] Robinson commented: "Notwithstanding all possible care, the Compositors will commit many gross Mistakes, and will not correct half the Errata made on the Sides. They are an ungovernable race of men; however it is as correct as most Books of the kind."[48] Huddesford confirmed Robinson's complaints in his letter to Emanuel Mendes da Costa about Lhuyd's work:

... [the text] is in some places obscure and difficult to the greatest proficients in this study. The descriptions are not distinct, the new coined terms, &c. render it very difficult to be understood. As I found there great impediments to my progress in the undertaking, I concluded they must be so to other young Readers, for whose service I also intended the second Edition; but though I was sensible how much it wanted it, I was diffident of correction. Mr Lhuyd's knowledge I dreaded to question—and of his carelessness I had no suspicion; but I am now sensible that he was sometimes deficient in the former, and very often guilty of the latter. [49]

Although Da Costa also "had a quantity of Lhuyd's original papers," (two portfolios of his correspondence) which he put at Huddesford's disposal, Lhuyd's original fossil specimens were "much damaged by the confusion in which they lay."[50] In his edition of 1760, Huddesford thus attempted to right

46 Bodl. MS Lister 36, f. 215r.

47 MacGregor, "Huddesford," 59.

48 Bodl. MS Eng. hist. c. 11, f. 89. The letter was sent 23 August 1698.

49 "William Huddesford to Emanuel Mendes Da Costa, 25 July 1758," in Nichols, *Illustrations*, 4: 459–460.

50 Nichols, *Illustrations*, 4: 457 and 459.

some omissions. For instance, Lhuyd claimed he had been given *buccina* and *pecten* shells specimens *et cretaceis Richmondius*. Da Costa pointed out to Huddesford that this was an erroneous observation, as there is no chalk in the area around Richmond, a mistake that should be amended to *ex argillaceis Richmondianis*, or from the clay pits in the area.[51]

Huddesford even had some of the plates retouched or re-engraved to make them adhere to updated taxonomy, correcting for instance a figure legend for specimen 1656 in the table of *Ichthyospondyli* and omitting a figure of "piped waxen veins," otherwise known as *lapis syringoides* from table CLI of *Lapidi Crystallini* (Figs. 4.2a & 4.2b).

FIGURE 4.2A *Edward Lhuyd* (LHWYD), Lithophylacii Britannici Ichnographia (*London, 1699*).

51 Nichols, *Illustrations*, 4: 458–59.

FIGURE 4.2B *Edward Lhuyd* (*LHWYD*), Lithophylacii Britannici Ichnographia (*London, 1760*).
COURTESY OF HISTORY OF SCIENCE COLLECTIONS, UNIVERSITY OF OKLAHOMA LIBRARIES, © THE BOARD OF REGENTS OF THE UNIVERSITY OF OKLAHOMA.

Sixteenth-century natural philosophers thought these were Ludi Helmonti, prepared as a remedy for the 'stone disease' then sweeping through Europe, first described in J. B. van Helmont's *De lithiasi* (1644).[52] In contrast, seventeenth-century naturalists like Lhuyd, Nehemiah Grew and John Woodward thought these were "*stalagmites cereus, tubulorum in quibusdam calculis marinis cavitates occupans*" or aggregated marine worms, caught up into masses in the time of the Deluge. Although Woodward noted they had the texture of wood, it was not known until the time of Huddesford that they really were petrified wood,

52 AM Alfonso-Goldfarb, PM Rattansi, and MH Ferraz, "Seventeenth-century 'treasure' found in Royal Society Archives: the Ludus helmontii and the stone disease," *Notes and Records of the Royal Society* 68,3 (2014): 227–43.

often filled with iron pyrite in large masses or in the form of slender twigs and branches. As a result, they were no longer classified as crystalline, and one of the copperplates in the ephemera was retouched to eliminate their figure. There seems to be a good deal of rubbing evident around the area on the recto of the plate; this evidence indicates alteration of the title using a burnisher, the copper leveled out with a scraper. In a process akin to *repoussage*, the plate was also beaten out from the back with a punch or small hammer to knock out the old title and achieve a smooth surface that could be cut again; *repoussage* was a common technique of alteration among engravers.[53] Making corrections such as these may have been behind Huddesford's keeping the Lhuyd prints and plates of fossils in the Lister ephemera aside. Huddesford republished Lhuyd's work for similar reasons, as a means to preserve the Ashmolean's reputation, and to update and repurpose a valuable work of natural history.

5 Physical Remains: The Shell Specimens

When he catalogued the Sloane Shell collection, Guy Wilkins, the curator of mollusca at the Natural History Museum, first noticed the existence of the original specimens that were used by the Listers to create the *Historiae* in the NHM collections. They were part of the original collection of Sir Hans Sloane. When Sloane went to Jamaica in 1687, Lister asked him to bring back specimens not only of shells but also of what he termed "naked snails" or slugs; in turn Sloane permitted Lister and his daughter to borrow specimens from his collection to illustrate. Lister also borrowed specimens from the virtuoso and collector William Courten or Charleton (1642–1702), dedicating his *Historiae* to him, and proposed him as a candidate for fellowship of the Royal Society.[54] Courten had a public museum of curiosities in a suite of ten rooms in the Temple, London, including artwork, specimens of flora and fauna, and archaeological objects, "celebrated as one of the finest cabinets of natural and artificial rarities in Europe."[55] Just as he bought the collection of James Petiver (see the chapter by Arnold Hunt), Sloane bought Courten's collection, including Courten's shells that the Listers illustrated in their book. Courten also kept an inventory of what he bought from 1689–1702, extant in MS Sloane 3961; several of the shells that Lister borrowed from Courten are in the inventory with

53 Mei-Ying Sung, *William Blake and the Art of Engraving* (London: Pickering and Chatto, 2009), 13.

54 Thomas Birch, *The History of the Royal Society* (London: A. Millar, 1756–7), 4: 326–7.

55 Carol Gibson-Wood, "Classification and Value in a Seventeenth-Century Museum," *Journal of the History of Collections* 9,1 (1997): 61.

their geographic and commercial origins. Specimen exchange and collection involved far-reaching networks: traders, apothecaries, physicians, naturalists and collectors all populated a vast intellectual geography to create the conchological collections of Sloane that indirectly made Lister's work possible.

As an example, the first shell in the NHM collection with a clear provenance is *Borus oblongus*, a buccina from Surinam, a large ventricose shell, with a spine shorter than the body-whorl and of four volutions. It is unusual and was very collectable as it had especially large white eggs about the size of a blackbird's.[56] In his inventory, William Courten recorded in an entry for 4 July 1689 having paid £1. 5s for the specimen, described as a "Large \viviparous/ snaile of Surinam with one of ye eggs the small snails come out of 3 of xxx f. g. they are found at ye ~~foot~~ bottom of ye Hedges about an inch under ground ye yolke of ye egg is of a kind of glewy substance."[57]

The shell is referred to again in a letter from Martin Lister to Edward Lhuyd of 8 April 1690. Lister wrote "The governour of *Sirinam* presented Mr Charleton [Courten] the other day with a land snail not bigger than a hens Egg. yet it layes eggs with *hard* shells as bigg a sparrows egg full; & the yong ones that are hatched of them, are as large again as the egg that holds them. The egg is finelie strieated length wayes."[58] The puzzle of the size of the egg continued to be a matter of strenuous discussion, as Lhuyd in turn reported to the naturalist John Ray, "Dr. Lister acquaints me that Mr. Charlton has lately received a land-snail from Surinam, not bigger than a hen's egg, which yet lays eggs as big as those of a sparrow; and the snails that are hatched of them are, he says, twice as large as the eggs."[59] Ray considered the specimen "very remarkable. But how the young Snayl hatched of the egges sould [should?] be so big of the eggs, I understand not." [60]

The accompanying fragile eggs and young snail drawn by Lister's daughters also are, not surprisingly, missing in the NHM collection, perhaps already

56 Lewis B. Brown, "Notes on the Land and Freshwater Snails of Barbados," *Journal of Conchology* 10, 9, (1902): 68.

57 BL MS Sloane 3961, f. 27v.

58 Bodl. Ashmole 1816, F. 79R.

59 The location of the original manuscript for the Lhuyd letter is unknown, but there are printed versions in: William Derham, ed. *Philosophical Letters Between the Late John Ray and Several of His Correspondents, Natives and Foreigners. To which are Added Those of Francis Willughby* (London: Jenys, 1718), 224; Edwin Lankester, ed., *The Correspondence of John Ray* (London: The Ray Society, 1848), 212–213; R.T. Gunther, ed., *Early Science in Oxford: Life and Letters of Edward Lhuyd* (Oxford: for the subscribers, 1945), 14: 99–100.

60 John Ray to Edward Lhuyd, 7 May 1690, Bodl. MS. Eng. hist. c. 11, f. 45.; Printed in R. T. Gunther, ed. *Further Correspondence of John Ray*, 206–207.

destroyed when Sloane bought Courten's collection; Sloane recorded his purchase in his inventory as 1895: *Buccinum admodum crassum ingens laeviter purpurascens a Surinam viparum*, noting its placement in table 23 in Lister's *Historiae.* [61] "Lister's figure of the young shell is a little larger than the egg figured on the same plate, and it is quite possible that some of the eggs sent to Courten hatched out, and increased the size of their shells in transit."[62] Lister also recorded that he received another specimen of the same type "idem cum proxime" to the Buccina with the egg, illustrating it in the next table in the *Historiae.* One of his draft notebooks for the *Historiae* reveals this copy of the specimen was also borrowed from Courten's museum. Lister subsequently gave his engraving of Courten's shells to him in thanks; Courten recorded in his inventory in April 1690, "1 Sheet of the ovum Testaceum a Surinam." [63]

Not only does having the original specimens present help us to understand their intellectual and material geographies, but also assists us in comprehending the techniques that Susanna and Anna Lister used to portray the specimens, and in turn, developing standards of classification and identification of species. For example, when Susanna and Anna Lister had an actual specimen to illustrate, they portrayed the shells in a one-to-one scale for ready identification. In the case of a shell (*Patella granularis*) from the genus *patella*, or a true limpet, the shell can be laid flat on the page, and it seems that his daughters traced around its periphery to portray its margins accurately in the final engraving. It is possible to place the shell down on the drawing and get a perfect match (Fig. 4.3).

Sloane's manuscript inventory reveals that the shell, which Lister borrowed from Sloane to illustrate, was originally "From Guinea by Mr Staphorst would have asperities [unevenness of surface or roughness]."[64] The patella is indeed quite ridged. "Staphorst" may have been Nicholas Staphorst, who was the Society of "Apothecaries' fourth Chemical Operator and author of *Officina Chymica Londinensis, sive exacta notitia medicamentorum spagyricorum* (1685, 1698)." As William Poole has indicated, "Sloane had lodged with Staphorst in Walter Lane [in London] while he studied chemistry at the Apothecaries Hall; Staphorst was presumably his instructor as well as housemate."[65] It is

---

61 NHM Sloane Inventories, f. 275r., shelfmark 50.L.6.

62 Wilkins, *A Catalogue and Historical Account*, 22.

63 BL MS Sloane 3961, f. 40v.

64 NHM Sloane Inventories, f. 116 r., shelfmark 50.L.6.

65 BL Additional MS. 421, f. 2r. "... [Sloane] came into England, and liv'd in a House adjoining to the Laboratory of Apothecaries Hall with Mr Staphorst, the Chemist, who had learn'd that art under Mr. Stahl his Kinsman." William Poole, "A Fragment of the Library of Theodore Haak (1605–1690)," *Electronic British Library Journal* (2007): 12, accessed 12 October 2013, http://www.bl.uk/eblj/2007articles/pdf/ebljarticle62007.pdf

FIGURE 4.3 Patella Granularis. *Martin Lister Collection, Division of Invertebrates, Department of Life Sciences, Natural History Museum, London.*
PHOTOGRAPH BY THE AUTHOR.

also possible that Mr Staphorst was Nicholas's brother, Barthold Staphorst, a merchant of Rotterdam; In a letter to Sloane, Pieter Hotton (Petrus Houttuyn 1648–1709), the professor of Botany and Medicine at Leiden University, advised Sloane that Nicholaus Staphorst could get books or specimens through his brother Barthold.[66]

We also see the same 1:1 technique utilized in the Lister's portrayal of an NHM specimen, the scallop shell, *Chlamys squamosa* (formerly *Ostrea squamosa* Gmelin, 1791). This shell is the lectotype, a biological specimen selected to serve as a definitive "type" example of a species, as Lister's *Historiae* is the earliest record of an identified specimen.[67] Lister described it in marginal notes in the Linnean Library's draft copy of the *Historiae* it as the "toothless under shell of a Scallop with a flat rib; it is smooth and curiously marbled with a white and dark hair colour." Anna Lister portrayed the markings on the surface of the shell absolutely accurately, paying special attention to the number of ribs, and equality or inequality of the valves and ears of the shells, which are the type characteristics for dividing the species of Pecten into groups.[68]

Anna and Susanna Lister also used tricks of artists' perspective to bring out type characteristics. *Melo aetheopica* has a distinctive *umbilicus*, the origin from which the whorls of the shell grew. However, looking down upon the shell hides this feature that is of great use in classification. As a result, Susanna Lister traced its outline to obtain the general shape and then portrayed it as tilted it upwards to reveal the *umbilicus* (Fig. 4.4).

FIGURE 4.4
Melo Aethiopica. *Martin Lister Collection, Division of Invertebrates, Department of Life Sciences, Natural History Museum, London.*
PHOTOGRAPH BY THE AUTHOR.

---

66 BL Sloane MS 4038, ff. 193–194, 296–97. Letters from Peter Hotton to Sir Hans Sloane, 22 July 1701 and 30 January 1702.

67 H.H. Dijkstra, "A contribution to the knowledge of the pectinacean Mollusca (*Bivalvia: Propeamussiidae, Entoliidae, Pectinidae*) from the Indonesian Archipelago," *Zoologische Verhandelingen Leiden* 271, 24 (1991): 32.

68 Wilkins, *A Catalogue and Historical Account*, 15.

Her use of perspective construction was thus was not "strictly correct" but opportunistic, entirely in keeping with what Martin Kemp has demonstrated in his work concerning the historical uses of perspective construction.[69] Her artistic judgment went beyond copying the shell, to featuring it as a taxonomic specimen of use in identification. In a letter of Rev. George Ashby to Huddesford, Ashby noted much the same thing, stating the copy of the *Historiae* that St. John's College, Cambridge had in its possession, "has many imperfect shells, which look like sketches only; but perhaps they are compleat and were intended to shew the remarkable part of the shell, as the *Hinge*."[70]

Lister's daughters also simply exaggerated type characteristics in their illustrations. *Neverita* (*Polynices*) *duplicata*, the shark eye, is a moon snail from the Bay of Campeche in the Gulf of Mexico; its circumference is often not completely smooth, and there can be two slightly angular protuberances, which generally can only be felt by holding the specimen. These protuberances are most pronounced in the illustration of the shell, its angular margin more felt that seen, but definitely perceptible (Fig. 4.5).

FIGURE 4.5
Neverita (Polynices) Duplicata. *Martin Lister Collection, Division of Invertebrates, Department of Life Sciences, Natural History Museum, London.*
PHOTOGRAPH BY THE AUTHOR.

69 See for instance, Martin Kemp, *Visualizations: The Nature Book of Art and Science* (Oxford: Oxford University Press, 2000).

70 Bodl. MS Ashmole 1822, f. 231r.

Although in early modern natural philosophy, we tend to see a Platonic and Aristotelian privileging of sight as the noblest of senses, the training, in Robert Hooke's words, "of a sincere Hand and a faithful eye" to accomplish illustration is accomplished by observing and doing, an interplay between sight and touch.[71] "Ideas of touch became increasingly visual during the Renaissance, giving rise to what Sander Gilman calls the "fantasy of 'seeing' the sense of touch, already commonplace in the seventeenth century."[72] It is also important to remember that Lister in his medical practice performed a type of embodied empiricism; just as early modern anatomical illustrations translated the dissector's manual touch into a visual touch for the viewer and readers of anatomy books, Lister's *Historiae Conchyliorum* did not just put his daughters handiwork in front of the reader's eyes.[73] The images of the shells developed a tactile dynamic of their own; seeing became a form of touching, of holding the specimens to classify them.

6 Conclusion

In a letter of 16 May 1694, John Place, the physician of the Grand Duke of Tuscany, told Lister of "the Great Duke's singular satisfaction ... especially with your ingenious, and elaborate booke [the *Historiae*]. I told him that the figures were the work of your daughters, which surprises him extremely." Place then wrote, "I believe he will present you with a parsell of his Florence wine."[74] The detailed illustrations and classifications of molluscs performed by Lister and his daughters meant the *Historiae* established a new standard for conchology, so that the work was in constant use by natural historians and taxonomists of the seventeenth, eighteenth and nineteenth centuries. The *Historiae's* plates and principles of classification were utilized by (among others) the explorer, botanist and entomologist James Petiver (1663–1718); Scottish physician and antiquarian Sir Robert Sibbald (1641–1722); Sir Hans Sloane, founder of

71 Robert Hooke, "Preface," *Micrographia* (London: John Martyn, 1665), 4.

72 Mark Michael Smith and Tristan Palmer, *Sensing the Past: Seeing, Hearing, Smelling, Tasting, and Touching in History* (Berkeley: University of California Press, 2008), 99.

73 Bettina Mathes, "As Long as a Swan's Neck?: The Significance of the Enlarged Clitoris in Early Modern Anatomy," in *Sensible Flesh: On Touch in Early Modern Culture*, ed. Elizabeth D. Harvey (Philadelphia: Pennsylvania State Press, 2003), 117. See also Joe Moshenka, *Feeling Pleasures: The Sense of Touch in Renaissance England* (Oxford: Oxford University Press, 2014).

74 Bodl. MS Lister 3, ff. 211–218.

the British Library; Belgian humanist and natural historian Carolus Langius (1670–1741); John Morton, the eighteenth-century natural historian of Northamptonshire, and Linnaeus. Thanks to Huddesford's efforts, his 1770 edition of the *Historiae* was in use until the nineteenth century.

In addition to these intellectual afterlives of Lister's work, this selective analysis of the significance and archival afterlives of the ephemera and the shell specimens revealed the employment of an embodied empiricism in the determination of type characteristics of species. The ephemera also demonstrated the migration of knowledge of nature, which occurred from one medium to another, from object to drawing to printed image. The archival remains of the Lister and Lhuyd ephemera are also witnesses to the important and often hidden role of intellectual networks, artisanal work, archival provenance and the history and bibliography in scientific book production in the early modern era.

CHAPTER 5

Playing Archival Politics with Hans Sloane, Edward Lhuyd, and John Woodward

Elizabeth Yale

1 Introduction

Writing to the physician Martin Lister in December 1699, the ethnographer, linguist, antiquary, and naturalist Edward Lhuyd made himself quite clear: "I intreat you not to publish anything out of my Letters in the *Transactions*, at any time." Written while Lhuyd was on a natural historical and antiquarian progress through Wales, Ireland, Scotland, Cornwall, and Brittany, letters like this, along with his and his research assistants' notebooks, were the primary records of his journey. He was not wholly opposed to publishing them, or some version of them—but he wanted to have some control over the process. He wrote on to Lister, "But if I chance to hint at any <thing> which you may judge worthy of publique view, I shall on your Intimation therof either send you a more particular account or offer some reason why I would not have it printed."[1] He sought to block his colleagues from automatically transmitting letters of

I would like to thank Vera Keller, Anna Marie Roos, Nasser Zakariya, Sachiko Kusukawa, and the external reviewer for their feedback on earlier versions of this essay and to acknowledge the Andrew W. Mellon Fellowship of Scholars in Critical Bibliography at Rare Book School, The Huntington Library, and The Royal Society for grants that supported this research and the "Archival Afterlives" conference in June 2015.

BL = The British Library, London
Bodl. = The Bodleian Library, Oxford
EMLO = Early Modern Letters Online (emlo.bodleian.ox.ac.uk)
JBO = Journal Book Original
Notes and Records = *Notes and Records of the Royal Society of London*
RS = Royal Society

1 Lhuyd to Lister, 15 December 1699, Bodl. MS Lister 36, 242–243, trans. Brynley F. Roberts and Helen Watt, *EMLO*. On the natural philosophical process of making and collecting hints, Kate Bennett, "John Aubrey, Hint-Keeper: Life-Writing and the Encouragement of Natural Philosophy in the pre-Newtonian Seventeenth Century," *The Seventeenth Century* 22 (2007): 358–80.

 | DOI 10.1163/9789004324305_007

interest to *Philosophical Transactions*, then under Hans Sloane's editorship, as they otherwise might.

Yet, despite Lhuyd's warnings, a selection of his letters was printed in *Philosophical Transactions* without his consultation—after his death. Their publication was superintended by Hans Sloane, serving as the Society's secretary. As is well known, the physician Hans Sloane was a collector.[2] He collected naturalia: rocks, shells, fossils, plants, insects. He collected artificialia: Afro-Caribbean musical instruments, shoes, and paper samples from around the globe. He collected books, buying up libraries from the estates of deceased physicians, competing (not always successfully) with the earls of Oxford and their librarian Humfrey Wanley for medieval religious, medical, and poetic manuscripts.[3] He collected people, assembling a global network of contacts to supply his collecting habit.[4] His collections overflowed his Bloomsbury home, and he expanded to a manor house in the London suburb of Chelsea to house them.

And he collected the papers of his contemporaries. It is this collecting activity, and how it intersected with Sloane's work in the Royal Society, that I explore in this essay. Even as he began to consider the historical potential of his own correspondence, becoming more careful about preserving and cataloguing it within his broader collections, he began to collect correspondence and papers from his contemporaries.[5] Some papers came to him as a gift. Others he bought at auction, such as some papers of Robert Hooke's or those of Mayerne discussed by Vera Keller elsewhere in this volume. [6] He inherited the papers—which included correspondence, memorandum books, and annotated sheets of specimens, of his friend the apothecary James Petiver, as Arnold Hunt explores, also in this volume.[7]

With these papers, Sloane played archival politics. He collected these papers, at least in part, in order to deploy them in his day-to-day life as virtuoso,

2 Alison Walker, Arthur MacGregor, and Michael Hunter, eds., *From Books to Bezoars: Sir Hans Sloane and his Collections* (London: The British Library, 2012); James Delbourgo, *Collecting the World: Hans Sloane and the Origins of the British Museum* (Cambridge: Harvard University Press, 2017).

3 Arnold Hunt, "Sloane as a Collector of Manuscripts," in *From Books to Bezoars*, 199.

4 James Delbourgo, "Listing People," *Isis* 103 (2012): 735–742; Delbourgo, "Collecting Hans Sloane," in *From Books to Bezoars*, 9–23.

5 Hunt, "Sloane as a Collector of Manuscripts," 203.

6 Sloane, BL Sloane MS 3972 B, 373v; Giles Mandelbrote, "Sloane's Purchases at the Sale of Robert Hooke's Library," Giles Mandelbrote and Barry Taylor (eds.) *Libraries within the Library* (London: The British Library, 2009), 110.

7 See Hunt, "Under Sloane's Shadow: the Archive of James Petiver," this volume.

naturalist, and physician. As secretary of the Royal Society, he used them as a means of accruing both personal and institutional power, of building and maintaining his position and relationships within the London natural philosophical community.[8] Sloane's use of these papers, then, offers insight into the intersection of personal collecting with the Royal Society's functioning as an institution, as well as the development of *Philosophical Transactions* as a learned journal. Through his tenure as secretary, Sloane (and others) drew on the papers of the dead, bringing them in for discussion in the Society's weekly meetings and publishing them in *Philosophical Transactions* and in independent volumes of correspondence and posthumous papers. Sloane built his archive for the sake of its afterlife, for his—and others'—continued use and transformation of these materials.

I focus specifically on Sloane's collection and use of Edward Lhuyd's correspondence and his use and manipulation of the Society's meeting minutes, the key record of the face-to-face business transacted in regular meetings of the fellowship and in meetings of its governing Council. Lhuyd, the second curator of the Ashmolean Museum, died in 1709, leaving behind a trove of letters and other papers in Oxford. Letters he wrote to others were, of course, scattered across the British Isles. Sloane collected some of this material and deployed it in Royal Society meetings and published it in *Philosophical Transactions*, which he personally bankrolled as secretary.[9] Sloane did so in the context of his ongoing battle with physician John Woodward for control of the society's direction and the direction of *Philosophical Transactions*, which played out during Sloane's tenure as Secretary. In life, Lhuyd and Woodward had disliked each other, a personal antipathy compounded by scientific disagreements over

8 On "archival politics," see Laurie Nussdorfer, *Brokers of Public Trust: Notaries in Early Modern Rome* (Baltimore: Johns Hopkins University Press, 2009); Ann Laura Stoler, *Along the Archival Grain: Epistemic Anxieties and Colonial Common Sense* (Princeton: Princeton University Press, 2009); Kathryn Burns, *Into the Archive: Writing and Power in Colonial Peru* (Durham: Duke University Press, 2010); Filippo de Vivo, "Ordering the Archive in Early Modern Venice (1400–1650)," *Archival Science*, 10 (2010), 231–48; Nicholas Popper, "From Abbey to Archive: Managing Texts and Records in Early Modern England," *Archival Science*, 10 (201): 249–266; Arndt Brendecke, "'*Arca, archivillo, archivo*': The Keeping, Use and Status of Historical Documents about the Spanish *Conquista*," *Archival Science*, 10 (2010), 267–83; Jacob Soll, *The Information Master: Jean-Baptiste Colbert's Secret State Intelligence System* (Ann Arbor: University of Michigan Press, 2011); Kirsten Weld, *Paper Cadavers: The Archives of Dictatorship in Guatemala* (Durham: Duke University Press, 2014).

9 Noah Moxham, "Fit for Print: Developing an Institutional Model of Scientific Periodical Publishing in England, 1665-ca. 1714," *Notes and Records* 69 (2015): 246.

the origins of fossils.[10] In collecting and publishing Lhuyd's correspondence, Sloane enrolled Lhuyd as an ally from beyond the grave—a role he would have, perhaps, been happy to play. And, as we will see, in manipulating the production and preservation of the meeting minutes of the Royal Society's Council, Sloane controlled the historical record of his own feud with Woodward.

On a deeper level, Sloane constructed an archival politics of knowledge grounded in an antiquarian logic. For Sloane, natural philosophical insights, experiments, and observations, as well as practical applications in medicine and industry, lay not only scattered on the surface of the present, but also buried in the archive, in papers gathered and accumulated over decades, if not centuries. Sloane used his archive as a source of natural philosophical learning as well as institutional and personal power, through it consolidating his position as Royal Society secretary and then president.

2 Archival Afterlives: Discussing and Publishing the Papers of the Dead at the Royal Society

When Lhuyd wrote to Lister, enjoining him not to publish his letters without first consulting him, he was on a five-year research journey around Britain and Ireland. He wrote from the village of Bathgate, along the Edinburgh-Glasgow road, after a period of some months where, traveling in the Highlands, he was unable to send or receive mail—he rambled "through countreys so retir'd, that they affoarded neither post nor carrier."[11] From 1697–1701, Edward Lhuyd left his post as curator of the Ashmolean Museum, where he presided over a collection that including antiquities, natural artifacts, as well as an extensive library of printed books, pamphlets, and contemporary manuscripts, to travel through Wales, Scotland, Ireland, Cornwall, and Brittany. He sought to reconstruct the human and natural history of Britain and Ireland. His study self-consciously excluded the English, instead searching for the common threads of history, landscape, and language linking Britain's other inhabitants. Lhuyd

10 For Lhuyd and Woodward's personal and intellectual disagreements, see Joseph Levine, *Dr Woodward's Shield: History, Science, and Satire in Augustan England* (Berkeley: University of California Press, 1977), 80–85; Rhoda Rappaport, *When Geologists were Historians, 1665–1750*, (Ithaca: Cornell University Press, 1997), 161–162; on the competing theories of fossil origins in the 1690s and early 1700s, see Rappaport, 136–172.

11 Lhuyd to Lister, 15 December 1699, Bodl. MS Lister 36, 242–243, *EMLO*, trans. Roberts and Watt.

was the first linguist, for example, to apply the word "Celtic" jointly to Welsh, Irish Gaelic, Scots Gaelic, Cornish, and Breton.[12] Using a standardized form of Welsh orthography as a phonetic alphabet, and a core list of vocabulary words derived from John Ray's *Dictionariolum Trilinguam*, a Latin-Greek-English dictionary, he systematically recorded spoken language around the British Isles and Ireland. He used this data to classify these languages into what are now known as "P-Celtic" and "Q-Celtic" based on systematic differences in their pronunciation. These family groupings still stand today.[13]

After Lhuyd's death, Sloane obtained Lhuyd's correspondence from others who knew of his interest in it. In 1713, the cleric John Morton gave him letters he had received from Lhuyd, along with stones and fossils Morton had collected, noting that they would "be a present <not un>acceptable to you, and might be of use to the publick."[14] They would certainly be of interest to Sloane: he collected them along with others Lhuyd wrote to Walter Charleton and Richard Richardson, as well as extracts from Lhuyd's correspondence with John Ray and Tancred Robinson. The latter extracts, annotated by an unknown hand, discuss competing theories of fossil formation, including John Woodward's theory that, after the Deluge, fossils had settled in layers in the earth according to their "specific Gravity." (The annotator described this notion as "a meer amusement: and a confused notion.")[15] Some of these letters also passed through the hands of Richard Waller, Sloane's co-secretary at the Royal Society, who marked them "Extracted by R. Waller," indicating they were reviewed and considered through the Royal Society's institutional channels, probably after they were given to Sloane.[16]

During Sloane's tenure as secretary (1693–1713), the letters and papers of the dead made frequent appearances at the Royal Society's weekly meetings.

12 For Lhuyd's research plan, see Lhuyd, *A Design of a British Dictionary, Historical and Geographical; With an Essay, Entituled, Archaeologia Britannica: and a Natural History of Wales* (Oxford: s.n., 1695); Elizabeth Yale, *Sociable Knowledge: Natural History and the Nation in Early Modern Britain* (Philadelphia: University of Pennsylvania Press, 2016); on his life, see Richard Ellis, "Some Incidents in the Life of Edward Lhuyd," in *Early Science in Oxford*, ed. Robert Gunther (Oxford: Printed for the Subscribers, 1945), 14:1–51; Arthur MacGregor, "Edward Lhuyd, Museum Keeper," *Welsh History Review* 25 (2010): 51–74.

13 David Cram, "Edward Lhuyd's *Archaeologia Britannica*: Method and Madness in Early Modern Comparative Philology," *Welsh Historical Review* 25 (2010): 75–96.

14 Hunt, "Sloane as a Collector of Manuscripts," 203; Morton to Sloane, 12 October 1713, BL Sloane MS 4043, 189v.

15 BL MS Sloane 4062, 278v.

16 Lhuyd to Morton, 18 October 1694, BL Sloane MS 4062, 256v; On Waller, see Sachiko Kusukawa, "Picturing Knowledge in the Early Royal Society: The Examples of Richard Waller and Henry Hunt," *Notes and Records* 65 (2011): 273–294.

Living fellows—first and foremost, though not alone among them, Sloane—brought the papers of the dead forward much as they might a contribution of their own, or a letter from a living correspondent, reading them aloud for the gathered company to discuss, evaluate, and respond to. As with contributions from living fellows and correspondents, the Society responded to these papers in a number of different ways. In some cases, they recommended that the papers of the dead be published—shorter pieces might appear in *Philosophical Transactions*, longer pieces could be published independently. On other occasions, they advised that the papers or manuscript be entered in the Society's library; that further experiments and observations be done to test and explore the author's ideas; or, most basically (and most dismissively), that the individuals who had brought forward the papers be thanked. Following Robert Hooke's death, Richard Waller regularly presented items from his papers.[17] Hooke's executors gave some of Hooke's papers directly to the Royal Society, who gave them to Waller, while others, including a set of drawings of fossils, passed through the book market into the hands of Hans Sloane.[18] Sloane regularly introduced discussion of both the general status of, as well as specific items from, the papers of John Ray, Sir Thomas and Edward Browne, and Edward Lhuyd.[19] The society was interested more broadly in the survival of scholarly papers and manuscripts (or, at least the texts they harbored), as well: over the years, topics of conversation at weekly meetings included the state of the Cottonian Library (a report was read by Humfrey Wanley); the manuscripts and papers of Tycho Brahe and Johannes Kepler; the Oriental books and papers of the Oxford professor Thomas Hyde; and the posthumous works of microscopist Marcello Malpighi.[20] In some cases, these discussions led to attempts (some successful, such as with Malpighi's

17 Felicity Henderson, "Robert Hooke's Archive," *Script and Print* 33 (2009): 92–108. Noah Moxham argues further that Waller structured his life of Hooke around the Society's archive, which Hooke participated in creating and filling. See Moxham, "An Experimental 'Life' for an experimental life: Richard Waller's biography of Robert Hooke (1705)," *British Journal for the History of Science* 49 (2016): 27–51.

18 RS JBO 11, 16–20; Michael Hunter, "Hooke's Possessions at his Death: A Hitherto Uknown Inventory," *Robert Hooke, New Studies*, ed. Michael Hunter and Simon Schaffer (Woodbridge, Suffolk: The Boydell Press, 1989), 287–294; Sachiko Kusukawa, "Drawings of Fossils by Robert Hooke and Richard Waller," *Notes and Records* 67 (2013): 123–138.

19 See repeated mentions in RS JBO 11, covering the years 1702 to 1714.

20 For Kepler's papers, see RS JBO 15, 8 November 1733, 328–330; for Brahe's papers, see JBO 11, 8 January 1706, 107; on the Cottonian library, JBO 11, 4 July 1703, 28; on Hyde, see RS JBO 11, 10 May 1704, 50.

papers) to print the papers in question or to bring them into the Royal Society Library.[21]

Hans Sloane presented Lhuyd's letters, written to the physicians Richard Richardson and Tancred Robinson, at Royal Society meetings in 1711–1713.[22] He published them in *Philosophical Transactions* in 1712 and 1713. Lhuyd's letters came to Sloane's hands as waves of interest surged around the collection of papers Lhuyd left behind at the Ashmolean at his death in 1709.[23] Lhuyd died in debt to his printer and to the university, and the university decided to sell his books and papers to cover his debts. From 1709 to 1715, as the manuscripts "lay in state," various figures, including the Harleys (via Wanley, now their secretary and librarian), the antiquary John Anstis, and Lhuyd's Welsh patron Thomas Mansel competed over their disposition.[24] Wanley was particularly interested in the *Red Book of Hergest,* one of the most important manuscript collections of medieval Welsh prose and poetry, but it turned out Lhuyd had only been borrowing it from Jesus College at the time of his death.[25] In the end, the collection went to Sir Thomas Sebright, a book and manuscript collector who had studied at Jesus College (this was not Sebright's first rodeo with the Harleys).[26]

Some collectors, Wanley and the Earl of Oxford included, were especially interested in Lhuyd's linguistic collections, the materials he had prepared towards a comparative dictionary of Welsh, Irish Gaelic, Scots Gaelic, Cornish, and Breton, which he had published as the first (and, in the event, only) volume of *Archaeologia Britannica* in 1707. (Lhuyd's own interest in the *Red Book of Hergest* had to do with tracing the historical development of Welsh.) Sloane, on the other hand, collected and published Lhuyd's natural historical and antiquarian letters, focusing in particular on his discussions of plant and animal life, fossils, and Celtic artifacts.

---

21 RS JBO 9, 22 April 1696, p. 227; JBO 9, 1 July 1696, 240; JBO 9, 22 July 1696, 241, 244; JBO 10, October 28, 1696, 3; Marcello Malpighi, *Opera Posthuma* (London: A. & J. Churchill, 1697).

22 RS JBO 11, 29 November 1711, 252–253; JBO 11, 17 January 1711/12, 265; JBO 11, 8 January 1712/3, 320–321; JBO 11, 29 January 1712/3, 327–328; JBO 11, 5 February, 1712/3, 330–331; JBO 11, 12 February 1712/13, 332–333; JBO 11, 5 March 1712/3, 337–339; JBO 11, 9 April 1713, 346–347.

23 Eiluned Rees and Gwyn Walters, "The Dispersion of the manuscripts of Edward Lhuyd," *Welsh History Review* 7 (1974): 148–178.

24 Rees and Walters, 150–151; quote on 150.

25 Humfrey Wanley, *The Diary of Humfrey Wanley*, ed. C.E. Wright and Ruth C. Wright (London: Bibliographical Society, 1966), 2 May 1715, vol. 1, 10.

26 Rees and Walters, 152.

At about the same time, Sloane also began sharing with the Society extracts from the correspondence and papers of the physician Sir Thomas Browne.[27] Sloane's use of these materials coincided with the auction of the library of Sir Thomas's son Edward Browne which incorporated his father's, suggesting a possible channel for Sloane's access to these papers.[28] The material Sloane presented largely consisted of letters to Browne rather than by him. These papers conveyed a range of antiquarian and natural historical particulars, and included images, as well. Presented to the Society in February 1711/12, correspondence with the Bishop of Iceland related information on "the Country, its Productions, its Migratory Birds, its Animals … not mentioned by any Writer."[29] Letters from William Dugdale discussed the draining of the Lincolnshire fens, and in particular, evidence, both material and textual, that the marshy fens had once been solid, dry, land, and that the Romans "had formerly been concerned in the same Works."[30] In May, Sloane presented from Sir Thomas Browne's papers a drawing of a Hungarian tomb, along with copies of the Latin inscriptions found there. The letters and extracts that Sloane chose to present covered roughly the same range of topics as those he presented from Lhuyd's correspondence, yet they dated from half a century earlier—roughly, the 1650s and 1660s.

The papers of Robert Hooke proved uniquely productive, inspiring further discussion and experimentation for over ten years after Hooke's death. Though Richard Waller took primary responsibility for editing these materials and presenting them at Royal Society meetings, Sloane took an interest in them, as well. Waller was surprised when he was unable to find a set of drawings of figured stones and petrified shells amongst Hooke's papers. But they turned up in Sloane's library, and "by the Favour of Dr Sloane, into whose Hands they happily fell," Waller was able to include engravings of these drawings in the *Posthumous Works*.[31] But these were not the only papers of Hooke's to fall happily into Sloane's hands: Sloane's personal library catalog lists three volumes of maps from Hooke's collection; "Dr Hooks letters, loose papers &c … gathered & bound up together"; some "Miscellaneous Collections of Mathematicks," partly in Hooke's hand; a manuscript on using magnetic variation to determine longitude taken "out of the library of Mr Robert Hooke"; and, finally, one

27 RS JBO 11, 7 February 1711/12, 268; 14 February 1711/12, 270; 21 February 1711/12, 272; 22 May 1712, 293.

28 Mandelbrote, 102.

29 RS JBO 11, 7 February 1711/12, 268.

30 RS JBO 11, 21 February 1711/12, 272.

31 Waller, *Posthumous Works*, 281. On Sloane and Hooke's books and manuscripts, see Mandelbrote, 98–145.

of the manuscripts of Hooke's diary.[32] That Waller was surprised to find the drawings missing suggests that he believed he had received all of the "personal" papers from Hooke's executors. However, there was some slippage in the process of transfer. It was not as neat, or as complete, as Waller would have hoped. Although Sloane probably acquired some manuscript material through his purchases at the auction of Hooke's library (sheets and slips tucked in to the books escaped the notice of Hooke's executors), this was not his only route to Hooke's papers. He collected other materials privately soon after Hooke's death, and he continued to purchase Hooke's papers through other library sales through the first half of the eighteenth century. "Dr Hooks letters, loose papers &c", for example, were "apparently acquired about three years" after Hooke's death, while the "Miscellaneous Collections of Mathematicks" were purchased at a library auction in 1739.[33]

3 Lhuyd's Letters as Visual and Textual Resources for Antiquarian Studies and Natural History

In the letters of his that Sloane presented and published in *Philosophical Transactions*, Lhuyd highlighted things he collected in his Celtic travels, as well as the social and natural contexts in which he found them. Lhuyd used visually descriptive language; he and his research team made drawings of individual specimens and artifacts, while also including textual information describing a specimen's color and size. He identified specimens by pointing readers to images resembling them in other books. His techniques here jibe with a number of approaches identified in recent studies of the visual culture of early modern natural history and antiquities, such as those by Alexander Wragge-Morley, Stephanie Moser, and Dániel Margócsy: the use of visually descriptive language, the categorization of "small finds" across sites, and the use of reference books to identify specimens when the physical specimen or a drawing could not be transported.[34]

32 Sloane, *Catalogues of his MSS*, BL Sloane MS 3972 B, 48v, 59r, 373v, 381v; Mandelbrote, 110.

33 Sloane, BL Sloane MS 3972 B, 373v; Mandelbrote, 110.

34 Alexander Wragge-Morley, "The Work of Verbal Picturing for John Ray and Some of His Contemporaries," *Intellectual History Review* 20 (2010): 165–79; Stephanie Moser, "Making Expert Knowledge Through the Image: Connections between Antiquarian and Early Modern Scientific Illustration," *Isis* 105 (2014): 58–99; Dániel Margócsy. "Refer to Folio and Number: Encyclopedias, the Exchange of Curiosities, and Practices of Identification before Linnaeus," *Journal of the History of Ideas* 71 (2010): 63–89.

What many of these techniques had in common was the isolation of the individual specimen, and comparison of that specimen to others that were similar to it: creating classes of specimens, abstracted from their local contexts, across space and time. So, for example, in a letter written from Monmouthshire in June 1697 (extracted in the *Philosophical Transactions* in the spring of 1712, fifteen years later), Lhuyd described a visit to a coalworks in Brecknockshire, where the mines were cut horizontally into the mountain, rather than dug vertically. Lhuyd found there a stone, a "valuable Curiosity." He offered a detailed visual description—the stone was made of a "substance like those they make Lime of; of a compress'd Cylinder Form; and as it were cut off even at each End: About 8 Inches long, and 3 in breadth: Its *Superficies* adorn'd with equidistant Dimples … and in each Dimple a small Circle; and in the Center of each Circle a little Stud like a Pins head"—and a reference comparing it to "Dr Plot's *Lepidotes*," a specimen engraved in Plot's *Natural History of Oxfordshire*, also taken from a coal mine, though in Cornwall. The stone Lhuyd had found was remarkable, he thought, in that it was "not referable to any thing I can think of, either in the Animal or Vegetable Kingdom," except to stones collected by other naturalists.[35]

Yet, Lhuyd's letters also offer a broader visual description of the places in which he wandered, situating the individual specimens in the sites in which he found them and giving his correspondents a fuller picture of the geological and human landscape of Wales. He described the high lonely lakes, deep in the Welsh mountains, where he found previously undescribed plants, and the coal mines where he found fossils. He meditated on the appearance of the rocks at the tops of Welsh mountains: "There is no Brimstone or Pumice-stones on the Tops of our Mountains, nor any thing else that I suspect to have been the Effects of Vulcano's. What seemed to me most strange, were waste confus'd Stones, and (to appearance) fragments of Rocks, standing on the Surface of the Earth, not only in wide Plains, but on the Summits also of the highest Mountains."[36] Lhuyd described the dry wells that stood at the tops of some of those mountains: they could be found on mountains on which "stood anciently Castles or Forts." [37]

35 Lhuyd, "Some farther Observations relating to the Natural History of Wales. In a Letter from Mr. Edw. Lhuyd to Dr Tancred Robinson, F.R.S.," *Philosophical Transactions* 27 (1712): 468.

36 Lhuyd, "A Letter from the Late Mr. Edward Lhuyd, Keeper of the Ashmolean Museum in Oxford, to Dr Tancred Robinson, F.R.S.," *Philosophical Transactions* 27 (1712): 465.

37 Lhuyd, "A Letter from the Late Mr. Edward Lhwyd, Keeper of the Ashmolean Museum in Oxford, to Dr Tancred Robinson, F.R.S.," *Philosophical Transactions* 27 (1712): 465.

As natural specimens were described in natural and human contexts (the high lake, the coal mine), so Lhuyd set antiquities within cultural contexts, allowing readers not only to visualize individual objects, but the human milieu in which they were found. Lhuyd did this in a description of a series of small stone finds in Scotland, in a letter written from Linlithgow (a town about 20 miles west of Edinburgh) in December 1699. The letter was printed in *Philosophical Transactions* in 1713, the last of a series printed together, tracking his travels from 1696 through 1699. In this letter, he described a series of small finds, stones and glass beads of particular forms, unto which the people he met with in his travels ascribed various kinds of magical powers. There were small glass beads, known as "snake-buttons" in the Scottish Highlands and "Adder-beads" in Wales. In both places, locals told a story that the beads were made by snakes, as Lhuyd wrote in his 1695 additions to William Camden's *Britannia*:

> "About Midsummer-Eve ...'tis usual for snakes to meet in companies, and that by joyning heads together and hissing, a kind of Bubble is form'd like a ring about the head of one of them, which the rest by continual hissing blow on till it comes off at the tail, and then it immediately hardens, and resembles a glass ring; which whoever finds (as some old women and children are perswaded) shall prosper in all his undertakings."[38]

Lhuyd believed these objects were of human manufacture, but in Scotland, he wrote, "not only the Vulgar, but even Gentlemen of good Education throughout all *Scotland*, are fully perswaded the Snakes make them."[39] Flint arrowheads, which Lhuyd compared to those found among the Indians of New England, were believed to fall out of the air, shot by Fairies.[40] As with natural objects, while he situated objects in their cultural and natural contexts, Lhuyd also assimilated stone tools to a larger class of similar objects. The arrowheads showed similarities not only to those found elsewhere in Britain, but in North America, as well. Lhuyd was not alone in drawing links between Native American and British material cultures. His fellow antiquaries made speculative connections between observed Native American practices and historical features of the British landscape. John Aubrey, for example, noted that natives in Virginia burned the forests to clear land "to cultivate the soile with Maiz,

38 Lhuyd, in "Additions to Denbighshire," in William Camden, *Britannia* (London: F. Collins, for A. Swalle and A. & J. Churchill, 1695), 683.

39 Lhuyd to Richardson, 17 December 1699, *Philosophical Transactions* 28 (1713): 98; Lhuyd, "Additions to Denbighshire," in Camden, *Britannia* (1695), 683–684.

40 Lhuyd to Richardson, 17 December 1699, *Philosophical Transactions* 28 (1713): 99.

and Potato-rootes.... Who knows, but Salibsury plaines, etc: might be made long time ago, after this manner and for the same reason?"[41] Perhaps the high downs of Britain and the open meadows and plains of North America developed through similar processes of human cultivation.

4 Editorial Transactions: From Personal Correspondence to Printed Philosophical Letter

What did it take to transform personal correspondence into a printed philosophical letter? Sloane (or possibly Waller, who extracted at least some of Lhuyd's letters for the Society) edited out Lhuyd's pleasantries, as well as his day-to-day accounting of his movements. In the published version of his letter to Tancred Robinson, dated 14 September 1696 and published in volume 27 of *Philosophical Transactions*, these opening lines are dropped: "Since my coming to Wales I have been generally upon the ramble in the day time and in company at night; so that I have had but little leasure for correspondence."[42] Not only pleasantries were edited out; the mundane work of managing as extensive a research project as Lhuyd's—crowd-funded and crowd-sourced through his correspondents—went missing, too. In his 22 December 1696 letter to Robinson, written during a brief rest in Oxford, he thanked Robinson for his assistance with gaining contacts among the Cornish gentry, and asked for his further help in distributing his natural historical and antiquarian questionnaire, a key research instrument, amongst them. Lhuyd distributed roughly 4,000 copies of his questionnaire. It allowed him to pinpoint artifacts, sites, and manuscripts that he wished to seek out, as well as knowledgeable people who might further assist him as he traveled.[43] Lhuyd typically distributed the questionnaire through his correspondents, as he did here, asking them to spread the sheets out among their contacts.[44] In the published version, all that remained of this letter were a few notes about a flock of scarlet birds observed in a hemp field in Pembrokeshire.[45]

41 John Aubrey, *Naturall Historie of Wiltshire*, Bodl. MS Aubrey 1, 18v. See also M. Goodrum, "Recovering the Vestiges of Primeval Europe: Archaeology and the Significance of Stone Implements, 1750–1800," *Journal of the History of Ideas*, 72 (2011): 51–74, 54.

42 Lhuyd to Robinson, Swansea, Glamorgan, 14 September 1696, RS LBO 14, 285–288.

43 Yale, *Sociable Knowledge*, 193–201.

44 Lhuyd to [Tancred Robinson?], Oxford, 22 December 1696, RS LBO 14, 289–290.

45 Lhuyd to Robinson, Oxford, 22 December 1696, *Philosophical Transactions* 27 (1710–1712), 466.

Some changes were more thorough-going. Before publication, Lhuyd's letters were edited so as to emphasize their visual, descriptive character. Causal speculation was reduced or removed. In the unpublished version of Lhuyd's September 1696 letter to Robinson, he ventured an opinion as to what the rough, rocky fields at the tops of the Welsh mountains might support in terms of geological theory. He thought it argued against a dissolution that covered the surface of the earth: how could such a dissolution "create solitary Rocks on the surface of the Earth, without a Mould to cast it into for supporting of the sides"?[46] This was dropped from the print version. In another case, Lhuyd's speculated that an "Ear of Rye" found in the side of an infant in Merionythshire could be supporting evidence for his theory that fossils germinated in the earth from seeds. This, too, was edited out.[47]

In the process of editing out the business of natural history, Sloane, working with Waller, edited out controversy and personal animus, removing Lhuyd's comments against the natural philosopher John Woodward. In a letter of 24 November 1696, written to Richard Richardson and published in 1713, Lhuyd made reference to a book, Thomas Robinson's *New Observations on the Natural History of This World of Matter* that "dealt very freely" with Woodward's theories of the formation of the earth.[48] More direct criticism of Woodward—reaching beyond his scholarship to his personal conduct—was dropped from a letter to Richard Richardson, dated 18 April 1699. Lhuyd's *Lithophylacium Britannicum*, his pocket guide to British fossils, was then in press. Since Lhuyd was absent from London, Tancred Robinson superintended publication for him. Lhuyd and Richardson had agreed to print some letters of Richardson's, compressed into one letter and translated into Latin, in Lhuyd's book.[49] In preparing the text for the press, Tancred Robinson inserted Woodward's name into what had been a veiled criticism of his theories: Richardson challenged Woodward to find fossil shells in the quarries in Richardson's North Yorkshire neighborhood—they should have been there, according to Woodward's theory that fossils had settled in sedimentary layers during the Noachian deluge, but

46 Lhuyd to Robinson, Swansea, Glamorgan, 14 September 1696, RS LBO 14, 285–288, trans. by Watt and Roberts, *EMLO*.

47 Lhuyd to Richard Richardson, 19 June 1698, RS LBO 14, 349–353, trans. Watt and Roberts, *EMLO*.

48 Lhuyd to Richardson, 24 November 1696, RS LBO 14, 337–338, trans. Watt and Roberts, *EMLO*.

49 Transcribers' note, Richard Richardson to Edward Lhuyd, 16 June 1698, trans. Watt and Roberts, *EMLO*; Lhuyd, *Lithophylacii Britannici Ichnographia* (London: ex officina M.C., 1699), 106–111.

Richardson was unable to find any. The section dropped from Lhuyd's 18 April 1699 letter to Richardson was Lhuyd's apology to Richardson for the inclusion of Woodward's name: "I was likewise concernd at the inserting any person's name in your Letter: but truly that Gentlmans carriage has been such, towards Mr Ray, Dr Lister, Dr Robinson & others, that no Gentlman of worth need scruple the public contradicting him in a plain matters of Fact wherin any persons of a Thousand may be Arbitrators: and I find that notwithstanding his \<great choler &/> Indignation few Naturalists spare him when he comes in their way."[50] According to Lhuyd, Woodward's private conduct excused public criticism of his theories, attached to his name. Sloane, in preparing Lhuyd's correspondence for publication, didn't necessarily disagree with this point; however, he thought it best not to air in print the personal animus that accompanied Lhuyd's reasoned scientific critique.

5 Sloane's Archival Politics

Sloane published Lhuyd's letters in *Philosophical Transactions* following the resolution of his own longstanding feud with John Woodward. Woodward and Sloane had different visions of the Royal Society. In Woodward's uncharitable view, Sloane focused too much on irrelevant minutiae, filling *Philosophical Transactions* with descriptive accounts of unphilosophical phenomena—an approach parodied, beyond the bounds of the Royal Society, in William King's 1700 *Transactioneer*, which attacked Sloane and the apothecary James Petiver.[51] Rumor, perhaps advanced by Sloane and Petiver, said that Woodward and his Royal Society ally John Harris had assisted King in his satire. Woodward denied it, but, on the other hand, he did believe Sloane's editorship was a scandal to the Royal Society's reputation. He sought to focus the Royal Society more tightly on the search for causal knowledge.[52]

Their dispute played out in person, at meetings of the Royal Society and its governing council, and in print. In May 1706, Council reprimanded Woodward and reminded him of the rules of right conduct: "unjust reflections" spoken against the Society or any of its members during meetings would be considered,

50 Lhuyd to Richardson, 18 April 1699, BL Sloane MS 4062, 300, trans. Watt and Roberts, EMLO; *Philosophical Transactions* 28 (1713): 275–276.

51 See Barbara M. Benedict, "From Benefactor to Entrepreneur: Sloane's Literary Reputation 1685–1800," in *From Books to Bezoars*, 34–37.

52 Levine, 85–86; Moxham, "Fit for Print," 243–244. For the outlines of the feud between Woodward and Sloane, see Levine, 85–92.

according to the Society's statutes, as grounds for ejection.[53] The reprimand suggests Woodward's "reflections" could be construed as violating gentlemanly norms, whether they were against Sloane or others. Just before Lhuyd's death, Sloane solicited from him a review of Johann Jacob Scheuchzer's *Itinera Alpina*, a survey of Alpine natural history, and published it in *Philosophical Transactions* in the form of a letter to Richardson. In this review, Lhuyd took a gratuitous swipe at Woodward's fossil theories, noting that Sheuchzer seemed to support Woodward's idea "of an Atomical Dissolution of all Things" at the Deluge, though this theory had been "sufficiently exploded" by Lhuyd and others.[54] Whatever Woodward may have said in person about fellows, including Sloane, this reflected publicly against Woodward's theories. Things came to a head in 1710. At a meeting of the full fellowship, Sloane read an account in which he discussed bezoar stones as gallstones and "a cause of colic"—a theory with which Woodward strongly disagreed.[55] They exchanged hard words—Sloane grimaced and laughed at Woodward, and Woodward accused Sloane of being ignorant of anatomy.

What words, precisely, did Sloane and Woodward exchange? What faces were made? What insults lobbed? It might seem that the answers to these questions were relatively straightforward, but it took further council meetings to pin them down. This process was recorded in the pages of the rough minute book, the paper book in which minutes were taken during Council meetings. The final judgment, both of what was said and the actions that were taken in response, is recorded in the clean copy of the Council minutes. On March 29, Council took up the business of Sloane and Woodward's contré-temps. Woodward attempted to manipulate the Council process in his favour; he requested that the matter be put off until his ally John Harris, a member of Council in 1709–1710, could be present. However, he was unsuccessful in delaying the proceedings. According to the minute book, "The Words spoken of Dr Sloane by Dr Woodward were; Speak sense or English and we shall understand you. If you understood Anatomy you would know better: or to that purpose." It was affirmed, as well, "that Dr Sloane made Grimaces, with a Laughter, and holding up his hands at Dr Woodward, before the reflecting Words above-mentioned were spoken." But did Sloane's grimaces provoke or excuse Woodward's reflecting words? With Sloane and Woodward out of the room, Council voted that Woodward's words were "reflecting" but that Sloane's grimaces were not "sufficient provocation for

53 RS, Council Minutes, vol. II, 1682–1727, 132.

54 Lhuyd, "A letter from Mr. Edward Lhuyd," *Philosophical Transactions*, 26 (1709), 157.

55 Levine, *Dr Woodward's Shield*, 91.

the above-mentioned Reflections."[56] On May 3, the Council retroactively laid in place statutes outlining the procedure for ejecting a fellow from Council and banning him from regular meetings of the full fellowship for speaking "words reflecting upon any other Member."[57] They then proceeded to reinforce the archival recording of these statues by requiring that "new Statutes" should be "reported to the Society in a convenient time" and a full copy of the statues should be recorded on vellum and deposited in the library for the fellows' consultation.[58] The use of vellum suggests the permanence and formality invested in this process. On May 24, Council performed a formal attempt at reconciliation as a final prelude to kicking Woodward out. Sloane "declared that he meant no Affront to Dr Woodward" by his grimaces, laughter, and gestures. The assembled company judged that this should be sufficient apology to Woodward, and asked whether he meant to accept the apology and offer his own to Sloane in return. Woodward, refusing "to beg [Sloane's] pardon for the reflecting words he spoke" was ejected. The episode was closed by a motion of Council, carried in the affirmative, to thank "Dr Sloane for his pains and fidelity in serving the Society as Secretary."[59]

In the clean copy of the minute book, all of this is recorded in a smooth, even italic hand. The story plays out sequentially and neatly. Even in the clean copy of the minute book, though, the entire process, with its additions to the statutes and the formal thanks to Sloane for his pains, appears rigged against Woodward. His agitated, angry comments disrupted the smooth, gentlemanly surface of the Royal Society's scientific waters, and many had felt him to be disruptive for some fifteen years. The institutional archive recorded a history of his disruptions, in the form of the reprimand against "unjust reflections" issued in 1706. The dominant forces on the Council, backing Sloane, manipulated institutional procedure against Woodward.

More of that agitation comes through in the rough minute book, the notes taken as Council worked out what Woodward had said, what Sloane had done, and what actions were to be taken in response. These minutes are preserved as Sloane Manuscript 3342, "Minutes of the Royal Society, 1699–1712," having passed in to the British Library's Sloane manuscripts collection from Sloane's library. The minutes are bound out of order, with the notes for the meetings of May 3, 10, and 17th preceeding those for March 29th, at which Council initially discussed Sloane and Woodward's altercation and settled on the precise

56 RS, Council Minutes, vol. II, 1682–1727, 166.
57 RS, Council Minutes, vol. II, 1682–1727, 167.
58 RS, Council Minutes, vol. II, 1682–1727, 168.
59 RS, Council Minutes, vol. II, 1682–1727, 169–170.

wording of Woodward's "reflecting words".[60] Council discussed two alternate wordings. The first was, "That nobody who understood anatomy would say so." The clerk—likely John Thorpe, a physician ally of Sloane's—crossed this out in favor of "If you understood Anatomy you would know better; or to that purpose", the wording used in the clean minute book.[61] In the course of recording, Woodward's statement was generalized; Council refused to pin the words down exactly, adding the phrase "or to that purpose." In the rough minutes, the motion asking whether Sloane's actions were "sufficient provocation" for Woodward's "reflecting words," described those actions as "such a grimace." Here, the clerk (or the assembled Council members) appeared to get ahead of themselves, referring to Sloane's actions as "such a grimace" before the grimace had been described fully in writing. The clerk crossed this out, expanded his description of Sloane's response to Woodward, and attributed it to a particular witness: "Mr Clavel affirmed that Dr Sloan made grimaces with a laughter & holding up his hands ~~before~~ at Dr Woodward before the reflecting words above mentioned were spoken." In this sentence, the clerk began to write "before the reflecting words," but then crossed out the initial "before" in order to note that Sloane held his hands up at Woodward. The wording here suggests the importance of specifying the details of Sloane's gestures, laughter, and grimaces, as well as of clearly identifying Woodward as the target of those gestures. We also see the importance of sourcing the recorded version of Sloane's grimace and laughter to a specific witness who was not Sloane himself. These details worked out, the clerk then recopied the text of the motion asking whether Sloane had provoked Woodward.[62] The clerk recopied this rough version into an intermediate copy, minus the crossings-out, in the rough minute book: from this he copied the final version for the clean minute book.[63] This may have happened in the Royal Society's rooms in Gresham College. By statutes adopted by Council in 1686, Thorpe, as clerk, was required to remain unmarried and live in Gresham College, where he was supposed to carry out his work.[64] Given his closeness to Sloane, however, and the fact that the rough minute book ended up in Sloane's archive, it is also possible Thorpe copied the minutes at Sloane's Bloomsbury home. Society

60 BL Sloane MS 3342, 68–78.

61 BL Sloane MS 3342, 76r; Henry Lyons, *The Royal Society 1660–1940: A History of its Administration under its Charters* (Cambridge: At the University Press, 1944), 143–144.

62 BL Sloane MS 3342, 76r.

63 BL Sloane MS 3342, 77r-v.

64 Thomas Birch, *The History of the Royal Society of London* (London: Printed for A. Millar in the Strand, 1757), vol. 4, 453–454.

records—such as journal books and correspondence—had been known to wander before.[65]

The olive branch that Sloane offered to Woodward was perhaps most carefully worded of all, as the rough minute book reveals. Sloane formally attempted to reconcile with Woodward before Council on May 24. After hearing Sloane's declaration that he "did not intend by any Grimaces to affront Dr Woodward," Council voted on whether that statement should be sufficient satisfaction to Woodward. Should Sloane also be required to express an apology? According to the minute book, they debated "whether the words ~~[& he is sorry that his gestures were so interpreted] shall be added after the words~~ [affront Dr Woodward]."[66] The language of apology is crossed through in the rough minutes, housed at the British Library, and does not appear in the fair copy of the minutes, housed at the Royal Society Library. Officially, Sloane offered only the minimal statement that he had meant no affront to Woodward.

6 Conclusion

For Sloane, paper was power. Sloane and Woodward's feud, worked out through letters passed from person to person and edited for publication as *Philosophical Transactions* articles, through their face-to-face maneuverings, and on the archival page, illustrates this principle concretely. Sloane not only sought to control archival records and manuscripts as sources of medical and natural philosophical knowledge, as with his collection and publication of Lhuyd's letters. He also sought such records as sources of institutional power. When Sloane was elected President of the Royal Society in 1727, following the death of Isaac Newton, his second official act, after settling a point of order about the use of the ceremonial mace during the Society's meetings, was to take in hand the Society's record keeping. According to the fair copy of Council minutes,

> The President observed that some years since a Copy of the Journals and other Books was ordered to be made out, but was never yet finished. The chief design of which was to have had two Copies, that one might be lodged in some different place for a Security to preserve Copies of the

65 Mordechai Feingold, "Of Records and Grandeur: the Archive of the Royal Society," in Michael Hunter, ed., *The Archives of the Scientific Revolution: The Formation and Exchange of Ideas in Seventeenth-century Europe* (Woodbridge: Boydell Press, 1998), 175.

66 BL Sloane MS 3342, 72r.

> Books against any accident of fire or otherwise, whereby they might be destroyed were there but one Copy of them. therefore He judged it might be fit to renew this Order. whereupon it was Ordered
>
> That the Copies of the Journals and other Books which were begun to be made be continued to the present time and when they shall be finished, That the said Copies be for the future lodged in the hands of the President, the President at the same time giving an Obligation under his hand to oblige himself, his Executors and Administrators to deliver the same to the Society at the Expiration of his Office.[67]

In this order, Sloane's concern is for the survival of the Royal Society's records: he orders off-site backups of the records to protect against the possibility that "any accident of fire or otherwise" might destroy them. Yet it is also about control: the copies are to be lodged in his house in Chelsea, built expressly to hold his collections. During his presidency, Sloane further required that Cromwell Mortimer, secretary and editor of *Philosophical Transactions* from 1728 to 1750, move to Chelsea to be closer to him.[68] Letters published in *Philosophical Transactions* from Sloane's secretaryship forward, including Lhuyd's letters, can be found in both Sloane's archive and in the Royal Society's.[69] In fact, the rough minute book in which Council recorded the process of working out precisely what Woodward had said to Sloane, whether Sloane provoked Woodward, and whether Woodward should be ejected survives only amongst Sloane's manuscripts. It is the only volume of rough minutes not housed in the Royal Society Library. Sloane allowed only the official, polished version of their dispute to survive in the institutional archives. *Philosophical Transactions* was officially Sloane's project as editor, bankrolled by him, though the Society had input into its contents, but the Council minutes more properly belonged to the Society. Whereas the distribution of letters published in *Philosophical Transactions* in Sloane's archive reflects Sloane's control over the journal, his capture of the rough minutes represents a more serious breach of the Society's archive. Perhaps preferring that the messy business of working out the details—the crossings-out and rephrasings that occurred as Council wrote, revised, and polished its communal voice—be less visible, he assimilated the rough minute book into his own collection. Sloane was concerned not only with manipulating the verbal, face to face processes in which Council determined his and Woodward's fate; he was also keen to

---

67 RS, Council Minutes, vol. II, 1682–1727, 303.

68 Moxham, "Fit for Print," 253.

69 Moxham, "Fit for Print," 253 and 260, n. 71.

secure the written record, leaving in the institutional archive only the gleaming surfaces of decisions made. In his editing of *Philosophical Transactions*, as well as in his daily interactions with the Royal Society record-keeping apparatus, Sloane manipulated the boundary between private and public archive, and deployed his collections strategically as a means of reinforcing his position within the Royal Society.

In this, Sloane's editing of the archive is not entirely unlike the editing of Lhuyd's letters from his Celtic travels, in which the logic of surfaces also asserted itself. Observations remain intact; but the business of doing natural history, of sorting out the books and tracking the progress of the distribution of research questionnaires, was deleted. No evidence of Lhuyd's personal grudge against Woodward remained. It's not surprising that Sloane deleted references to these kinds of business and personal matters, especially when they were upwards of fifteen years out of date. What's remarkable is that Sloane edited the archive as one might edit letters for publication, preferring to leave in the public archive only the polished official record.

With Woodward effectively vanquished, Sloane deleted any reference to Woodward from Lhuyd's letters, though he had happily published Lhuyd's review of Scheuchzer's *Iter Alpina*, with its side swipe at Woodward, in 1709. The dispute finished, there was no need to attack Woodward by name. Whether or not direct criticism appeared in the letters, those who needed to know would know that, given Lhuyd and Woodward's personal and intellectual disagreements—and that, in life, Lhuyd had been on Sloane's side—in publishing these letters, Sloane attacked Woodward.

Sloane collected for posterity, yes; one sees this in the moment he began writing his own correspondence into his library catalogues in the 1730s.[70] But he also collected to build institutional and personal power in the present. James Delbourgo has articulated the notion that, for Sloane and his suppliers, "collecting things meant collecting people."[71] Sloane collected his suppliers, the contacts who brought him things, and the suppliers collected Sloane, building a relationship with him that allowed them to then build contacts with other collectors to whom they could sell or give materials.

For Sloane, collecting the papers of his contemporaries was perhaps the purest form of this sort of collection of people. As Alison Walker argues in this volume, Sloane applied a similar logic in his purchases of printed books, seeking out the libraries of physicians who annotated their books. In doing so, Sloane

70 Hunt, "Sloane as a Collector of Manuscripts," 203.
71 Delbourgo, "Collecting Hans Sloane," 17.

brought his contacts, and their relationship with him, into his archive. There, those papers and the relationships that they carried could be further deployed as matter for Royal Society meetings, in print in *Philosophical Transactions*, and in his own publications.[72] Sloane could maintain his network of allies, marshaling them on his behalf, even from beyond the grave.

72 As in his later publication of physician Luke Rugeley's remedy for sore eyes, sourced from papers he had purchased from Rugeley's associates. Discussed by Arnold Hunt, "Sloane as a Collector of Manuscripts," 202–203; and Alison Walker, in this volume.

CHAPTER 6

Under Sloane's Shadow: The Archive of James Petiver

Arnold Hunt

In July 1710, the German traveller Zacharias Conrad von Uffenbach called on the botanist James Petiver at his house in London, bearing a letter of introduction and a gift of some fossils from the Frankfurt physician Johann Georg Kisner. Petiver was a member of the Royal Society, with an international reputation as a collector of natural curiosities, but the visit proved to be a great disappointment. "We expected to see a paragon of learning and refinement," Uffenbach wrote in his diary, "but he was quite deficient in both. For he appeared to be wretched both in looks and actions, and he had no parts, speaking very poor and deficient Latin and scarcely able to string a few words together." And Petiver's celebrated natural history collections turned out to be as disappointing as the man himself. "As soon as he gets any object of the least value he immediately has printed a short and insipid description of it, dedicating it to any person with whom he has some slight acquaintance; and then he takes a present for it. Everything is kept in true English fashion in prodigious confusion in one wretched cabinet and in boxes."[1]

Uffenbach was not alone in remarking on the disorganised state of Petiver's collections. On Petiver's death in 1718, Thomas Hearne wrote that he was "a very curious Man, particularly in Plants, but his things were in no Method."[2] Sir Hans Sloane, who bought Petiver's specimens and incorporated them into his own vast museum, paid tribute to him in the introduction to the second volume of his *Voyage to Jamaica* (1725) as having collected a greater quantity of natural curiosities than anyone before him. Unfortunately, Sloane went on, "he did not take equal Care to keep them, but put them into heaps, with sometimes small labels of Paper, where they were many of them injured by Dust, Insects, Rain, &c."[3] This sometimes led Petiver into making embarrassing mistakes, as in the case of the butterfly known as Albin's Hampstead Eye, supposedly

1 W.H. Quarrell and Margaret Mare, eds., *London in 1710: From the Travels of Zacharias Conrad von Uffenbach* (London: Faber & Faber, 1934), 126–7.

2 Thomas Hearne, *Remarks and Collections* (Oxford: Oxford Historical Society, 1885–1921), 6: 255.

3 Hans Sloane, *A Voyage to the Islands Madera, Barbados, Nieves, St Christophers, and Jamaica*, vol. (London, 1725), 2: IV.

 DOI 10.1163/9789004324305_008

discovered by Eleazar Albin on Hampstead Heath and described for the first time by Petiver in his *Papilionum Britanniae Icones* (1717). In fact, this was an Indo-Pacific butterfly that could not have been collected in England, and the likelihood is that Petiver received it from one of his overseas correspondents and accidentally mislabelled it.[4] Taking their cue from Uffenbach, Hearne and Sloane, modern studies of Petiver have reinforced his reputation as a sloppy and unsystematic collector. Raymond Stearns believed that much valuable information was lost because of Petiver's "careless and disorderly habits," while David Allen's verdict in the *Oxford DNB* was that "for Petiver, acquisition was virtually all: he had scant inclination to document or arrange his collections with the care that they deserved."[5]

Yet this view of Petiver as a man without method cannot be the whole story. No botanist of this period, and certainly no collector, could avoid taking sides in the debates over botanical method ignited by John Ray's *Methodus Plantarum Nova* (1682) and Joseph Tournefort's *Eléments de Botanique* (1694).[6] Petiver, a staunch supporter of Ray's system, was an active participant in these debates. On hearing that his friend Patrick Blair was planning to publish a synopsis of Tournefort's method, Petiver informed him that "although it be accurately done and you have taken great pains in it, yet it will meet with very few buyers" and urged him to adopt Ray's method instead: "I do not find that Dr Tourneforts Method in that Book has gained any great esteem amongst our Brotherhood or the Physitians and therefore yours may have the same Fate."[7] In common with many other early modern virtuosi, Petiver was also a fervent believer in detailed and extensive note-taking as a means of assembling and storing information for future retrieval.[8] His surviving archive is full of such notebooks, including a long series of books of letters sent which he labelled his "Adversaria," another long series of prescription books, and many

4 R.I. Vane-Wright and W. John Tennent, "Whatever happened to Albin's Hampstead Eye?," *Entomologists' Gazette*, 58 (2007): 205–18.

5 Raymond Phineas Stearns, *Science in the British Colonies of America* (Urbana: University of Illinois Press, 1970), 200; D.E. Allen, "Petiver, James (c.1665–1718)," *Oxford Dictionary of National Biography* (Oxford: Oxford University Press, 2004). http://www.oxforddnb.com/view/article/22041.

6 For a discussion of Ray and Tournefort and their respective methods, see Julian Martin, "Sauvages's Nosology: Medical Enlightenment in Montpellier," in *The Medical Enlightenment of the Eighteenth Century*, ed. Andrew Cunningham and Roger French (Cambridge: Cambridge University Press, 1990), 111–37 (esp. 118–20).

7 Petiver to Blair, 25 Jan 1708–9 and 6 Jan 1709–10, BL MS Sloane 3337, ff.30–32 and 68–9.

8 On early modern notetaking, see Richard Yeo, *Notebooks, English Virtuosi, and Early Modern Science* (Chicago: University of Chicago Press, 2014).

other volumes of collectanea and commonplace books. This forms a striking contrast with Sloane's archive, which consists mainly of loose papers and correspondence, with relatively few bound notebooks. Sloane may well have kept similar runs of letter-books and medical casebooks, but if so, he does not seem to have deemed them worthy of permanent preservation.

Petiver thus presents us with a paradox: a man who took great interest in methods of botanical classification, whose archive allows us to reconstruct his research methods in exceptional detail, but who was perceived by many of his contemporaries as singularly unmethodical. It should be remembered that he was active at a time when the "order of things" was still very far from settled, and when, as his correspondence shows, it was not unusual for botanists and naturalists to adopt their own idiosyncratic systems of ordering and classifying nature. John Banister, for example, in a letter to Petiver outlining his plans for a natural history of Virginia, explained that he did not propose to follow "so nice and particular a Method as that of the Learned Mr Ray" but to divide his book into four sections corresponding to the four elements. Under the heading of fire, he would describe the frequency of thunderstorms and "the Indians way of fire-hunting, sweating &c"; under air, the climate and some of the common diseases; under water, the "great Rivers" and "what quantity of Ships trade hither yearly"; under earth, "the Nature and felicitie of the soile, also somewhat of the Nature, Customes, and qualitie of the Natives and of the Trade we have with them."[9] Arbitrary as these divisions seem to us, they evidently appealed to Banister as a way of integrating natural and human phenomena into a single system of classification. In the same spirit, we should consider the possibility that Petiver appears unmethodical to us because his methods of knowledge production were so different from our own.

It is a central contention of this chapter that Petiver's reputation is inextricably linked to the afterlife of his archive. Because his collections were swallowed up by Sloane's, it is hard to visualise them as an independent body of material; and because Sloane's collections have been heavily reorganised and are now divided between the three "Sloane institutions" (the British Museum, the British Library and the Natural History Museum), it is even harder to see how Petiver's archive might originally have been organised and articulated. The terms "archive" and "collection" can be used interchangeably here, because Petiver himself would have perceived no distinction between the two. His paper archive of letters, drawings and notebooks in the British Library was intended to be used in conjunction with the plant and insect specimens in the

9 Banister to Petiver, undated, BL MS Sloane 3321, f. 7.

Natural History Museum (to say nothing of the living archive in the Chelsea Physic Garden). His herbarium is itself a paper archive, containing many hundreds of Petiver's original labels (the "small labels of paper" that Sloane referred to) as well as many other lists, memoranda and items of correspondence. It is therefore necessary to read across current institutional boundaries in order to reconstruct the integrity of his collection. Only by doing so is it possible to bring Petiver out from under Sloane's shadow and reconsider him on his own terms.

1 Petiver as Collector

In his own lifetime, Petiver's collection was better known and arguably more celebrated than Sloane's. His name was familiar not just to botanists and naturalists in Britain and Europe but to a global network of agents, travellers and collectors across the Near East, Africa and the New World; John Ray called him "a man of greater correspondence in Africa, India and America than anyone I know of besides."[10] His celebrity was further magnified by his printed catalogues, the *Museum Petiverianum* (1695–1703) and the *Gazophylacium Naturae et Artis* (1702–11), which publicised and illustrated his new acquisitions. Michael Valentini's *Musei Museorum* (1714), a compilation of selected highlights from contemporary cabinet-collections, included a list of no less than 610 items from the *Museum Petiverianum*, putting Petiver in the company of some of the most distinguished virtuosi in Europe.[11] Sloane, by contrast, never issued a printed catalogue, so that knowledge of his collection was limited to a smaller circle of friends, acquaintances, and those visitors to London who could obtain a personal introduction to see his museum.

This extraordinary one-man operation was conducted by Petiver from his apothecary's shop at the White Cross in Aldersgate Street. How did he do it? He was clearly successful in his business, remarking of the apothecary's trade, a little smugly, that "though the profit falls short of the vulgar saying, yet there is a very handsome maintenance to be obtained from it."[12] In 1702 he claimed to have been offered, and refused, one hundred guineas to take an apprentice, and when he drew up his last will in 1717 he was financially secure enough to make cash bequests to family members totalling £550. But as an apothecary he was not

10 Ray to Edward Lhwyd, 11 June 1701, Bodl. MS Eng. hist. c.11, ff. 63–4.

11 Michael Bernhard Valentini, *Musei Museorum* (Frankfurt am Main, 1714), vol. 2: 43–52 (second pagination).

12 Petiver to William Bennett, [after 10 May 1704], BL MS Sloane 4064, f. 6.

in the same league, socially or financially, as the great physician-collectors like Mead and Sloane. In a letter to John Ray he explained that "the daily attendance I am obliged to give in my Vocation" prevented him from writing to his friends as often as he would have liked.[13] Moreover, as he lamented to another of his correspondents, the French botanist Sébastien Vaillant, his means as a collector were limited: "I could yearly procure from most parts of the World things more Curious were there some annual Encouragement given to defray the Expences I am at, which are not a little to a Person of my private Circumstances."[14]

Petiver was therefore obliged to build up his collections through an elaborate process of bargaining and exchange which he termed "retaliation." A typical example comes from a letter to the German botanist Johann Philipp Breyne proposing an exchange of printed books or natural curiosities:

> We do not live in an Age that gives the mean incouragement that true Philosophers only expect, viz a smal competency beyond starving: however I am willing where money is not to be had to exchange what I do for a like vallue in other books of that kind which Curious Persons have often duplicates of or at least can easily procure, & if you will [not] or cannot do it I am willing to traffick with the Booksellers in your parts on the same Account, viz. Barter or Exchange for your Fathers acceptible works, your own or any others published in Dantzic or neer you, or otherwise even for Collections of Naturall Things themselves or other Curiosities if you vallue them not too dear.[15]

This process of exchange encompassed an enormous variety of items. From Madras, the East India Company surgeon Samuel Brown sent opium to Petiver to sell for him in London.[16] From South Carolina, Hannah Williams sent an Indian tobacco pipe, a wild bees' nest, a collection of shells and "the head of a strange Bird," to which Petiver responded with "some Hysterick pills and others for your head and stomack, with some Packetts of News and a printed Account of the Shells you last sent me."[17] Other correspondents in North America were

13 Petiver to Ray, 1 April 1701, BL MS Sloane 3334, f. 23.

14 Petiver to Vaillant, 21 April 1701, BL MS Sloane 3334, ff. 30–2.

15 Petiver to Breyne, [April 1706], BL MS Sloane 3335, f. 9.

16 Petiver to Brown, 15 April 1697, BL MS Sloane 3332, ff. 271–4.

17 Petiver to Williams, 17 Nov 1706, BL MS Sloane 3335, ff. 39–40. For Petiver's correspondence with Hannah Williams, see Beatrice Scheer Smith, "Hannah English Williams: America's First Woman Natural History Collector," *South Carolina Historical Magazine*, 87: 2 (1986): 83–92.

supplied with pots of balsam in exchange for plants and animal skins. The shop in Aldersgate Street was thus at the centre of a constant two-way traffic, with books, medicines and specimens flowing out as well as in.

This makes Petiver's correspondence a uniquely valuable historical source, both on the practical logistics of conveying fragile and perishable objects over very long distances and on the economy of exchange among early modern collectors. Petiver was quite explicit about the *quid pro quo* that this entailed. In a letter to Vaillant thanking him for a gift of plants, he assured him: "you shall be amply recompensed for them with the choicest of my dry specimens ... I will take the first Opportunity of sending them, that I may oblige you to continue your so well begun Favours to me."[18] Another correspondent was given an album of specimens to present to the French botanist Antoine de Jussieu, but instructed not to hand it over until Jussieu had offered something in return:

> When you come to Paris I desire you will give Dr Jussieu his Packett with my service & shew him the Hortus Siccus (but part not with it out of your hands,) & let him know if such a Collection will be acceptible to him, I will send it, as soon as I have had a return for what he has already received from me .. You know how desirous I am to keep a Correspondence with so able a Botanist, if he pleases to incourage it by a mutuall & frequent Commutation.[19]

Petiver also played on the vanity of his correspondents by emphasising the fame they could obtain by being the first to report a new specimen:

> It will be your immortall Honour to be the first discoverer of such Plants Shells Insects & Serpents as you shall send before another, which will cause your name to [be] eternized in Golden Letters which will remain to future Ages when you shall be no more. I must tell you its noe small Honour to be taken notice of by so illustrious a Body as the Royall Society besides the advantage that you may in time reap by it, therefore let me advise you to be the first in the discoveries by sending a Collection of Plants Shells & Insects with a small Cask of Snakes &c to me as soon as this comes to your hands to prevent others from being before you who are employed for the same purpose.[20]

18 Petiver to Vaillant, 20 Feb 1700–1, BL MS Sloane 3334, f. 33.

19 Petiver to Dr Urquhart, 30 April 1717, BL MS Sloane 3339, f. 322.

20 Petiver to Edward Bartar, 16 Feb 1696–7, BL MS Sloane 3332, ff. 247–8.

As James Delbourgo has perceptively observed, Petiver was above all a list-maker, and his collecting was a project of "enlistment" in the literal sense of the word, in which people as well as objects were co-opted into lists and catalogues of one sort or another.[21]

To some of his contemporaries, Petiver's collection seemed to reflect the mindset of the tradesman rather than the disinterested pursuit of knowledge. He never fully mastered the epistolary conventions of early modern *commerce de lettres*, constantly pestering his correspondents for new specimens or chiding them for not packing and preserving them properly. He was not embarrassed to offer money for specimens or even, in his *Gazophylacium Naturae*, to promise that anyone could have a plate dedicated to them on payment of a guinea. Here again the contrast with Sloane is instructive: when Sloane's correspondents offered him items for his collection, they frequently chose to disguise these as gifts, even when they evidently hoped for payment in return. But the significance of Petiver's correspondence lies precisely in the fact that it exposes, sometimes with more frankness than finesse, the hard commercial calculations which made early modern collecting possible. When Robert Stevens, in South Carolina, was approached by Petiver offering to purchase specimens, he replied coolly that "what he takes in Collecting of Seeds and Plants, is only to please himself, and oblige one or two particular friends." Yet he added, meaningfully, that if he were to make "a Trade of it," he would require "greater offers" than the sum Petiver had mentioned.[22] His objection, in other words, was not so much to the social solecism of offering money for specimens as to the fact that Petiver had not offered enough.

Petiver's association with Sloane reveals a similar combination of friendship and self-interest. Sloane, five years Petiver's senior, was instrumental in advancing the younger man's career and getting him elected to the Royal Society. Their friendship flourished despite the long history of conflict between the College of Physicians and the Society of Apothecaries, which came to a head in 1696 with the founding of the London Dispensary.[23] Petiver, like other apothecaries, hoped for more independence from the physicians and argued that it was in an apothecary's interests to acquire a working knowledge of practical physic, as "it will qualifie him to give to his own Patients till acquaintance

21 James Delbourgo, "Listing People," *Isis*, 103 (2012): 735–42.

22 Robert Ellis to Petiver, 25 April 1704, BL MS Sloane 4064, f. 2.

23 On the Dispensary affair, see Frank H. Ellis, "The Background of the London Dispensary," *Journal of the History of Medicine*, 20 (1965): 197–212. An anti-Dispensary tract, "A modest defence of the Apothecaries against the dispensary," is copied into one of Petiver's letter-books, BL MS Sloane 3334, ff. 19–22.

and obligations with Physitians further enable him, for I have found it very necessary so to improve myself that I might live without dependance." This was the point at issue in the Rose Case of 1704, the test case which officially broke the physicians' monopoly by giving apothecaries the right to diagnose and prescribe.[24] Sloane's views on the Rose Case are not clear: in private he may have sympathised with the apothecaries, but in public he stood four-square with his colleagues in the College of Physicians.[25] However, both he and Petiver both tried to repair relations between the two professions and saw them as united by a common reliance on botanical knowledge, which Petiver described as "a Science absolutely necessary towards the accomplishing all Physitians Apothecaries Surgeons Chymists &c in the knowledge of Physick."[26]

Petiver made himself useful to Sloane in other ways: for example, by acting as Sloane's proxy when the botanist Leonard Plukenet attacked him for errors in his catalogue of Jamaican plants. Even by the standards of early modern controversy, Plukenet's attack on Sloane stands out for its ferocity: he declared that Sloane's work was filled "not merely with falsehood, but rashness and stupidity" and scornfully quoted Cicero, "any man can err, but only a fool persists in his error."[27] Yet it was Petiver who took up the challenge, in a series of papers published in the *Philosophical Transactions* ostensibly written as a commentary on Samuel Browne's account of East Indian plants. As he explained to William Sherard:

> I have ventured to publish another Vollum of Dr Browns Indian plants in the last Transactions & did design (as you may see in the first sheet) to take very little notice of our angry adversary, but finding so many Errors that it would be injustice to the world not in some Measure to Correct

24 Petiver to William Bennett, [May 1704], BL MS Sloane 4064, f. 6. Harold J. Cook, "The Rose Case Reconsidered: Physicians, Apothecaries, and the Law in Augustan England," *Journal of the History of Medicine*, 45 (1990): 527–55. Cook argues that many apothecaries were already practising physic and that the judgement in the Rose Case simply reaffirmed the status quo.

25 Cook notes that William Rose appealed to Sloane for support, which may mean that Sloane was thought to be sympathetic to the Apothecaries' case: see Cook, "The Rose Case," 539.

26 Petiver to "Mr S.D." (Samuel Dale) on Ray's method of English plants, BL MS Sloane 3337, f. 20.

27 Plukenet, *Almagesti Botanici Mantissa* (London, 1700), 39: "in quo non falsa modo, sed inscitiae et temeritatis plena cuncta congessit. Cujusvis hominis est errare (inquit Cicero) nullius nisi insipientis in errore perseverare." Petiver's copy of the book, which later belonged to Sloane, is now BL 441.g.10.

> them, I was obleiged to take cognizance of them & him, which if he can disprove, I expect to be severely lasht, tho at present he seems very moody .. and his Zeal to Botany is now grown so cold that he only wishes the Expences he has been at were again in his Pockett.[28]

Most observers seem to have judged Petiver the victor of this "botanical war."[29] John Ray wrote to him: "You have discovered so many oversights and mistakes in Dr Plukenets works, that I fear he may have led me into some errours, who followed him as a most exact Botanist without due examination."[30] At the same time, Petiver's friend David Krieg warned him candidly that he was doing his own reputation no good by engaging in controversy: "pray do not give any occasion to fall out with other people, as you have done [with] Dr Plukenet, for those quarrels hinder allways by some way or other the improvement of that studie."[31] Petiver's reply to Plukenet was thus of the greatest value to Sloane in allowing him to vindicate his intellectual reputation without having to get his hands dirty in the arena of public debate.

This forms something of a pattern in Petiver's relations with Sloane. In their dispute with John Woodward it was, once again, Sloane who remained above the fray and Petiver who acted as his fixer. Writing to Hugh Jones in Maryland, Petiver earnestly warned him not to collaborate with Woodward or allow any of his specimens to fall into Woodward's hands:

> I understand our good friend Dr Woodward is mightily disgruntled that you should send petrifactions to me and Mr Lhwyd .. This I thought fit to acquaint you withall that forewarned, you may know who are your Friends & that Dr Woodwards Anger will not I hope frighten you from sending by all Opportunities to me & the rest of your Friends: And to obleige you to it let me tell you he hath (as he repeated he would spoil my Correspondence) already sent a Person over with a stipend & sallary from the Arch Bishop of Canterbury on purpose to collect Plants, Shells, Insects & particularly Fossills & form"d Stones or Petrifactions, so that

28 Petiver to Sherard, 19 Feb 1700–1, BL MS Sloane 3334, ff. 16–17. Petiver, "An Account of Mr Sam. Brown his Fifth Book of East India Plants," *Philosophical Transactions*, 22 (1700–1): 1007–29.

29 The phrase is Tancred Robinson's, from a letter to Martin Lister, 17 Oct 1699: "Dr Plukenet ... reflects very severely upon Dr Sloan and Mr Pettiver, so we are like to have a Bellum Botanicum." Bodl. MS Lister 36, ff. 235–6.

30 Ray to Petiver, 4 April 1701, BL MS Sloane 4063, ff. 77–8.

31 Krieg to Petiver, 30 March 1702, BL MS Sloane 4063, ff. 149–50.

> I begg you would double your diligence & send all you can of every of these that you may have the honour of their first discoverer & he only what's left.[32]

Sloane and Petiver were convinced that Woodward was behind the publication of the satirical pamphlet *The Transactioneer* (1700).[33] Woodward denied it, but even as he did so, recapitulated the pamphlet's portrayal of Sloane and Petiver as a tight little mafia running the Royal Society for the benefit of themselves and their cronies. "The matter is this," wrote Woodward, "Dr Sloane and his friend Mr Petiver cause it to be spread abroad that I am the Author .. They do not directly charge me with it, that is not their way, but they do the thing as effectually by insinuating in their Clubs and meetings."[34] It is a rare glimpse of Sloane behind the scenes, revealing a less affable Sloane than the carefully crafted public image might suggest.

The *Transactioneer*—now attributed to the satirist William King, though possibly a collaborative effort written with the help of some of Sloane's enemies in the Royal Society—presented a well-informed caricature of Petiver as Sloane's client and factotum, the "Philosophick Sancho" to Sloane's Don Quixote. "Sir, he and I are all one," Sloane was made to say of Petiver. "You must know we club Notions, laying them up in a kind of Joynt Stock, and have all things in common .. By my good will I would never be without him."[35] The allusion to Sloane and Petiver as partners in a joint-stock enterprise may have been a dig at Petiver's social position, suggesting that, for all his pretensions, he remained a shopkeeper for whom philosophical ideas were merely articles of trade. However, it may also be a more direct reference to their collaboration as collectors. Petiver's correspondence shows that he and Sloane sometimes acted jointly in buying and exchanging specimens: for example, he wrote to Jean Rudolphe Lavater in 1708 that he had seen "a Letter from you to Dr Sloan our worthy Friend, by which I understand you are in the House of a Curious Collector of Rarities and if we could suddenly see a Catalogue of them as you promised Dr Sloan, either he or I would endeavour to make the Gentleman a reciprocall return for whatever he has duplicates of or after seeing his

32 Petiver to Jones, [Nov 1697], BL MS Sloane 3333, ff. 91–3.

33 On the *Transactioneer* affair, see Delbourgo, "Listing People," and Richard Coulton, "'The Darling of the *Temple-Coffee-House Club*': Science, Sociability and Satire in Early Eighteenth-Century London," *Journal for Eighteenth-Century Studies*, 35: 1 (2012): 43–65.

34 "Dr Woodwards Letter," 28 Feb 1699, copied in BL MS Sloane 3334, ff. 58–9.

35 *The Transactioneer, with some of his Philosophical Fancies; in Two Dialogues* (London, 1700), F1v (p. 34).

Catalogue wee shall desire."[36] On occasion Sloane seems to have stepped in to buy material that Petiver could not afford. In their dealings with Maria Sybilla Merian it was Petiver who conducted most of the correspondence, but it was Sloane who eventually bought Merian's drawings for the sum of 200 guineas on Petiver's recommendation.[37]

Petiver's most important errand on Sloane's behalf was his journey to Holland in 1711—the only time he ever travelled outside England—to bid at the auction of Paul Hermann's collection in Leiden.[38] Before his departure for Holland, Petiver drew up a will in which he bequeathed to Sloane "all my Collections of Naturall Things whatsoever, as well Duplicates as single Samples, and of all my Manuscripts relating thereto, excluding my Printed Bookes: on consideration of your cancelling my Bond" and making further bequests of £500, including £100 to the Royal Society "for the Discovery and Collecting Naturall Productions" and £100 to the Chelsea Physic Garden. The reference to the cancellation of a bond is intriguing, as it suggests that Sloane may have loaned Petiver money to support his collecting activities.[39] Petiver's final will, drawn up in August 1717, eight months before his death, left his residual estate to his sister, Jane Woodcock, and made no mention either of Sloane or of his collection.[40] However, as Sloane later wrote, "he always intended if he died before me, that his Collections should come into my Hands," and Sloane reportedly paid the sum of £4000 to acquire the whole collection, either shortly before Petiver's death or soon afterwards.[41]

This was the largest single acquisition that Sloane ever made, and one of the most important. It was a collection of collections, containing not just Petiver's own archive and specimens but a number of other discrete collections formed by his correspondents and suppliers, including James Cuninghame's plants and botanical drawings from China, and George Joseph Kamel's plants and drawings from the Philippines. The relationship between Petiver and Sloane has often

36 Petiver to Lavater, 30 Jan 1708, BL MS Sloane 3336, f. 24.

37 Petiver to J.P. Breyne, [April 1706], BL MS Sloane 3335, f. 9. On Sloane's purchase of Merian's drawings, see also Ella Reitsma, *Maria Sibylla Merian and Daughters: Women of Art and Science* (Los Angeles: J. Paul Getty Museum, 2008), 203.

38 On Petiver's purchases at the Hermann auction, see his letter to Sloane, 29 June 1711, BL MS Sloane 4042, f. 305 (copied in Petiver's letter-book, BL MS Sloane 3337, ff. 160–1, with some textual differences). Petiver's copy of the sale catalogue, annotated by him with prices and names of buyers, and with related documents bound in, is BL 1044.c.4.

39 Petiver's will is contained in a letter to Sloane, 7 June 1711, BL MS Sloane 4042, ff. 295–6.

40 Petiver's will, National Archives, PROB 11/563.

41 Sloane, *Voyage to Jamaica* (1725), 2: IV. The sum of £4000 is given by Hearne (n.2 above) and in Richard Pulteney, *Historical and Biographical Sketches of the Progress of Botany in England* (London, 1790), 2: 32.

been presented as a study in contrasts: Petiver the industrious accumulator, Sloane the systematic collector, as if the natural fate of Petiver's collection was to be absorbed into Sloane's museum where it could be properly catalogued and organised for the first time. But in terms of their collecting activities, Petiver and Sloane were joined at the hip. Much of what now passes as the "Sloane collection" is in fact Petiver's creation; indeed, many of Sloane's catalogue descriptions were taken straight from Petiver. As the next section will show, Petiver was by no means without a system for organising his collection, but the dynamic nature of his collecting, with letters, printed papers and objects in constant circulation, means that his indexing system can only be partially captured in a static archive. It is this, I suggest, that accounts for his reputation as an unsystematic collector.

2 Petiver as Archivist

Petiver's massive collection of natural curiosities required the creation of an equally massive paper archive to keep track of it. Petiver understood this very well, and, like many other early modern naturalists, took a lot of trouble to develop efficient methods of information storage and retrieval. In a letter of 1696 he explained his system for recording insect specimens, which resembled the system pioneered by Francis Bacon earlier in the century in using two separate notebooks: one for rough jottings, which Bacon likened to a merchant's waste book, and the other for more digested observations, which Bacon likened to a merchant's ledger.[42] The key to the archive was the numbering system, which allowed for easy cross-reference between the paper book and the physical specimen. Equally important was the use of blank space so that new material could be inserted at a later stage:

> Whatever comes in my way, I put down into a small Paper booke I keepe for that purpose under a Series of Numbers even as they come to hand without any Method beginning with No. 1. 2. & so goe on, I allow 2 Insects to each Octavo page in my Adversaria or Note Booke first giving some rough name that may easily convey the remembrance of him to my memory & after that I leave a certain space to add the Synonymous names of Authors as they may come in my way. Then I remark the Place & time I first observed them att, with what else may be remarkable in them & a short description of them. Each Insect I pinn in my Box under

42 On Bacon's method, see Angus Vine, "Commercial Commonplacing: Francis Bacon, the Waste Book, and the Ledger," *English Manuscript Studies*, 16 (2010): 1–33.

> its peculiar Class with the same number I have made my remarks. Then I have a Seperate Catalogue of them in a sheet of Paper, allowing halfe or a whole page or more according to the largness of the Class as for Example
>
> 1. Apum Vesparum & Bombylum Catalogus.
> 2. Muscarum Catalogus. & so on, to each Bee, Fly, Beetle &c adding the Number under which it is in the History as I have done to the end of the names of those I have already mentioned & such I shall send you. This Method till you can find a better I would have you follow.[43]

The "small Paper booke" described here is still preserved among the Sloane manuscripts, with the entries laid out two to a page and numbered 1 to 500.[44] It includes detailed field notes on Petiver's insect specimens, many of them collected on his rambles around Islington, Hampstead and "Caen Wood" (Kenwood). In some cases Petiver did not even need to leave his house to add new insects to his collection: one butterfly was caught by his servant at his kitchen window, another by Petiver himself "on the steps of my street door." Of one small moth he noted: "these are frequent most part of the summer in my parlour, proceeding as I suppose from some dead Birds kept there."[45] Petiver's notes leave no doubt of his dedication to the accurate recording of specimen data; if this has not been widely recognised as it might be, it is because his notebooks in the British Library are now physically separated from the specimens in the Natural History Museum to which they refer.

Petiver also took great interest in systems of botanical classification. In a paper to the Royal Society in 1699, he argued that plants with the same physical characteristics often had the same medicinal effects: for example, umbelliferous herbs such as parsley, caraway and cumin were effective in treating colic and flatulence, while vegetables such as radishes and turnips, which Ray had classed together as "flore tetrapetalo uniformi" in respect of their four-petalled flowers, were powerful diuretics.[46] As an apothecary, Petiver was well aware that this information could have commercial implications. As he pointed out, rosemary and lavender were generally valued for their flowers, but his system of classification suggested that, like other umbelliferous plants, their chief effect

43 Petiver to John Scampton, 4 June 1696, BL MS Sloane 3332, ff. 212–5.

44 Petiver, "Historia Universalis Insectorum," BL MS Sloane 2347.

45 BL MS Sloane 2347, ff. 34r, 52v, 60r.

46 Petiver, "some Attempts made to prove that Herbs of the same Make or Class for the generality, have the like Vertue and Tendency to work the same Effects," *Philosophical Transactions*, 21 (1699): 289–94.

lay in their leaves and husks. If so, then, he predicted, the market value of these plants would rise: "I fear they will quickly sell the Husks as dear as the Flowers, if they find a great vend or a frequent demand for them."[47] But his paper was also, in a sense, his professional manifesto, reflecting his long-held conviction that the botanical knowledge of apothecaries like himself could be directly applied to the improvement of physic. "I doubt not," he told Patrick Blair, "but if you take the Classicall Method into your Curious Consideration" ("classical" in this sense meaning classed or classified), "you will find the Medicinall subservient to it, and perhaps many Plants which grow in such abundant plenty with us to be of more Vertue or Use than we have hitherto found out or conceived, and are not yet reckoned amongst our Materia Medica."[48]

This makes it implausible to suggest that Petiver was solely interested in collecting for its own sake, not in organising his collection. Among the many services he rendered to Sloane was that of putting his herbarium into order. When Sloane purchased Plukenet's dried plants after the latter's death in 1706, he found them "very confused and without Names and references" and called on Petiver to help. "I have taken some pains about them," Petiver wrote to Sherard, "and as I knew most of his Correspondents and the greatest part of his Collection coming from the same hands I had mine, I find I can go a great length towards the elucidating of them."[49] In the same spirit, Petiver urged Sir George Wheler to preserve his important collection of plants from Greece and Asia Minor by "letting them be put into his own Method and pasted all in one Book together," a task which Petiver offered to do for him "and to add some Notes and Synonyms to them by which they would be the better known." Petiver was shrewdly aware that good classification could significantly enhance both the intellectual and financial value of a herbarium. The collection of the great sixteenth-century botanist Leonhard Rauwolff, he pointed out, had been bought by the University of Leiden, "where they Vallue them at the Price they gave for all Dr Isaack Vossius his Library in which they were & paid for them £2500 as I am informed." Wheler, he suggested, could do the same by allowing his collection to be put in order, "and then wherever he leaves them, they may without prejudice be viewed and remain a lasting Monument of his great pains & Travails."[50]

For Petiver, collecting was not simply about the accumulation of objects; it was a continuous process of listing, recording and labelling. His obsessive list-making is exemplified in a volume from Sloane's collection (now Sloane MS

47 Petiver, "Some Attempts," 291.

48 Petiver to Blair, 6 Jan 1710, BL MS Sloane 3337, ff. 68–9.

49 Petiver to Sherard, 20 Sept 1710, BL MS Sloane 3337, ff. 85–6.

50 Petiver to John Covell, 22 May 1717, BL MS Sloane 3340, ff. 323–4.

3331) described in Sloane's catalogue as "loose papers of Mr Petivers pasted on brown paper." At first glance this gives the impression of absolute chaos: hundreds of notes in Petiver's crabbed hand, scribbled on tiny scraps of paper, often literally on the backs of envelopes. On closer examination, however, order starts to emerge. The majority of the loose papers consist of lists of specimens sent to, or received from, Petiver's numerous correspondents, often identified with reference to the numbered descriptions and tables of engravings in his two major publications, the *Museum Petiverianum* and *Gazophylacium Naturae*.[51] One intriguing survival is a list headed "Collectanea Quotidiana," recording new accessions to Petiver's collection in order of acquisition. This may have been intended as part of a larger notebook, modelled on the entry-books that Petiver would have used in his apothecary's business to keep track of day-to-day transactions. The surviving fragment is dated September to November 1717, only a few months before Petiver's death, and reflects the extraordinary diversity of his collection as well as the hectic pace of acquisition: it begins with a list of crabs, shells and sponges sent by James Campbell from Minorca, followed by a list of animal specimens sent by Henry Barham from Jamaica, including a crocodile, an iguana, a galliwasp and a porcupine, all received on a single day, 26 September 1717.[52]

Alongside the collecting and listing of specimens, Petiver was also involved in collecting and commissioning images. This was partly because drawings were less perishable than plants or insects, and easier to transmit across long distances, but also because Petiver needed a constant supply of images to reproduce in his *Gazophylacium Naturae*. In 1697 he employed David Krieg to draw the insects in his collection, describing him as "a most ingenious person" who "hath lodged at my howse about a year in which time he hath painted most of our English Insects and severall other things admirably well."[53] In 1717 he wrote to the Dutch collector Levinus Vincent asking for the loan of some of his specimens and drawings so that they could be copied and engraved; the letter is of interest both for Petiver's remarks on the copying of imperfect specimens and for his candid admission that it was sometimes necessary to improve on nature rather than copying the original "exactly as it lies":

> I could wish for the future what you send me the Paintings of, should be only your most rare, & such you have no duplicates of, for the Charge

51 For a typical example, see the list of "Insects sent Mr Seba," 6 Feb 1712–13, BL MS Sloane 3331, f. 313.

52 "Collectanea Quotidiana," BL MS Sloane 3331, f. 636.

53 Petiver to Hugh Jones, [Nov 1697], BL MS Sloane 3333, ff. 91–3.

> would be the same as painting the most common, which are easier to send over in their own Specimens & if they are not so intire as they should be, we can make up the imperfection of one side or part by the other that is not so, and rectifie what your Painters I find, often err in, viz. of laying the under wing of some above the upper, and sometimes not dividing the one from the other as in Fig. 4. or making one side unequall & differing from the other as in Fig. 4, 5, 6 & 8, they exactly copying the insect as it lies, which may be prevented by sending over the Patterns themselves of such you can spare.[54]

Petiver also acquired natural history images from a number of his overseas correspondents. From Madras, the East India Company surgeon Edward Bulkley sent him drawings of Indian birds; from the Cape, the naval surgeon Martin Dolneus sent him drawings of African plants; while in Virginia, the naturalist John Banister was said by Petiver to have "a very happy hand at designing Plants Shells and Insects the which I have seen curiously performed by him, he frequently sending over what he there observed."[55] Many of the images from Petiver's collection can be found today in Sloane's albums of natural history drawings, identifiable as Petiver's because of the references in his hand to "MP" (*Museum Petiverianum*) and "GN" (*Gazophylacium Naturae*) where they were described and reproduced.

Petiver's publications, particularly the *Gazophylacium Naturae*, drew a decidedly mixed response from some of his correspondents, who were disappointed by the poor quality of the engravings and the brevity of the descriptive text accompanying them. James Bobart, the keeper of the Oxford Botanic Garden, told him bluntly: "I dare not deliver these books to the subscribers without the key to them, or without some account of their contents." One of those subscribers, the Oxford academic Dixon Colby, complained that "such pictures" served no purpose except "to please Children or amuse fooles, being very unserviceable to any Philosopher, without names," while even the mild-mannered John Ray confided to Sloane: "I doe not well like the Cuts of Mr

54 Petiver to Vincent, [March 1717], BL MS Sloane 3340, f. 316.

55 Petiver to Dr Lewis, 2 Feb 1696–7, BL MS Sloane 3332, ff. 254–8. On the important role of ships' surgeons in supplying Petiver with material, see Kathleen S. Murphy, "Collecting Slave Traders: James Petiver, Natural History, and the British Slave Trade," *William and Mary Quarterly*, 3rd series, 70: 4 (2013): 637–70, and Anna Winterbottom, "Medicine and Botany in the Making of Madras, 1680–1720," in *The East India Company and the Natural World*, ed. Vinita Damodaran, Anna Winterbottom and Alan Lester (Basingstoke: Palgrave Macmillan, 2015), 35–57.

Petivers Gazophylacium, they are not so elegant and polite as I could wish mine might be."[56] When the first instalment of the *Gazophylacium* came out in 1702, Petiver sent several copies to David Krieg in Paris in the hope of gaining more subscribers, but without success; Krieg showed the plates to a distinguished group of botanists, including Sébastien Vaillant, Claude Joseph Geoffroy and Charles Plumier, but failed to persuade any of them to subscribe, Plumier remarking unenthusiastically that "it was pity, that they was not well engraved and anatomised, especially the plants." Krieg only made matters worse by his clumsy efforts to excuse Petiver on the grounds that he might have been working from imperfect specimens and, in any case, did not have the artistic ability to draw them himself.[57]

What then should we make of these publications on which Petiver lavished so much effort and expense? The *Museum* and *Gazophylacium* are best understood, I suggest, not as independent works of reference but as extensions of Petiver's collecting activities. The printed text could be cut up to produce labels, which could then be attached to specimens in his collection or duplicates exchanged with other collectors. An inventory of Petiver's miscellaneous papers, drawn up by Sloane after Petiver's death, lists bundles of unbound sheets from the *Museum* and *Gazophylacium* "cutt for labells," and many of these spare labels survive today, loosely tucked into the volumes of Petiver's herbarium in the Natural History Museum.[58] The plates in the *Gazophylacium* also served as desiderata lists, which could be sent to Petiver's overseas correspondents as a handy visual guide to the specimens he wanted them to collect. Two examples survive among Petiver's papers (Fig. 6.1): one of a starfish, annotated "A Sea starr, pray send me all the varieties you meet with," the other of a lizard, annotated "A sort of Lizard which you have many kinds of, which may be sent in a Bottle of Rum or Brandy."[59] While Petiver's subscribers were understandably keen to assure themselves that their copies of the *Gazophylacium* were complete with all the text and plates, Petiver took little interest in creating what bibliographers term an "ideal copy"; to him, his publications were

56 Bobart to Petiver, 6 March 1708–9, BL MS Sloane 3321, f. 243; Ray to Sloane, 22 Aug 1704, BL MS Sloane 4039, ff. 350–1.

57 Krieg to Petiver, 30 March 1702, BL MS Sloane 4063, ff. 149–50.

58 For Sloane's list of printed papers in Petiver's archive, see BL MS Sloane 1968, f. 171. For typical examples of loose labels in Petiver's herbarium, see Natural History Museum HS 150, f. 73v (10 loose labels for "Muscus Saxatilis Foeniculaceus Anglicus") and f. 118 (3 loose labels for "soft Crested Grass").

59 BL MS Sloane 1968, f. 213.

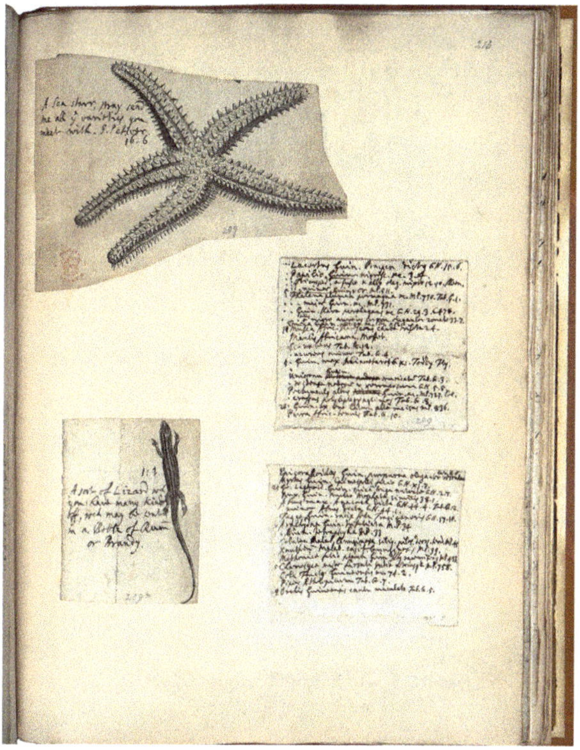

FIGURE 6.1 *A typical group of Petiver's loose notes, with annotated engravings of a lizard and a starfish, and two lists of specimens.*
BRITISH LIBRARY, SLOANE MS 1968, F. 213.

objects to be cut up and disassembled in an ongoing process of commerce and exchange.

The *Museum* and *Gazophylacium* were also used by Petiver as indexes and finding-aids to his collection. One of his working copies of the *Museum* (now BL 972.g.2) is marked up with cross-references to the *Gazophylacium* and a range of other botanical works, including Ray's *Historia Plantarum*, the *Hortus Malabaricus* and Sloane's *Voyage to Jamaica*. Many of his natural history drawings are marked up in the same way with cross-references to his own publications. To take only one example from many: Sloane's album of insect drawings (Sloane MS 5271) includes a drawing of a dragonfly-like insect with mottled wings (Fig. 6.2), with no identifying information other than a reference in Petiver's hand to the *Gazophylacium*. Following up this reference leads one to an engraving of the drawing, in reverse (Fig. 6.3), with a caption identifying it as "Libella Turcica major, alias *Locustae*," noting that the spots on the wings are "very dark, near to black" (obvious in the drawing but less

FIGURE 6.2 *Drawing of the antlion* Palpares libelluloides, *possibly by David Krieg, with Petiver's reference to the* Gazophylacium.
BRITISH LIBRARY, SLOANE MS 5271, F. 172.

clear in the engraving), and giving the source of acquisition: "Caught about Aleppo, by Mr Samuel Daniel, Surgeon."[60] Petiver did not need to note any of this identifying data on the drawing, as the reference to the *Gazophylacium* gave him all the information he needed to retrieve it. In other words, the *Gazophylacium* seems to have been intended by Petiver primarily for his own reference, to be used in conjunction with the specimens, drawings and notes in his collection.

These indexing arrangements were continued and adapted by Sloane after he acquired Petiver's collections. One of the copies of the *Gazophylacium* now in the British Library (443.i.1) is marked up with Sloane's catalogue numbers, showing that Sloane was still using it as a finding-aid to Petiver's collection of butterflies. Another example shows how the *Museum* and *Gazophylacium*

60 The insect is a species of antlion, called *Myrmelion libelluloides* by Linnaeus and now known as *Palpares libelluloides*.

FIGURE 6.3 *Engraving of the same image, known to Petiver as 'Libella Turcica major', from the* Gazophylacium.
BRITISH LIBRARY, 443.I.1 (1).

functioned as what we might call hyperlinks, allowing Petiver, and Sloane after him, to connect the different parts of the collection. Petiver owned both a drawing and a specimen of the beautiful moth from Jamaica, now sadly extinct, which Sloane described as "one of the most elegant Sorts of Butterflies I ever saw" and which was later named *Urania sloanus* in his honour.[61] The drawing, now in the British Library (Fig. 6.4), is annotated by Petiver in his usual fashion, "MP 509." The specimen, now in the Natural History Museum, is catalogued in Sloane's hand with a description copied out of the *Museum Petiverianum*, "Papilio caudatus Jamaicensis nigrescens, utrinque ex viridi aureo splendide striatus" ("long-tailed black butterfly from Jamaica, both wings splendidly striped in green and gold"). But one still has to turn back to the *Museum* to find Petiver's note on the provenance of this specimen: "Mr Patrick Rattray

61 Sloane, *Voyage to Jamaica*, 2: 216 and plate 239.

FIGURE 6.4 *Drawing of the Jamaican moth* Urania sloanus, *with Petiver's reference to the* Museum Petiverianum.
BRITISH LIBRARY, SLOANE MS 5271, F. 101.

brought me this some time since from Jamaica."[62] After three hundred years, the *Museum* and *Gazophylacium* are still, as they were for Petiver and Sloane, indispensable tools for navigating across Petiver's collection and identifying his sources of supply.

Petiver's publications illustrate the point recently made by Daniel Margócsy that early modern taxonomy was "not a simple quest after the

62 Petiver, *Museum Petiverianum*, D1v (p. 50).

FIGURE 6.5 Gazophylacium Naturae et Artis, *plate 14, showing the juxtaposition of natural and artificial curiosities.* BRITISH LIBRARY, 443.I.1 (1).

order of nature" but "a method for facilitating identification in the commerce of curiosities."[63] This explains why Petiver was not chiefly concerned with detailed descriptions or highly accurate illustrations. His catalogues served other purposes: as finding-aids for his own collection, as a public acknowledgement of his suppliers, and as a means of selecting and identifying duplicate specimens which could then be used for trade and exchange. In 1696

63 Daniel Margócsy, "'Refer to Folio and Number': Encyclopedias, the Exchange of Curiosities, and Practices of Identification before Linnaeus," *Journal of the History of Ideas*, 71: 1 (2010): 81.

his correspondent Richard Wheeler sent him a sample of the Norwegian moss now known as *Splachrum rubrum*, popularly known as umbrella moss or sometimes moosedung moss because it grows on reindeer dung. Petiver was delighted with the specimen, calling it "one of the finest discoveries that has lately been made"; in a flight of fancy, he christened it bongrace moss "because it hath a Scarlet cap like an Umbrella or Childs Bongrace," and wrote to Wheeler: "pray order those you employ to get Herbs Shells & Insects for me to gather me 40 or so fair Specimens of it, for I design to have a figure of it engraven on a Copper plate which I will send you."[64] Petiver asked for forty specimens not only to ensure that he would have a perfect example to copy, but also to send to other collectors. By having it engraved for the *Museum Petiverianum*, he was able to create a cycle of exchange in which the specimens sent to London were reciprocated by the printed image making the return journey to Norway. In short, Petiver's archive was not a closed collection, but part of a network of long-distance communication to which his printed publications served as keys.

3 Petiver's Afterlives

The amalgamation of Petiver's collection with Sloane's was not a simple or straightforward process. In 1727 his nephew John Woodcock wrote to Sloane to inform him that several important items, including "a large book of very valuable Paintings which I find were painted in Holland and several small books of Exotick Butterflies and severall kinds of Animalls," had only just been recovered from another collector who had borrowed them shortly before Petiver's death.[65] As for Petiver's correspondence, it was not until the 1730s that Sloane incorporated this into his library, at the same time as he accessioned his own archive of incoming letters. This, as I have argued elsewhere, was a pivotal moment in Sloane's self-fashioning as a public figure: the moment when, in a manner of speaking, he added himself to his collection.[66] But by this time Petiver's papers had got inextricably mixed up

64 Petiver to Wheeler, 26 March and 18 July 1696, BL MS Sloane 3332, ff, 200–1 and 223. Petiver, *Museum Petiverianum*, no. 70, "Muscus Norwegicus umbraculo ruberrimo insignito," A6r (p. 11).

65 Woodcock to Sloane, 4 April 1727, BL MS Sloane 4048, f. 273.

66 Arnold Hunt, "Sloane as a Collector of Manuscripts," in *From Books to Bezoars: Sir Hans Sloane and His Collections*, ed. Michael Hunter, Arthur MacGregor, and Alison Walker (London: British Library Publishing, 2012), 203.

with Sloane's, as shown by one entry in Sloane's accession register: "Letters to Mr Petiver and some to my selfe."[67] And despite considerable rearrangement over the years, inextricably mixed up they remain, with several volumes of letters to Petiver labelled on the spine "Letters to Sir Hans Sloane."

A preliminary listing of Petiver's herbarium was carried out by William Sherard, whose catalogue survives among Sloane's papers.[68] However, Sherard later fell out with Sloane over what he claimed was "a breach of his promise" in refusing him access to Petiver's collection.[69] According to Sherard, this was because Sloane wanted priority in using the plants for his *Voyage to Jamaica*. According to Sloane, it was because Petiver's plants had been left in "great confusion" and could not be made available to anyone until they had been properly labelled and pasted into volumes. As he wrote to their mutual friend, Richard Richardson:

> I told him that I was then printing my second volume of the Naturall history of Jamaica which is now near finished, the plates of which had been graved this 20 years & that I intended after that was published to take my Collections of dry"d plants to putt in order, that they were 200 volumes, some of which were very large, that Mr Petivers collections he had seen in great confusion but that if the labels & references which lay loose were not kept with them that neither he nor any body else could ever putt them to rights .. Now you may judge by this account what I can do more, for I shall be allways ready to assist him.[70]

Even after the second volume of the *Voyage to Jamaica* was finally published in 1725, Sloane stuck to his insistence that Petiver's collections were not in a fit state to be seen by anyone else:

> I have taken as much Care as I can to bring his Collections and Papers out of the Confusion I found them in, and will take farther Care, that what he hath gather"d together, by very great and undefatigable Industry, shall not

67 BL MS Sloane 3972B, f. 306v.

68 "A Catalogue of the late Mr James Petiver's Collections" (unattributed, but in Sherard's hand), BL MS Sloane 1968, f. 170.

69 Sherard to Richardson, 14 Oct 1721, Bodl. MS Radcliffe Trust c.4, ff. 53–4, printed in *Extracts from the Literary and Scientific Correspondence of Richard Richardson*, ed. Dawson Turner (Yarmouth, 1835), 175–7.

70 Sloane to Richardson, 9 March 1720–1, Bodl. MS Radcliffe Trust c.4, ff. 4–5, printed in *Extracts*, 161–4.

> be lost, but preserved and published for the good of the Publick, doing right to his Memory, and my own Reputation.[71]

It is true that Petiver may well have left his herbarium in an unfinished state, though Sherard asserted that he had seen it "in order and together," and his listing of the collection shows that it was organised geographically and bound into volumes, very much as it remains today.[72] But it was important for Sloane to stress the disorganisation of Petiver's collection in order to justify himself. As his remark in the *Voyage to Jamaica* makes clear, it was not just Petiver's memory that was at stake, but his own reputation for liberality in sharing his collections with others.

This, then, was how Petiver would be remembered: as a great collector, and in Sherard's words, a man of "indefatigable industry in procuring and graving natural history," but not equal to the task of organising and systematising his collection.[73] His own publications, particularly the *Gazophylacium*, tended to reinforce this impression. As the title made clear, this was a collection of natural and artificial curiosities—*Gazophylacium Naturae et Artis*—and Petiver saw nothing incongruous in juxtaposing the two: in one plate, for example, the new coinage for Queen Anne was pictured alongside a selection of English butterflies, a fern from the Philippines, lichen from Norway, a lizard from America and a scarab beetle from Borneo (Fig. 6.5). It was a purposely eclectic selection, designed to evoke wonder and curiosity in the reader and to mimic the experience of viewing a private museum or cabinet. In this, Petiver was hardly unique. As Kay Dian Kriz has shown, Sloane did exactly the same thing in his *Voyage to Jamaica*, where natural and artificial objects were frequently placed together on the same page, inviting the viewer to read them emblematically as well as scientifically. This juxtaposition, Kriz argues, "promoted the notion that such collections represented the world in microcosm as well as suggesting that human artifacts were linked to and emulations of divine creativity."[74] But with the development of biological classification in the eighteenth century, this began to seem hopelessly unsystematic and unscientific, contributing to Petiver's reputation as a haphazard accumulator of objects.

---

71 Sloane, *Voyage to Jamaica*, 2: v.

72 Sherard to Richardson, 14 Oct 1721, Bodl. MS Radcliffe Trust c.4, ff. 53–4. However, a remark in the same letter implies that at least some of Petiver's plants were loose in the volumes: "they will be eat up (if they are not already) in a little time, I mean Petivers, for Plukenets are pasted on."

73 Sherard to Petiver, 25 March 1709, BL MS Sloane 4064, ff. 194–5.

74 Kay Dian Kriz, "Curiosities, Commodities, and Transplanted Bodies in Hans Sloane's 'Natural History of Jamaica'," *William and Mary Quarterly*, 3rd series, 57, 1 (2000): 55.

Petiver also used his publications to position himself as a broker of intellectual credit by acknowledging his correspondents in print and sometimes even naming new species after them. George Lewis, East India Company chaplain at Fort St George in Madras, was rewarded for a consignment of African plants by having an entire genus, Lewisanus, named in his honour.[75] Again, Petiver was not unique in this. As the Edinburgh botanist Charles Preston observed, he was merely following the example of Joseph Tournefort in "baptizing your plants after the names of your friends and benefactors (which is not amisse)." However, as Preston pointed out, Tournefort had provided detailed descriptions of his new species, whereas "you are pleased only to give us the bare names and leave us to seek for the Character some where else. As for instance how is it possible for me or any man else, who has not seen these plants or Specimens, to distinguish Lewisanus capensis foliis capillaceis from other capillaceous plants that grow in the same country, or even in other places, without some particular Character?"[76] Even in Petiver's own lifetime, the increasing precision of botanical taxonomy left publications like the *Museum Petiverianum* looking distinctly old-fashioned. And Linnaeus later scoffed at Petiver's lack of discrimination in naming plants after his friends: "I am astounded when I see with what boldness, with what rashness *Petiver*, to say nothing of others, thrust priceless gifts, too brilliant and valuable for the uneducated, on florists, monks, relations, friends and the like."[77] In Linnaeus's opinion, this only succeeded in bringing botanical nomenclature into disrepute and devaluing what ought to be the highest honour that any botanist could receive. The *Gazophylacium* earned Linnaeus's disapproval too, as one of a number of books whose copperplate engravings made them too expensive for the average student of botany to afford. Linnaeus's aim was to make such books obsolete by basing his taxonomic method on exact verbal description, thus doing away with the need for costly engraved illustrations.[78]

Needless to say, there was a social dimension to this critique. Petiver's humble social origins and lack of university education were widely commented on by his contemporaries. Samuel Molyneux, secretary to the Prince of Wales (the future King George II), visited Petiver's museum in 1713 and observed: "This Gentleman is a very laborious Collector of Nature and I believe

75 J.E. Dandy, *The Sloane Herbarium* (London: British Museum, 1958), 155.

76 Preston to Petiver, 8 Sept 1701, BL MS Sloane 4063, f. 117.

77 Carl Linnaeus, *Critica Botanica*, trans. Sir Arthur Hort and M.L. Green (London, Ray Society, 1938), 54.

78 John L. Heller, "Linnaeus on Sumptuous Books," *Taxon*, 25: 1 (1976): 33–52.

if his Education had been suitable to his Genius and Inclination he might have been very useful to the Enquireing world." Indeed, Petiver himself lamented "my not being allowed Academicall Learning, whereby I might have been better capacitated to serve both my own Country and others."[79] This finds a curious echo in modern scholarship, where Petiver is often portrayed as a social outsider in the polite world of eighteenth-century collecting. Raymond Stearns, the first scholar to make a detailed study of Petiver's papers, described him as "vain, excessively ambitious and occasionally dull" (also, Stearns speculated in a footnote, possibly homosexual), and remarked, a little unkindly, that Petiver's correspondents were "often more interesting and sometimes more significant" than Petiver himself.[80] More recently, Marjorie Swann has argued that Petiver consciously used his collecting as a means of self-fashioning and self-advancement, "coordinating the labour of others and using his control of this work force to enhance his own social status."[81]

No one would deny the social dimension of early modern collecting, and Swann is quite right to observe that for Petiver, collecting objects was also a way of collecting people. But in trying to reconstruct Petiver's social world—a world in which, as Swann puts it, "social interactions [were] predicated on collecting"—there is the risk of fusing the social and the intellectual, and treating Petiver's intellectual shortcomings as a function of his social inferiority. I have argued in this chapter that Petiver was far more interested in method and organisation than modern scholars have been prepared to allow. Some visitors to his collection perceived this: Molyneux, for example, despite his rather patronising remarks on Petiver's lack of education, was impressed by "the method of his Collections which is in all its parts digested according to the different parts of the world so that you have the Plants and Insects of Europe by themselves and so of the rest." He was particularly gratified to see that Petiver gave priority to British botany, "and very properly, for I think all Collectors of this kind should have a particular regard to their own native Country."[82] But this method of arrangement would soon come to be regarded as no proper method at all. When Linnaeus visited Sloane's collection in 1736, he remarked

79 *The London Letters of Samuel Molyneux, 1712–13*, ed. Ann Saunders, with Paul Holden and Sheila O'Connell (London: London Topographical Society, 2011), 120. Petiver to Sir Charles Holt, 2 June 1713, BL MS Sloane 3339, f. 11.

80 Raymond Phineas Stearns, "James Petiver, Promoter of Natural Science, c.1663–1718," *Proceedings of the American Antiquarian Society*, 62 (1953): 245 and 247.

81 Marjorie Swann, *Curiosities and Texts: The Culture of Collecting in Early Modern England* (Philadelphia: University of Pennsylvania Press, 2001), 90–96.

82 *London Letters of Samuel Molyneux*, 120.

contemptuously that the herbarium was in "great disorder," a judgement not merely on Petiver or Sloane but on a whole system of botanical classification that Linnaeus was now proposing to sweep away.[83] If we fail to perceive method in Petiver's collecting, it may be a reflection of our own inability to think ourselves back into a pre-Linnaean world.

---

83 Dandy, *Sloane Herbarium*, 12. Linnaeus's remark occurs in a letter to Olaf Celsius, 30 Nov 1736: see Jackson, "The visit of Carl Linnaeus to England in 1736," *Svenska Linnesallskapets Arsskrift*, 9 (1926), 1–11.

CHAPTER 7

Collecting Knowledge: Annotated Material in the Library of Sir Hans Sloane

Alison Walker

1 Introduction: The Afterlife of Collections

The title of this volume emphasises that collections commonly undergo metamorphoses, or even rebirths: not static, they pass in whole or in part from one owner to the next, their significance variously changing according to use, location and context. An "archival afterlife" implies that this significance continues to be recognised by owners and users—whether individual or institutional—long after the creation of the collection. It does not necessarily mean, however, that all collections of documents, books or objects of long-term significance can be described by the specialised term of "archives." Collections of printed books rarely fall into this category, but in the case of heavily annotated material there can be a fine line between print and manuscript—one which even the original collector may find difficult to draw, or may consider unnecessary.[1] Within collections of printed material, items with marks or annotations made by owners or readers have a special kind of afterlife, reflecting the ways these marks are created and valued by their various owners, and often recording the transmission of knowledge from owner to owner.

Collections of printed material may show a variety of forms of marking and annotation. There may be reading notes, or simply marks on the page,

BL = British Library

BM = British Museum

ODNB = *Oxford Dictionary of National Biography* (Oxford: Oxford University Press, 2004–2016)

PROB = Records of the Prerogative Court of Canterbury

1 Sir Hans Sloane's manuscripts and printed material were numbered and stored separately, but a small number of heavily annotated items were classed as manuscripts, and continued to be so after accession by the British Museum. Some of these have now been moved to the printed collections. Sloane sometimes moved books from a manuscript number to a printed books number and vice versa; see www.bl.uk/catalogues/sloane, searching on Sloane number range MsA to MsD.

 | DOI 10.1163/9789004324305_009

indicating attention paid to the text. There may be references to other material read earlier, accounts of events or experiences, additions to, or comments on, the printed text, or expansions of the text in the light of the user's experience. In Sir Hans Sloane's library there are examples where empty space on the page, and blank interleaved pages, have been used by earlier owners to record case notes, medical remedies, or the location of botanical specimens found, and some where the books even hold the specimens themselves. All these additions to the printed text, and the collocation with similarly annotated material, can transmute print into archive.

Libraries are typically extremely rich repositories of knowledge, incorporating not only texts but also the products of the reader's experience. Often, however, library catalogues mask this, partly through the practice of simply indicating that a book contains "ms notes." There is far more to it than this. This essay will use a small number of case studies of material in Sloane's collection to show how "manuscript notes" can conceal a great variety of types of annotation, including some of those mentioned above. It will suggest that the accumulation and collocation of text and notes often creates a whole much larger and more significant than the parts, allowing information to become transmuted into knowledge.

2 Sir Hans Sloane: Collector of Knowledge

The aim of the collector in acquiring annotated copies may vary according to his scholarly or personal desires, the existing content of his collection, the opportunity offered by a brief period of availability of material or the unexpected availability of previously unknown material, and of course the content of the annotations. In the case of Sir Hans Sloane, all of these may have come into play at different times. Overall, however, he can be said to be collecting knowledge, that is, information, or facts, enriched, enhanced and given value by their context, and by their user's understanding. The acquisition of books annotated by previous owners who have already transformed information into knowledge creates multiple layers, as successive owners lay down their own evidence of usage, further adding to the significance of the original text, and building up a series of overlaid personal experiences of it.

His library was very large, and part of one of the most important scientific collections of the age. Built up over his long life (1660–1753) and career as a physician, his library functioned both as a collection in its own right, and as background and support for his collections of objects. The number of his volumes of books, manuscripts and prints were estimated in 1753 at 50,000.

Together with over 47,000 objects and specimens and approximately 23,000 coins, they became the foundation collection of the British Museum (BM).[2] Most of his library is still at the British Library (BL), but was never kept in the order in which he arranged it, and is now scattered throughout the older parts of the collection. Further dispersing the collection, several thousand volumes were disposed of as duplicates between 1769 and 1831. The Sloane Printed Books Project,[3] based at the British Library, is working to identify books from this dispersed library. Most of the items identified so far are located at the BL. The bibliographical database is enriched with copy-specific information, including evidence of previous ownership and the presence of manuscript notes. This provides a rich source of evidence for the afterlife of collections, and can be used to illustrate the survival, through sequences of ownership, of the scientific and medical knowledge which is the strength of Sloane's collection.

The number of annotated books in this collection has only been revealed by the copy-in-hand examination of his library undertaken by project staff since 2008.[4] Over 39000 bibliographical items from among Sloane's estimated 45000 printed volumes have now been identified. Over 1800 books contain annotations, of greater or lesser significance.[5] Some are significant either in their extent or by association with the person who wrote them or the person who owned the book. Others are by unidentified previous owners of the text, so remain anonymous. Others again are very slight and reveal only small traces of the writer's experience or intentions.

The selection of material from Sloane's library examined here comprises case studies which focus on annotated copies which do reveal something of

2 Arthur MacGregor, "The life, character and career of Sir Hans Sloane," in *Sir Hans Sloane, collector, scientist, antiquary, founding father of the British Museum*, ed. Arthur MacGregor (London: British Museum Press, 1994), 28–9.

3 www.bl.uk/catalogues/sloane. The database can be searched by previous owner for fuller details of items referred to in this article.

4 The various marks which identify Sloane's books are illustrated in Alison Walker, 'Sir Hans Sloane's printed books in the British Library; their identification and associations,' in *Libraries within the Library*, ed. Giles Mandelbrote and Barry Taylor (London: British Library, 2009), 89–97, and in the introduction to the database.

5 William H. Sherman, *Used Books: Marking Readers in Renaissance England* (Philadelphia: University of Pennsylvania Press, 2008), 5, notes that just over 20% of the STC collection at the Huntington Library contain manuscript notes. On the face of it, Sloane's collections would seem to be much less heavily annotated, but in certain subject areas there is a high concentration of books with manuscript notes. It is also likely that not all his annotated books have yet been identified.

the writer or owner, outlining the earlier life of some collections, and showing some of the ways in which Sloane valued and acquired this annotated material.

Sloane was a life-long collector. His earliest dated acquisitions of books, two pharmacopoeias and a work on botany, are from 1680, when he was a 20-year old student. In 1682 he was buying books in London, including a handful of medical titles at the sale of the library of Richard Smyth, sold at auction that year.[6] Between 1683 and 1684 he continued his studies in France, staying at Paris, Montpellier and Toulouse, buying books in each of these cities. Accessions to his library continued until his death in 1753. In his later years, he acquired a reputation for omnivorous or indiscriminate collecting, to the extent that he was viewed as a toyman or charlatan,[7] though it is clear from his later library catalogue volumes that he also continued to build on existing strengths and interests. However, his earlier collecting career, up to 1700, shows a great deal of focus on all aspects of medicine, natural history and science. There is evidence of his personal involvement in the selection and management of material: indeed, until 1700 he wrote the catalogues of his collections himself: later, he employed a number of amanuenses to catalogue his books.[8] While never excluding objects of great beauty or value, his principal interest in this early period was in collecting material which supported and expanded his work in science and medicine.

Most of the material described here was acquired early in Sloane's collecting career. It is natural that as a student and young practitioner he should have been attracted by other scientific and medical collections, and have made significant acquisitions from the libraries of his predecessors and peers in these fields. Among these are a number of books bearing marks of ownership and evidence of the collection's organisation, or textual annotations by their earlier owners. "Clean" copies would certainly have been available, had he preferred these, but in contrast to many collectors, he seems not to have sought only

6 Smyth, Richard, Secondary of the Poultry Compter, *Bibliotheca Smithiana: sive Catalogus librorum ... quos multo œræ [sic] sibi comparavit ... Richardus Smith Londinensis. Horum auctio habebitur Londini ... Maii die 15. 1682, etc.* [London, 1682]. BL 11906.e.29. Few MS. notes [of prices].

7 Barbara M. Benedict, "Collecting Trouble: Sir Hans Sloane's Literary Reputation in Eighteenth-Century Britain," *Eighteenth century Life* 36.2 (2012): 111–142; Barbara M. Benedict, "From benefactor to entrepreneur, Sloane's literary reputation 1685–1800," in *From Books to Bezoars: Sir Hans Sloane and his Collections*, ed. Alison Walker, Michael Hunter, Arthur MacGregor (London: British Library, 2012), 41–47.

8 Amy Blakeway, "The Library Catalogues of Sir Hans Sloane: Their Authors, Organization, and Functions," *The Electronic British Library Journal* (2011), Article 16, accessed April 17, 2017 http://www.bl.uk/eblj/2011articles/article16.html.

the pristine text, elaborately bound, preserved and possibly unread. Sloane's choices imply that he valued copies annotated by previous owners, especially in the areas of his own professional and scientific activities. In both Sloane's library and others of the period, certain subjects can be found to be quite commonly annotated, often in interleaved copies, reflecting the type of active reading such topics attracted. Medicine, botany and legal works are very frequently annotated. The acquisition of annotated copies does seem to have been a feature of Sloane's early career both as a physician and as a collector. Of the 1824 items posted by a search on the Sloane database for "ms notes", 90% were printed before 1700. Of course, not all were acquired before 1700, but it is certainly an indication that he was acquiring annotated material actively at this period.

The number of previous owners whose books ended up in Sloane's library, as identified by inscriptions in the books, is over 1100. Some of these owners are well-known either as scientists or as collectors. He acquired substantial parts of the libraries of Robert Hooke (1635–1703), natural philosopher; Francis Bernard (1628–1698), apothecary and physician; James Petiver (1665–1718), botanist; Leonard Plukenet (1642–1708), botanist; Edward Sherburne (1616–1692), translator and poet; Walter Charleton (1620–1707), physician and natural philosopher; and William Courten (1642–1702), naturalist and collector. Later acquisitions included considerable numbers of books from the library of Jean-Baptiste Colbert (1619–1683), French politician, whose books were sold in 1728, and Engelbert Kaempfer (1651–1716), German naturalist, physician and traveller.[9]

In other cases, only a few books from notable libraries reached Sloane, such as those of Sir Theodore Turquet de Mayerne (1573–1655), royal physician; Herman Boerhaave (1668 –1738), Dutch botanist and physician; John Dee (1527–1609), mathematician, astrologer, and antiquary; and Constantijn Huygens (1596–1687), Dutch poet and composer, father of the scientist Christiaan Huygens.

Naturally, not all people who leave interesting marks in their books are well-known, and conversely, significant owners such as those above often leave no more than their name in their books. Many annotators are anonymous, unidentified or unidentifiable. Their annotations can still be valuable examples of the ways in which people thought about and managed knowledge, and of the kind of knowledge that was valued by Sloane. Indeed, we might be glad

9 See Hamish Todd, "Kaempfer's cat," *Asian and African Studies Blog*, British Library, February 19, 2016. Accessed March 31, 2018. http://britishlibrary.typepad.co.uk/asian-and-african/2016/02/kaempfers-cat.html.

now if Sloane himself had written more in his books as it could have helped us to a better understanding of him as a scientist, but he only rarely wrote in his own books, usually to identify or remark upon an author, or to comment on the circumstances of his acquisition.

Medicine and botany were great strengths of Sloane's collection. As noted above, many practitioners in these fields did annotate their books, and it should not be surprising that Sloane acquired many. He owned many annotated copies of classical medical authors: his library contained at least 210 editions of Galen, of which 68 have manuscript notes; 274 of Hippocrates, 49 with notes; 26 editions of Dioscorides, 7 with notes; 17 editions of Celsus, 6 with notes; 226 editions of Aristotle, 32 with notes.[10] In the great majority of cases, the annotators have not been identified, either from names inscribed in the books, or by identification of the hand. He also had significant numbers of botanical works with manuscript notes relating to specimens. Among these the works of John Ray stand out: of 33 works held by Sloane, 20 have manuscript notes. A more general search of the database for works containing the word "plantarum" finds 228 books, of which 67 have notes. The majority of these annotators are not identified.

Despite the anonymity of so many writers, there is nevertheless a huge wealth of stories to be told from the afterlives within Sloane's collection. The selection of material discussed here illustrates some of the types of evidence it encapsulates. Firstly, evidence of ownership and inheritance, through a brief look at the afterlife of two medical collections; secondly, notes left by practitioners in a particular genre which Sloane actively collected, annotated pharmacopoeias, and finally a number of books owned by a medical practitioner which combine the two themes of inheritance and professional annotations.

3 Joseph Fenton's Library: A Ready-Made Working Collection

The first case study looks at the library of Joseph Fenton. Born circa 1565/70, he was a surgeon who practised at St Bartholomew's in London, where he was a colleague of William Harvey, until his death in 1634.[11] All his books, papers, and his surgical instruments were inherited by his grandson Joseph Colston or

10 The figures given here are derived from searches of the Sloane Printed Books database (www.bl.uk/catalogues/sloane) in March 2016. As the database grows or is amended, the numbers will change.

11 Norman Moore, *The History of St. Bartholomew's Hospital* (2 vols., London: C A Pearson, 1918), vol. 2, 615.

Coleston.[12] Fenton's will[13] expressly leaves his grandson not only his medical books but also those on grammar, philosophy and "other books whatsoever." He "pray[s] God to blesse and prosper him in his studies, for his mother's sake I am thus kinde to him." Colston continued the family's medical tradition, being admitted an Honorary Fellow of the College of Physicians in December, 1664. He died childless in 1675.[14]

Sloane seems to have acquired the bulk of Fenton's library in 1686, a few years after Colston's death, and despite the reference in Fenton's will to books of grammar and philosophy, his acquisitions were mostly medical. At this stage in Sloane's collecting career, he usually recorded the date and price of purchase in code on the title page or a preliminary leaf.[15] Many books bear Sloane's acquisition date 1686, along with Fenton's signature and/or motto, the definitive proof of his ownership.[16] Some, of course, have lost parts of the evidence of ownership and acquisition through the processes of binding and rebinding.[17] Nevertheless, over 340 books from Fenton's library can be traced in Sloane's collection. The exact means of Sloane's acquisition remains unknown: there is no record of an auction of the library, suggesting it may have been sold through a bookseller, or as a private purchase.

The date of Sloane's principal acquisitions from Fenton's collection can be traced through his first library catalogue, now BL Sloane MS 3995, a small notebook which starts with the heading "A catalogue of my books taken in February 1684/5 in London." He carefully recorded the cost of each item, with a running total at the foot of each page. By the end of the section covering 1685, 307 volumes are listed. On fol. 37v he started another list, headed "1686" and it is here that many books from Fenton's library are listed. By the time he stopped using this catalogue, in 1687, it listed 2700 books and volumes of manuscripts.[18]

12 David Pearson, "Joseph Fenton and his books," *Medical History* 47 (2003): 239–248.

13 National Archives, PROB 11/165/165, 12 February 1634.

14 See *Lives of the Fellows*, Royal College of Physicians, accessed April 17, 2017, http://munksroll.rcplondon.ac.uk/Biography/Details/936; *A Cambridge Alumni Database*, University of Cambridge, accessed April 17, 2017, http://venn.lib.cam.ac.uk/.

15 Illustrated in http://www.bl.uk/catalogues/sloane/Identifiers.aspx.

16 A representative title page with Fenton's signature and motto is illustrated in Pearson, "Joseph Fenton and his Books."

17 Sloane appears to have kept many items unbound, as evidenced by dirt still present on titlepages. The British Museum typically grouped and bound such material by subject, often with the loss of endleaves and evidence of provenance. Many have since been rebound at least once.

18 Around 1692 Sloane began a new catalogue, now BL Sloane MS 3972C,D, which continued in use unti his death in 1753. Latin medical books were listed separately in an interleaved copy of Joannes Antonides van der Linden, *Lindenius renovatus* (Norimbergæ, 1686).

In the period covered by Sloane MS 3995, his acquisitions were mostly in support of his studies in medicine and botany. The attraction of Fenton's library is clear: it offered a ready-made collection of medical classics, perfect for study and reference. Possibly too medical texts would have been a useful resource for his posting as physician to the Duke of Albemarle in Jamaica between 1686 and 1688, either to carry with him or to prepare for his duties. Sloane seems to have bought most of the library, but not all, since scattered Fenton books are to be found in other medical collections. Nor were all of Sloane's acquisitions bought in 1686: a few have purchase codes from the 1690s, probably acquired from other owners. 344 Fenton/Sloane books have now been identified at the British Library, and some 60 Fenton books in other collections. The library may have consisted of around 500 volumes.[19] The kinds of books Sloane acquired from Fenton's library continued to be strengths of his collecting throughout his career: the Greek, Roman and Arabic classic writers, such as Hippocrates, Celsus, Galen, Avicenna, and Rhasis, as well as later authors whose works became standards, including Du Chesne, Gesner, Riolan, and Fabricius.

Fenton did make notes in his books, certainly more than Sloane did, and in general they relate to his reading, being marginal notes recording an attentive reading of the text. The few exceptions, found on preliminary leaves rather than in the text, are usually references to other authors, or to sections in the text at hand. Unusually, there is an interesting list of his books on surgery, "A note of all the Bookes of Chyrurgerye that I have in my Studye,"[20] which is the nearest we have to a catalogue of Fenton's books. Unfortunately very little detail of the editions is given and only a few are identifiable.

The value for Sloane may have lain partly in Fenton's notes, which could have been a potential bibliographical resource, but principally it lay in the content of his library, an excellent collection of medical classics. Fenton had sought to ensure that his library had an afterlife in the hands of his grandson, and it acquired a later afterlife with Sloane in support of his professional education and as a core component of his medical library. Even further down the line, when Sloane reorganised his library in the 1690s he sent many duplicates, including Fenton copies, to the Bodleian Library. The reach of Fenton's legacy thus extends in time and scope far beyond what he could have foreseen.

---

19 Fenton used a system which placed books by size and numbered each shelf. The figure of 500 volumes is a rough estimate based on the apparent capacity of the shelves and the maximum number per shelf located so far. A more extensive article on Fenton's library is in preparation by the current author.

20 On a preliminary leaf of his copy of Eustachius Rudius, *De affectib. externarum corporis humani partium libri septem* (Venetiis, 1606). BL 548.l.3.

The authors which Sloane bought from Fenton continued to be a collecting interest throughout his life. He later recorded his ambition to update Johannes Antonides van der Linden's 1686 medical bibliography *Lindenius renovatus*,[21] by adding to the corpus texts and editions collected by him but not listed there.[22] The acquisition of a large library of medical literature early in Sloane's career could well have fuelled his interest in medical bibliography and appealed to his already established collecting habit. By 1725 he was able to note proudly in his annotated copy of *Lindenius renovatus* that he had effectively doubled the number of authors of medical works originally listed by van der Linden.[23] The afterlife of Fenton's library thus proved to be of some significance within the larger collection that Sloane built up, and may have influenced the direction taken by the collector.

4 Henry Stanley's Library: The Library of a Chemical Physician

Many of the same medical authors figure among Sloane's acquisitions from the library of Henry Stanley. Educated at Oxford and Padua, Stanley became a Fellow of the College of Physicians in 1649, and died in 1671.[24] It seems that a relatively small part of Stanley's library ended up in Sloane's collection. A few more of his

---

21 Joannes Antonides van der Linden, *Lindenius renovatus, sive J. van der L. ... de scriptis medicis libri duo: ... addita plurimorum authorum ... vitæ curriculorum succincta descriptione: ... continuati, ... amplificati, ... et ... purgati a G. A. Mercklino* (Norimbergæ, 1686).

22 This is explained in the introduction to the second volume of his *Voyage to Jamaica* (London, 1725): "In *van der Linden de Scriptis Medicis* put forth by *Mercklin* in 1687 [sic] are 3937 Authors, to which in my Library are added 3734, all such as have in Greek or Latin treated of the Medical Art, Natural History, Chymistry, Anatomy, &c. which are come to my knowledge; so that my Library consists of 7671 Greek and Latin Physical Authors, besides a great Number of Latin Books publish'd by the Authors already mentioned in *Mercklin's van der Linden*, and Editions, not taken Notice of by him, and other Physical Books publish'd in the *European* Languages. An Account of all these may perhaps be given hereafter by myself; or by some-body else from the Things themselves, and the Memoirs left with them, which I hope may be for the Benefit of Mankind."

23 BL 878.n.8, vol. 8, first unnumbered leaf. Sloane notes that van der Linden had listed 3937 works, and he had added a further 3734.

24 Munk's Roll describes him thus: "b.? d.24 February 1671 MD Padua(1637) MD Oxon(1641) LRCP(1646) FRCP(1649) Henry Stanley, MD, was the son of Henry Stanley of Chichester, gent, by his wife Ann, daughter of William Madgick of Southampton, gent, and was educated at New college, Oxford; but was a doctor of medicine of Padua of 10th July, 1637; incorporated on that degree at Oxford 2nd April, 1641. He was admitted a Licentiate of the College of Physicians 3rd July, 1646; a Candidate 22nd December, 1646; and a

books have been identified through the British Armorial Bindings database,[25] bearing various forms of Stanley's crest. The Victoria and Albert Museum has a few, as do the Bodleian Library, Oxford, and Cambridge University Library. Probably there are many more scattered, unrecognised, perhaps having lost the owner's binding mark, but since Stanley punctiliously recorded the bindings he commissioned, we may assume that most of the books in his library bore his crest. Stanley cared for the outward appearance of his books, and indeed shows all the characteristics of a book collector. He recorded his purchase and its binding, occasionally also noting the provenance of his copy: several were said to come from the library of a Dr John Moore. He noted whether the book was bought in sheets, its cost, and what he paid for the binding, or what he paid for embellishment to an existing binding, and remarked on improved editions. He noted, for example, of Bernardus de Gordonio, *Opus Lilium Medicinae*, 1574, "I have this in a later & better edition Frankfurt 1617."[26]

His library was almost certainly larger and probably more varied than Fenton's. Most of the books have an alphanumeric mark, almost certainly a shelfmark, in Stanley's hand, consisting of a letter and two numbers, e.g. M.7.5. It starts with folio volumes, such as B.2.2–3, an expensive 1619 edition of Aristotle for which Stanley paid £2 plus 3s 6d for "filleting," embellishment of the binding, and ends with octavos and duodecimos at the end of the alphabet. If it can be assumed that the letter represents a bay of bookshelves, the first number a shelf and the second a number on the shelf, there is a system which identifies the place of each book.[27] In itself, this type of tripartite numbering system: bay, shelf, place on shelf, is not uncommon- indeed, the British Museum Library followed the same type of scheme. It allows us to posit that this must have been a substantial library, though at present only 48 items from the collection have been identified.

Fellow 7th December, 1649. He was Censor in 1653 and 1669. Dying 24th February, 1671, he was buried at Little Gadesden, co Herts, where there is a monument with the following inscription: Henr: Stanley, MD celeberrimi utriusque Coll: Novi primum in Univ: Oxon: Medicorum deinde in civitate Lond: ob: 24 Feb: 1671. Dr Stanley bequeathed to the College fifty pounds, which was received 1st October, 1672."

25 *British Armorial Bindings*, University of Toronto Libraries, https://armorial.library.utoronto.ca/.

26 BL 773.b.18, blank preliminary leaf.

27 On the basis of the shelfmarking system, some calculations can be made about the size of the library, say 24 bays, each with up to 10 shelves, and up to 14 books (for the smaller categories) per shelf, which gives a maximum of 3360 volumes. The realistic total is likely to be fewer, given that folio and quarto books would require fewer shelves per bay and fewer items per shelf.

Stanley's texts are not heavily annotated, but his concern to record acquisition and provenance prefigures his later efforts to ensure that his library continued an afterlife. His nephew Dr Nicholas Stanley, also a medical practitioner, was willed not only his uncle's books and papers but the book presses and other library furniture, his doctor's gowns and cap, the drugs and medicinal preparations and all the equipment used in his "Repository," which seems to have been a laboratory in his house. Among the equipment were furnaces, vessels, instruments used for distillation and other chymical operations and pharmaceutical preparations, together with his mathematical and surgical instruments.[28] Such details tell us much about Stanley's medical style. Stanley must have been a wealthy man, and made other bequests to medical persons and organisations. He was certainly concerned that his nephew should continue his work as a chemical physician. Nicholas Stanley became a successful practitioner in Winchester, and died in 1687. His son, also Nicholas, at one time a Fellow of All Souls, Oxford, continued in practice at Winchester, and died in 1710. His grandson, a third Nicholas, graduated from Oxford in 1713. It seems likely that some of Henry Stanley's library stayed in the family, its final dispersal coming at auction in 1722.[29] However, the auction catalogue contains a relatively small proportion of medical works for a physician's library: only 257 out of 889 lots are medical books. Possibly this is the rump of the library, since most of Sloane's copies, nearly all medical books, were bought in 1697, and were presumably disposed of by Stanley's great-nephew.

Again, Sloane's purchases cover medical classics, Aristotle, Hippocrates, Riolan, Sebisch, Sennert, and a few more recent authors. He also acquired some non-medical books, including an edition of Rabelais' (himself a physician) *Gargantua and Pantagruel* of 1605. On the whole, though, Sloane's approach to Stanley's library follows the pattern of classics and working texts established in his purchases from Fenton's library, already perhaps inspired by his nascent ambitions to update Lindenius, and thus a "collector's choice."

These two case studies represent examples of medical professionals who were also collectors, organisers, who valued the literature of classical medicine, and sought to ensure the survival of their libraries by passing them to persons who would value and use them. This should sound familiar, because this is exactly what Sloane himself did, in ensuring his collection remained intact

28 National Archives PROB/11/338/383, 26 March 1672.

29 *Catalogus bibliothecæ viri eruditissimi H. Stanley, M.D., or a Catalogue of the Library of the learned Dr Henry Stanley, which will begin to be sold at Dick's Coffee house, 12th Feb, 1722/23,* (London, [1723]).

and available for study after his death.[30] The afterlife of their collections, and the preservation of the knowledge embodied in the context of the collection was of great importance to all three men.

5 Annotated Pharmacopoeias: Collecting Archives of Practice

Next, pharmacopoeias, a body of material collected extensively by Sloane, but acquired from a variety of different sources rather than a single previous owner, and typically containing a very different type of annotation. Pharmacopoeias were produced in many European cities from the late sixteenth and early seventeenth century as an important mechanism for the local and national regulation of medicinal products: Sloane had copies from Augsburg, Messina, Blois, Amsterdam, Leiden, Paris, Lille, Brussels, Utrecht, Stockholm, Bordeaux, Edinburgh, Strasbourg, and many others, including, of course, London.[31] It is no surprise that he should have owned so many of these, given his interest in collecting medical texts, nor is it surprising that many owners should have annotated their copies. Indeed, the standard format, consisting of short, well-spaced lists of medicinal ingredients and formulations, lent itself perfectly to the addition of notes about the content or the writer's medical practice. Many copies were interleaved, allowing even more notes to be added. Within Sloane's library, the types of publication most often found in interleaved format are pharmacopoeias, medical text books and botanical works, all commonly used to record the findings of professional life.

Sloane's copies of the London pharmacopoeia are notable for the amount of annotation they contain. The *Pharmacopoeia Londinensis* was first published by the College of Physicians in 1618. Of the seven editions between then and 1678, Sloane had at least 15 copies, of which 12 contain manuscript notes. Five of the 12 annotated copies are interleaved. Three are so heavily annotated that Sloane classed them as manuscripts.[32] Of these, two have notes by

30 Marjorie Caygill, "Sloane's will and the establishment of the British Museum," in Arthur MacGregor, ed., *Sir Hans Sloane Collector, Scientist, Antiquary* (London: British Museum, 1994), 45–68.

31 Many of Sloane's pharmacopoeias are shelved in close proximity at the BL. A search on the Sloane database for the truncated shelfmark 777* reveals 195 items of which a considerable number are pharmacopoeias. Others can be found by searching for 'Pharmacopoeia'. The approximate number of pharmacopoeias owned by Sloane is around 95.

32 BL 777.k.7, 1627 edition, has Sloane's numbers Ms 1872 (deleted) and Ms A 237. BL 777.k.5, 1618 edition, has Ms 305 (deleted), Ms A 238. BL 777.d.1 has Ms 487 (deleted) and Ms B 945.

the Huguenot Royal physician, Theodore Turquet de Mayerne. A copy of the 1650 edition has notes by Christopher Merrett, the first Harveian librarian of the College of Physicians and a founding member of the Royal Society.[33] Some volumes have annotations in more than one hand, the volume having been used by several practitioners. Other medical collections, naturally enough, also have interleaved and annotated pharmacopoeias, acquired from various sources over the years. The Royal College of Physicians in London, for example, has a considerable number, though on a brief examination they seem to be less heavily annotated than Sloane's.

The fact that Sloane had five annotated copies of the 1650 edition of the *Pharmacopoeia Londinensis* is persuasive evidence that his interest was at least as much in the notes as in the printed text, confirming the general finding that added information was a major attraction for Sloane across a wide range of material. In some cases, as with Merrett and Mayerne, they can be seen in the context of manuscript collections from the same owners also acquired by Sloane.[34] Heavily annotated printed books were barely distinguished from manuscripts. How and when Sloane acquired these various editions, and how he used them, is more opaque. Unfortunately, many have lost his numbering, through wear and rebinding, and it is therefore difficult to date his acquisitions[35] and understand how Sloane organised this type of publication. It is clear, however, that they were acquired over many years, and not from a single source. There is also an implication of choice. Even though Sloane's acquisitions were less precisely targeted after 1700 than before, it can be seen from his catalogues that there was very little buying of duplicate copies or editions. We must assume that Sloane made a deliberate choice in acquiring so many copies of the same text, and that the presence of annotations was an incentive.

The annotated copies of the 1650 *Pharmacopoeia Londinensis* can be used as a case study to demonstrate a certain number of clearly defined types of annotation. Chief among these is commentary on the formulations and their uses, either by reference to other sources or to experience of their use in practice. Thus in BL 777.i.14, an unidentified annotator has filled almost all the interleavings with notes on the formulations, including examples of their use,

33 BL 777.k.9.

34 For details of Sloane's considerable holdings of Mayerne's manuscripts, and the rather fewer number of Merrett's, see "British Library Catalogue, Archives and Manuscripts," accessed 5 December 5, 2016, searcharchives.bl.uk.

35 As the Sloane Printed Books Catalogue continues to grow, and further information from Sloane's manuscript catalogues is incorporated, it will become easier to map this type of information.

sometimes with references to printed sources of authority.[36] The great majority of the notes in this example are in Latin, but occasionally in English or Italian. There are frequent occurrences of additional formulations, and, less frequently, formulations credited to particular physicians, among whom is Theodore Turquet de Mayerne. BL 777.i.15, a copy without interleaving, shows a quite different form of annotation, consisting mainly of references alongside the formulations to printed sources or authorities for them.[37] It is less obvious what Sloane would have learned from these annotations since they consist in bibliographical references with very little in the way of comment or interpretation. BL 777.i.13 is also not interleaved, but annotated in the text and on the endleaves by one or more unidentified owners. The majority of the notes relate to the use of the various formulations, or give additional examples. No bibliographical references or authorities are offered, though a small number of formulations are credited to "Browne"[38] and one refers to Turquet de Mayerne.[39] BL 777.i.16, not interleaved, and again annotated by an unidentified owner, offers a more systematic set of additions to the text. The annotator has created an index of abbreviations explaining the benefits of the various formulations,[40] though this is used only in the Catalogus Simplicium. He has also created a set of codes to refer to his authorities. Not all can be deciphered, but they include: Wecker, Augustan, Cordus, Culpeper, Schroder, Morellus, Brugis, Hartman, Montagn: (possibly Montagnana), Fernelius, Bauderon and Mesue (Yuhanna Ibn Masawah). There is considerable overlap with the authorities used in the copies above.

BL 778.k.8, not interleaved, has a relatively small amount of annotation, adding the English names of plants, additional formulations, and suggestions for medical conditions in which they might be used. This unidentified annotator does not refer to printed sources to any great extent, only referring to two unguents attributed to Ruland.[41] BL 778.k.9, again not interleaved, shows

36 This annotator makes extensive use of the following sources, referencing them throughout: Nicholas Culpeper, Jean de Renou ('Renod'), Martin Ruland, Joseph du Chesne, Hanss Jacob Wecker, Schroeder (presumably the *Pharamacopoea Schrodero-Hoffmannia*) and Zwelfer (presumably the *Pharmacopoeia Augustana* edited by Joannes Zwelfer).

37 Many of the same sources occur here, for example Wecker, the *Pharmacopoeia Augustana*, Schroeder and Renou, but others also figure: Johann Tilemann, van Helmont, Thomas Sydenham, Guillaume Rondelet and William Vaughan's *Directions for Health*.

38 BL 777.i.13, e.g. on the preliminary leaf facing the title page.

39 BL 777.i.13, title page verso.

40 BL 777.i.162, preliminary leaves.

41 BL 778.k.8, 163.

two very different forms of annotation. It was owned by Christopher Merrett (1614–1695), who was admitted to Fellowship of the College in 1651 and quickly took an active role in its affairs. He would later become a founding member of the Royal Society. There are very few notes in the text, but on blank leaves at the end of the volume he notes the number of each type of remedy, headings which he thought were required, and lists additional or omitted medications. On blank leaves at the front, however, he notes the names of Fellows (possibly in 1659), and names from the election in 1672. At the end of the volume there is a short list of the Gulstonian lecturers. Merrett has used the volume not only for his medical practice but also as a convenient place to record events relating to the College.

Since Turquet de Mayerne is mentioned by several annotators as a contemporary authority it is worth looking at his own annotated copies of the *Pharmacopoeia Londinensis*. His copy of the 1618 edition,[42] not interleaved, contains numerous formulations for syrups, pills and powders, written in various inks, showing additions and amendments. His copy of the 1627 edition[43] is interleaved and includes many remedies used by him. In both copies the notes are neatly set out with headings and other emphases in red, and longer text in black. Only rarely does he cite sources or authorities: the remedies are principally a record of his own formulations. As Vera Keller discusses in this volume, Mayerne (1573–1655) was a pre-eminent physician, the most successful and expensive in London in his day, the source of knowledge quoted by others, and here we see him too using his copies of the *Pharmacopoeia* to record his own practice.[44]

The various forms of annotation in Sloane's copies of the 1650 *Pharmacopoeia Londinensis* exemplify forms which occur in other pharmacopoeias and in other medical texts: additional remedies and formulations, occasions for their use and references to other sources or practitioners who describe the remedies or have used them with success. The annotators use references to printed sources both to locate additional information and to give authority to the remedies cited. These references tell us what texts practitioners were reading and using, and emphasise that practice was both what we might term "evidence-based" and highly reliant on the authority of others, whether contemporary or classic. The pharmacopoeias are a particularly striking example of how knowledge evolves from the original text when combined with added information,

42 BL 777.k.5. Sloane number MS 1872 (deleted); MS A 236.

43 BL 777.k.7. Sloane number MS 305 (deleted); MS A 238.

44 Hugh Trevor-Roper, "Sir Theodore Turquet de Mayerne (1573–1655)," *ODNB*; see Keller, "Scarlet Letters," for further biographical and historiographical references.

the context of professional experience, and the recognised authority of medical authors. Sloane's choice of so many annotated copies of the same edition shows that he specifically valued the information and experience embodied in the notes.

6 Luke Rugeley's Library: Chemical Medicine and the Search for a Remedy for Sore Eyes

Finally, a collection of books which further illustrate these types of annotation, but with a specific context for Sloane, books from the library of Luke Rugeley.[45] Rugeley (1616–1696) was a well-known physician with a reputation for curing eye disease, son of the chemical physician Thomas Ridgeley or Rugeley, c.1576–1656. Luke Rugeley and Sloane were neighbours at opposite sides of Southampton Square (now Bloomsbury Square), both associated with the Royal College of Physicians. Sloane too had an interest in eye disease, and would have dearly liked to have the recipe for a particular remedy used by Rugeley. His failure in this was still rankling when he wrote *An Account of a Most Efficacious Medicine for Soreness, Weakness and several other Distempers of the Eyes*, London, 1745, and recounted his attempts. In the end, he says, he bought the remedy from a former employee of Rugeley's, and found it to be nothing particularly innovative. However, when Rugeley's library was sold in 1697, soon after his death, Sloane purchased at least 50 items, of which 17 are annotated, and it seems probable that the potential for finding the remedy among these was an attraction here.

The search was not successful. Indeed, Sloane must have realised quite quickly that the significant annotations were not by Luke Rugeley himself. He made some suggestions: Thomas Rugeley, Luke's father?[46] Or Jonathan Goddard, founder Fellow of the Royal Society?[47] Or, as suggested later by one of Sloane's assistants, Theodore Turquet de Mayerne?[48] Probably none of these, and there is as yet no firm identification of their author, but it does seem

45 For a more detailed account of Sloane's acquisitions from Rugeley's library, and a list of those identified, see Alison Walker, 'Sir Hans Sloane and the library of Dr Luke Rugeley', *The Library*, 7th series, 15.4 (December 2014): 384–409.

46 In Matthaeus Silvaticus, *Pandectae Medicinae* (Lugduni, 1541). BL 544.h.5.

47 In Pietro Rostinio, *Trattato di mal francese* (Vicenza, 1623). BL 1174.b.7.

48 In Hieronymus Braunschweig, *New volkommen Distillierbuch* (Frankfurt, 1597). BL 717.l.38.(1).

likely that the annotated books were brought together by Thomas Rugeley, and inherited by Luke.

Sloane's purchases at the Rugeley sale were very different from the type of book he was buying from Fenton and Stanley, even though Rugeley's library had many of the Greek, Roman and Arabic classic authors, as can be seen from the printed catalogue of the sale.[49] What Sloane chose to buy was something new to his library, mainly late sixteenth and early seventeenth century chemical and alchemical books from German-speaking countries, many in the German language. It was from Rugeley's collection that Sloane first acquired works of Bartholomaus Carrichter (1510–1567), Swiss physician and astrologer; Johann Wittich (1537–1596), German court physician; and an important work on distillation by Hieronymus Braunschweig (ca. 1450—ca. 1512), German surgeon, alchemist and botanist. The popular *Artzneybuch* of Oswald Gaebelkhover (1539–1616) and chemical works by Johann Popp, Johann Pharamund Rhumel and Fidejustus Reinneccerus all entered Sloane's collection from Rugeley's library. In all of these cases, the copies Sloane bought had manuscript notes.

Sloane would have found much to interest him in this collection, even if the remedy remained elusive. The person who annotated these books seems to have been in practice in London in the 1640's-1650's, as either a physician or an apothecary, but possibly also an empiric, and had some association with the Netherlands. The majority of the notes fall into two categories; commentary on materials (chemical and botanical) described in the text, and remedies which use these materials. They generally take the form of citations from authorities, either books or persons known to have used the remedies. Notes are generally in Latin or English, but occasionally in Dutch, and extracts from authorities are in Latin, English, French, German, Dutch, Italian and Spanish. Plant names are generally given in Latin or English, translated from the original language of the text. There is a special interest in cures for gout and gonorrhea, but his main emphasis is on the preparation of remedies. He speaks frequently of his experiences in distillation, and describes in some detail the best materials to use in remedies. Many printed sources, mostly Continental authors, are mentioned. Some English sources are included, for example Thomas Moffett; the anonymous *A rich store-house or treasury for the diseased*, London, 1596; Elizabeth Grey, *A choice manual of ... Physick*, 1653, and Didymus Montanus [Thomas Hill], *The gardyner's Labyrinth;* John Hall and George Starkey. The Netherlands

49 *A Catalogue of ... Books ... being the Library of Dr. Rugeley ... will be sold by auction ... on Tuesday the 19th instant ... By John Bullord, etc*, [London, 1697]. BL S.C.886.

connection is emphasised by the annotator's frequent references to one of his sources, cited several times with apparent pride, a manuscript of "Abraham van Hacke Belgo Empirici: post mortem autoris apud Bibliopolam venali, et per me coempto."[50] Van Hacke may be the Abraham van Hack of Norwich, Norfolk, who was licensed on 16 August 1620 to practise medicine and surgery in the province of Canterbury,[51] and whose family later lived in London.[52]

The books acquired from Rugeley's library serve to illustrate several aspects of the descendence, or afterlife, of collections and material from several collections: the putative manuscripts of Van Hack acquired and much quoted by the unnamed annotator, referred to in texts collected by Thomas Rugeley and left to his son Luke, later selected and studied by Sloane. Although Sloane's purchases ultimately failed to provide the information he sought, his choices have ensured the survival of a significant body of material which illustrates the interests and practice of a medical professional of the mid 17 th century.

7 Afterlives

These few examples of annotated works in Sloane's library illustrate several forms of afterlife. We see the survival of parts of a library through selection by a new owner or the survival of the greatest part of a collection through its incorporation in another collection. But more than simply ensuring survival, intentionally bringing together annotated books creates new associations, new lives and new uses in new hands. None of these processes is uncommon, but they are brought into focus because of the subsequent incorporation and long-term survival of this material in a major national collection.

It is worth trying to interpret and summarise Sloane's motives in collecting this material. After all, not all collectors sought annotated copies; indeed, the survival of such material depends on many factors, including fashion, market forces, the depredations of age, and sheer chance. For Sloane, there was firstly a need for information which might be used in a medical practice. The remedies added to medical texts and pharmacopoeias could be of immediate practical use. But accompanied by case studies and accounts of the use of these

50 BL 1033.l.10, blank leaves at end of volume.

51 Lambeth Palace Library Reg. Abbot 2 f. 189v.

52 Van Hack family members are recorded in London from 1634 to 1689, and in Norwich between 1690 and 1722. An Abraham Vanhack married Ellen Derrick in London in 1634, and the will of an Abraham Vanhack was proved in 1661, National Archives PCC 1661 f. 189.

remedies, they amount to more, as the added context and evaluation promote understanding, and effect the transformation of information into knowledge.[53]

There are further reasons too, which have much more to do with the processes of collecting. Texts enhanced by previous owners, particularly those known to Sloane either in person or by reputation, offered knowledge by association, created through the accumulation of the experience of others. In any collection, the bringing together of material creates a form of identity for the collector and allows the objects collected to become part of that new identity, yet it also creates contexts which may be unknown to the collector himself, perhaps emerging only by collocation with still other collections. Sloane was both collecting the knowledge of others, and through this process creating new forms of knowledge. His desire that his collection should stay together shows that he understood the fruitfulness of such collocation, and valued the interrelated meanings that emerge from material brought together.

The afterlife of Sloane's own collections has been long but not uneventful. His desire that the collection should remain together as a whole has been mentioned earlier. Most of his books and objects are now held by the British Museum, the British Library and the Natural History Museum. There have been exoduses, notably in the sales of duplicates in the early nineteenth century. There have been losses of vulnerable natural history specimens. There was some destruction during the second World War. However, most of the material is still known and available for study, but its context is not that of Sloane's day. His books and manuscripts, once housed in close proximity to the objects to which they closely relate, have lost that connection. Books and manuscripts are separate. His correspondence survives, but there is little in the way of financial or family documentation. The collection which reached the British Museum after Sloane's death is not an archive of his life, nor is it an archive of the collections which he acquired, but it has archival qualities in that it records ways in which users interacted with its content. It has now lived alongside material from other sources for over 350 years, and during that time it has always been part of a continual process of integration into something bigger than itself. Metamorphosing from its first life in Sloane's ownership, the collection as part of the British Museum and its daughter institutions has had, and continues to have, an extended afterlife of great influence and worldwide impact.

53 As suggested earlier, the line between print and manuscript becomes fuzzy here, and it should be noted that Sloane's manuscripts also offer a wide range of case studies and remedies.

CHAPTER 8

Collecting Genomics: Documenting Modern, Collaborative Science

Victoria Sloyan

1 Introduction

Scientific research has long functioned through collaboration, whether between the wealthy natural philosopher and his unnamed scientific servants, as in the case of Robert Boyle and his laboratory assistants, or between correspondents who exchanged natural historical specimens and information, crediting each other in their publications.[1] In the modern era it has become only more collaborative and multidisciplinary, particularly since the post-war rise of big science. And yet many archivists striving to capture and preserve scientific records rely on traditional collecting methods that focus on individuals and organisations in isolation. Whilst pragmatic considerations play a part in this, it is also due to the pervasive cultural myth that dominates our understanding of scientific practice; that of the lone scientific researcher experimenting and theorizing in isolation.

The Wellcome Library project Collecting Genomics has attempted to tackle this through its work collecting records produced in the lead up to and during the Human Genome Project (HGP). We adopted a documentation strategy-based approach that explored the surrounding landscape of the HGP and was specifically designed to capture collaborative working and the interconnected relationships that defined the project.

One consequence of Collecting Genomics was increased contact between archivists and scientists. For reasons outlined below, we archivists became involved in record preservation at an earlier stage of the record lifecycle: at the point at which records were semi-current rather than ready for deposit in an archive. This meant engaging with record creators (scientists) who in many cases were still in the middle of their careers. It quickly became apparent that

1 Steven Shapin, "The House of Experiment in Seventeenth-Century England," *Isis* 79 (1988): 395, 2016 http://www.jstor.org/stable/234672; Yale, *Sociable Knowledge: Natural History and the Nation in Early Modern Britain* (Philadelphia: University of Pennsylvania Press, 2016).

scientists have very wide-ranging ideas of what archives are. Our role became just as much about educating scientists and improving recordkeeping practices as it was about collecting archives. Increasing reliance on digital media in the scientific sphere also necessitated an early, hands-on approach to archival collecting, as digital records are much more vulnerable than paper to deterioration and loss of accessibility.

This paper sets out the Wellcome Library's alternative strategy for archival acquisition. It explores how we attempted to engage with the scientific community, how scientists' support is vital for the successful acquisition of records and how an understanding of a scientist's working environment aids the preservation of their records.

In the context of this paper the term "scientist" is used to refer to anyone engaged in any kind of scientific endeavour. The term "researcher" refers to archival users and those engaged in academic study.

2 Collecting Genomics Survey Methodology

The Wellcome Library is a reference library for the study of health in historical and cultural contexts. In January 2012 we launched a project to survey records in the UK related to the Human Genome Project (HGP) to establish what had survived and what was needed to ensure the long-term preservation of the records. This was the Human Genome Archive Project which ran throughout 2012 and 2013. Phase II, Collecting Genomics, is building on this work and began in January 2014. The Human Genome Project was an international endeavour to sequence the human genome in its entirety for the first time and release the data openly. It ran from the late 1980s until the early 2000s and involved over 16 institutions in more than 6 countries. A working draft of the sequence was published in February 2001 and a "gold standard" was announced in April 2003. The Wellcome Trust was one of the primary funders of the HGP, establishing the Sanger Institute to undertake a major portion of the sequencing work. We at the Wellcome Library wanted to ensure we captured a sense of the HGP collaboration in the archives we collected, but were concerned that traditional archival collecting approaches might not achieve this aim.

Traditional archival acquisition models usually focus on individual people or organisations. Acquisition is often passive, whereby collections are offered to an archive, most commonly by the creator or a close family member. This limits acquisition to collections from individuals who are already familiar with archives and have some sense of the historical significance of their material.

Generally, these individuals begin the process of deposit towards the end of their career, or even life, and so a collection only contains those records which have survived many years.

These issues can be overcome to an extent through proactive acquisition, whereby an archival institution identifies and targets specific archives to collect. Nevertheless, the focus often remains on a single entity and the records they have created or collected. In addition, both forms of archival acquisition often lead to the capture of collections belonging to those individuals or organisations with larger public profiles, as they are more likely to have a stronger sense of their legacy and archivists targeting collections in an unfamiliar field are more likely to identify these individuals. Whilst this ensures there is at least some record of scientific endeavour in a particular field, there is a danger that focus on individuals can distort the historical picture and fail to reflect the collaboration behind modern scientific projects such as the HGP. This focus on individuals also fails to recognise the wider context in which a person or organisation works and in which records are created. As one critic of traditional archival acquisition says, "the integrated nature of society's institutions and its recorded documentation must be reflected in archivists' efforts to document those institutions. Institutions do not stand alone, nor can their archives."[2] This statement is easily applicable to people as well as institutions.

Therefore, we investigated alternative approaches to archival surveying and collecting, including documentation strategy. This concept developed in the 1980s and is closely associated with the archivist Helen Willa Samuels, who defines documentation strategy thus:

> A documentation strategy is a plan formulated to assure the documentation of an ongoing issue, activity, or geographic area... The strategy is ordinarily designed, promoted and in part implemented by an ongoing mechanism involving records creators, administrators (including archivists), and users.[3]

The idea is to survey the environment which surrounds an issue or, in the case of the HGP, an event, identify what material exists and does not, identify what needs to be documented in order to fully understand that event and then ensure that documentation is preserved.[4] The broad principle of documentation strategy seemed to cater for our needs as it was devised with complex

2 Helen Willa Samuels, "Who Controls the Past," *American Archivist* 49(2) (1986): 112.
3 Willa Samuels, "Who Controls the Past," 115.
4 Willa Samuels, "Who Controls the Past," 120–121.

creating environments and wider context in mind. It is also suitable for capturing of context beyond the scientific arena to broader social, ethical, economic and political contexts. Furthermore, documentation strategy is also appropriate for situations where records are physically dispersed, rather than gathered in a central location.

We adopted documentation strategy as our guiding concept and took inspiration from the Minnesota Method for our implementation approach.[5] Detailed information on our strategy is available elsewhere and what follows here is a short summary. [6] Firstly, we defined the collecting area. Surveying the entire HGP environment was not feasible with our resources and we restricted ourselves to UK participation with a focus on the years 1977–2004. The early period covers the development of genomics and key work that led to the emergence of the HGP. The second step was to identify relevant collections that had already been deposited in archival repositories. This was followed by prioritising areas of surveying, as even restricting the survey to Britain was a daunting task. Three scientific areas that had particular significance for the HGP in Britain were prioritised: the Sanger Institute, the Medical Research Council Laboratory of Molecular Biology (LMB) and the Human Genome Mapping Project (HGMP).

The Wellcome Library project was not a strict implementation of documentation strategy. Rather, the fundamental concept was adopted and implementation methods were adapted to suit our needs. Given this was a retrospective project we could not involve record creators at the point of creation. Nor could our strategy be described as an on-going mechanism involving these people.[7] Nevertheless, we took to heart the idea of surveying the wider environment surrounding the HGP, identifying and attempting to document the events, institutions, individuals and outputs that led up to it, that enable it to happen and that were involved in it.

After taking into consideration the limits of the defined collecting area and the prioritised areas of collecting, and after ruling out those individuals who had already deposited their archive or were deceased, 90 individuals were contacted about potential archival material they may have had. 64 replied and 17 archival surveys were undertaken, leading to six collections being

5 Richard Pearce-Moses, *A Glossary of Archival and Records Terminology* (Chicago: Society of American Archivists, 2005), 253, http://files.archivists.org/pubs/free/SAA-Glossary-2005.pdf.

6 Jennifer Shaw, "Documenting genomics: Applying archive theory to preserving the records of the Human Genome Project," *Studies in History and Philosophy of Biological and Biomedical Sciences* 55 (2015): 3, doi: 10.1016/j.shpsc.2015.08.005.

7 Willa Samuels, "Who Controls the Past," 115.

deposited with the Wellcome Library and additional collections being deposited with other, more suitable repositories. Six may not seem a large number, but it should be noted that several people are still in the middle of their careers and not yet ready to deposit their records. More important at this stage was to make initial contact and raise awareness of the potential archival value of their records in order to ensure their survival. Of those collections that were surveyed but not deposited, this was mainly due to there being little archival value contained within them.

An important feature of the six deposited collections is that they come from a range of people who were at different stages of their career during the HGP and had different public profiles. We took in the collections of the Nobel Prize winner Sir John Sulston (Wellcome Library reference: PP/SUL) and genomics pioneer Michael Ashburner (Wellcome Library reference: PP/MIA). We also took in collections from people who have a lesser public profile but still made a significant contribution to the HGP and, crucially, have documents that reflect areas of the HGP not covered in the archives of the Sulston and Ashburner. For instance, Sulston was the first Director of the Sanger Institute and oversaw its contribution to the HGP, but did little of the actual sequencing work. In contrast, Matt Jones (Wellcome Library reference: GRL/MCJ) was at the forefront of subcloning work (a preparatory stage of genome sequencing) and his archive provides details on how this was undertaken. If the Library had only acquired the Sulston archive, researchers could gain a good understanding of the high level strategy and politics of the HGP but only a vague idea of how the science was carried out.

Whilst the six deposited collections all come from individuals, the Library did look at institutional archives. The LMB already has systems in place to preserve its organisational records and whilst it does not hold records of individual employees, it helps them find suitable repositories. The HGMP was funded by the Medical Research Council which has recordkeeping practices in place, though is not always successful in implementing them beyond head office. The Sanger Institute does not have well established recordkeeping practices in place and the implications of this are discussed further on.

3 Engaging with Scientists

One consequence of our survey work was that it brought us into much closer contact with scientists, many of whom had little understanding of the purpose of archival institutions and records. They had varying ideas of what constitutes an archival record and which of their own papers would be of interest to future researchers. We realised that in order to successfully engage scientists we had to clearly articulate what constitutes an archival record, what does not, what the Library was interested in taking and what the benefits to scientists were of depositing records. We also identified a number of risk factors that threaten the survival of scientific archive records.

The aim of the Human Genome Archive Project and Collecting Genomics was to collect records that document the HGP and the environment in which it was established and carried out. Therefore we were not only interested in genome sequencing work, but also in the other roles and duties of scientists. In modern science, a scientist is not just a researcher but often combines this role with that of teacher, consultant and administrator, to name but three.[8] All these roles need to be captured in an archive if future researchers are going to be able to discern the context in which a scientists worked and in which a specific project (like the HGP) occurred. Precise appraisal criteria are often very specific to each collection, nevertheless we attempted to devise some broad guidelines to help scientists start thinking about their records.

Laboratory Notebooks: A key record series in a scientist's collection and one of the most likely to be kept as they have longstanding value to scientists. The personal value to the creator is discussed later, but in terms of archival significance laboratory notebooks document the chronology and direction of research and record the development of new techniques and practices. A comprehensive set is best able to chart these developments and we found it necessary to state this to scientists as we learnt that some only value, and in some cases only keep, those notebooks that document successes and potentially patentable research. A related type of record is research files. These are valued for the same reasons as laboratory notebooks and contain similar information, though are not kept in such a structured way.

Conference and Meeting Papers: These show the development of scientific networks and document the communication of ideas. Personal notes made by scientists during these events show how they responded to presentations and new ideas. Moreover, many informal yet important meetings were held during conferences as scientists took advantage of the fact that key people from across the globe were physically together. Genomics, particularly in the period 1970–2000 was quite a niche field and many conferences were organised by relatively small organisations, many of whom did not retain key official records such as abstract books. Many scientists throw away their copies of these documents, thinking a

8 Kalpona Shankar, "Understanding the Record Keeping Practices of Scientists," *Technical Leaflet Series no.10, Mid-Atlantic Regional Archives Conference* (1999): 1, http://www.marac.info/assets/documents/marac_technical_leaflet_10.pdf.

master record is held elsewhere, therefore it was important for us to explain the value of their copies and stress that we were interested in official documentation as well as individual's notes and annotations.

Correspondence: Capturing correspondence serves a similar function to conference papers in that they document the transfer of information and communication of ideas. They also provide rich detail on how collaborative projects were executed. In many cases correspondence are more revealing than official minutes and reports.

Teaching Records: These document the development of a discipline and the "intellectual lineage"[9] of teachers and students. In addition, they show how scientists reconcile research with other time pressures and also demonstrate the impact research has on teaching and vice versa.

Administrative Records: The administrative aspect of an institution can have a huge impact on scientific work, particularly in terms of funding, overall research priorities and strategic direction. Administrative records include grant files and we stressed we were interested in failed as well as successful applications. Administrative records in personal archives can be particularly valuable if they come from institutions with poor recordkeeping practices as these personal records can be the only copies accessible to researchers.

Advisory Board and Consultancy Files: These show the wider application and influence that scientific research has on the wider community, whether in the commercial sector, scientific ethics or political policy.

Manuscripts: This includes drafts of a scientist's published outputs along with working papers, such as research and correspondence. The published paper is available through other sources, but this background information is often unique. It documents the development of a paper, and shows how collaboration is achieved when there are multiple authors.

It was also important to provide information on the sorts of material the Library was not interested in. Scientists often have limited storage capacity for paper records and it was important to help identify those papers which could be disposed of, rather than risk the accidental destruction of archival records. Such disposable material includes records that lack contextual information as this makes it very hard for others to interpret records accurately, especially researchers who may not have in-depth genomics expertise and rely on contextual information. This type of record commonly includes unlabelled and undocumented images and raw research data. We also advised against retaining ephemeral material such as routine travel information (train tickets, hotel booking forms etc.), circulated

9 Joan K. Haas, et al., *Appraising the Records of Modern Science and Technology: A Guide* (Massachusetts: Massachusetts Institute of Technology, 1985), 14.

papers from committees and journals and daily financial records other than project accounts.

Armed with our criteria, we felt prepared to attempt to engage scientists with archival preservation. The scientists we spoke to had different ideas about archives, but this is not a new problem. One archivist writing in 1970 recalls how a scientist believed only his conclusions mattered and destroyed all his working papers once his conclusions were published.[10] Forty years on a similar attitude remains prevalent, with many scientists believing only their published papers have historical significance.[11] This is partly due to the importance placed on published outputs by the scientific community as a means of securing funding for research projects. In addition, a published paper (along with data deposited in a data bank) is often a requirement imposed by funding bodies and as such research councils, university administrators and scientific institutions prioritise the creation and preservation of these records but have little interest in other records. Yet it is the background information that researchers find so compelling and want to see preserved.

As mentioned, most scientists had kept at least some of their laboratory notebooks. This is something advised by administrators, since a well maintained notebook can be used to defend intellectual property rights.[12] Many undergraduates are given advice on maintaining laboratory notebooks and retaining them for a substantial period.[13] Unsurprisingly, there is a more consistent and higher level of recordkeeping for records that are important to scientists during their career and are recognised as such by administrators and funding bodies. In contrast, there is a much lower survival rate for other records and survival is reliant on personal disposition. This situation is reinforced by poor

10 Roger H. Ellis, "The Historical Manuscripts of Science and Technology," *Journal of the Society of Archivists* 4(2) (1970): 90, doi: 10.1080/00379817009513944.

11 Georgina Ferry: "Scientific Heritage: Science Today, History Tomorrow", *Nature* 493 (January 3, 2013): 19, doi: 10.1038/493019a.

12 Roland Smith, "Keeping a Good Laboratory Record Book," last modified September 2005, accessed January 7, 2016, http://www3.imperial.ac.uk/pls/portallive/docs/1/7289716.PDF.

13 Howard Hughes Medical Institute and Burroughs Wellcome Fund, *Making the Right Moves. A Practical Guide to Scientific Management for Postdocs and New Faculty, Second Edition*, (Maryland: Howard Hughes Medical Institute, 2006):147, accessed January 7, 2016, https://www.hhmi.org/sites/default/files/Educational%20Materials/Lab%20Management/Making%20the%20Right%20Moves/moves2.pdf and Oxford University, "Guidelines for Keeping Laboratory Notebooks," accessed January 8, 2016 https://www.admin.ox.ac.uk/media/global/wwwadminoxacuk/localsites/researchdatamanagement/documents/labnotes_policy.pdf

institutional recordkeeping. A lack of recordkeeping culture in many scientific institutions (particularly biology ones) means scientists are not given training and do not understand the importance of good recordkeeping.[14]

Another risk factor for the survival of archives is the laboratory or office move. In recent years internal relocation is often part of a process of physically fitting more people into an organisation and paper record storage is one of the first areas to be squeezed. Not only do scientists have to make do with less storage, making at least some disposal necessary, but the act of packing up and moving can also lead to record destruction. When records sit for years in an office untouched, benign neglect can often lead to their survival. When a scientist is forced to confront their records, unless they have posterity in mind they are likely to dispose of those they no longer see any personal need for. One collection deposited in the Library survived precisely because of benign neglect. The creator had worked continuously for one institution for the last twenty-four years, had his own office and suffered very few office moves. Had he, it is highly likely most of his archive would not have survived.

External moves are also dangerous. Scientists moving to a new institution or retiring are obliged to sort out their records. In some cases records are inherited by colleagues and so collections are dispersed. In others, particularly in the case of senior staff with an office, the records can be left on site where they are at the mercy of the institution. When the scientist takes his or her records with them, this has the same risks as internal moves when a scientist has to confront their records. If a scientist moves several times during their career, as is not uncommon, this further increases the risk of destruction.

To conclude, many of the scientists we spoke to did not have a good understanding of the purpose and nature of archival records, at least not in a scientific context. The majority had maintained laboratory notebooks and published outputs but did not value their other records. This is primarily due to a lack of recordkeeping awareness, which in turn is due to the culture in which they operate. It is a culture that does not prioritise recordkeeping and does not necessarily make it easy for scientists to retain their records. Where comprehensive sets of archival records have survived, this was due to the efforts of the individual record creator combined with a low level of risk factors such as continuous laboratory moves. We sought to encourage greater awareness of the potential archival value of scientific records so that those scientists in early or mid-career would be more likely to retain current and future records until they reached the point at which deposit in an archive was appropriate.

14 Shaw, "Documenting genomics," 3.

The focus of this paper has so far been on paper records. Despite the dominance of computers in our daily lives, paper records are still generated. In any case, scientists' records can extend back to their early careers in the 1960s and 1970s when there was greater reliance on paper formats. However there is a significant proportion of born-digital records in the collections we acquired and these have their own challenges when attempting to collect them.

4 Dealing with Digital

The majority of scientific collections deposited in the Wellcome Library are hybrid archives: they contained both paper and digital records. Digital records do not face survival issues in the sense that they are not commonly deliberately destroyed in the way paper records are. But in another sense digital records are much more vulnerable. Paper often survives through benign neglect, but neglect of digital material often leads to degradation or inaccessibility due to software and hardware obsolescence.

The collections we acquired date from the 1960s to 2000s. Consequently digital records were deposited on a range of media including 5.25" floppy disks, 3.5" floppy disks and external hard drives. This storage media is fragile. It is easy to bend, scratch and break and once that happens the data is often unrecoverable. In addition, data on the disks can decay over time and files become altered or corrupted. Early intervention by archivists can ensure that records are transferred to more secure storage before any damage or decay occurs. If it is too early to deposit records, archivists can advise on suitable preservation actions to take in the medium-term. Early intervention has the additional benefit of fending off obsolescence issues. Passive collecting results in archivists acquiring material that is years old. Add to this the speed of technological change and obsolescence becomes a real problem for archivists because it diminishes the likelihood we have the right hardware to access and recover data. It is much better to work with potential depositors at an early stage of the record lifecycle to ensure born-digital records are migrated to newer storage media as these media are developed, rather than try to deal with obsolete media.

Obsolescence also affects software. A file may be recovered from old media but it cannot then be rendered because the necessary software application no longer exists. Preservation steps are being devised by the digital preservation community to ensure future accessibility of various file formats, but understandably the focus is on common formats, such as MS Office files and PDFs. The scientists we worked with use these common formats, but also use other niche formats and in some cases develop their own. For instance, two collections contain a number of FASTA Sequence files and BED Annotation Track Files, which hold nucleic acid and protein sequences and genome annotation track data respectively. Being very niche, these file formats are harder to identify. They often are not recognised by software profiling tools and identification has to be done manually. This is difficult enough as it is, but is made harder if the file format in question is obsolete. The Wellcome Library can only ensure the long-term preservation of the file formats we have identified and know we have. Therefore it is crucial that all formats are identified. Manual identification can be aided by the scientist as they will be familiar with the formats they worked with and are more likely to recognise niche formats.

One other complicating factor of born-digital records is quantity. Unlike paper records, there is often less pressure on storage and so scientists are not compelled to sort through and weed digital records. Such large quantities can be hard for the archivist to appraise, especially where obscure filenames are used. Therefore the archivist is more reliant on the creator's knowledge in order to be able to catalogue and describe the records. This knowledge is easier to acquire through proactive collecting, where the creator is more likely to be in a position to remember and assist.

In short, born-digital records benefit from early intervention by archivists. Consequently our documentation strategy-based approach not only enabled us to acquire a broad range of collections but also proved beneficial when dealing with born-digital records.

5 Conclusion

From the outset the Human Genome Archive Project and Collecting Genomics aimed to devise a surveying approach that would yield better results than traditional acquisition methods. Our approach captured records that reflect not just British contribution to the HGP, but also the surrounding scientific, political and social contexts that the project existed within. We acquired archives from a range of scientists connected to the HGP in different ways that when taken together provide a comprehensive account of the project. Our collecting approach was not only broader than usual but was also more proactive. This had benefits for the preservation of digital material, where early intervention can be crucial in maintaining accessibility. It also provided the opportunity for us to raise awareness of archives within the genetic and genomics community.

Whilst we acquired six archives through this project, we perhaps could have collected more had recordkeeping practices within the scientific community been better. We found that many scientists are not aware of their legacy or of the historical value of their records. Their thoughts are, perhaps understandably, focused on the present and their current research. Record retention, beyond laboratory notebooks and published papers is not a priority and as a result records are vulnerable to loss and destruction.

The biggest risk factor to survival is the lack of recordkeeping culture in life sciences research institutions. An unsympathetic institution can hinder individual attempts to retain records and an apathetic one will not encourage or enable scientists to implement good recordkeeping practices. Therefore, the improvement of institutional recordkeeping is the focus for the remainder of Collecting Genomics. It is hoped that real and lasting improvements can be made, resulting in more records surviving for the benefit of future researchers.

Afterword

Michael Hunter

In reflecting on the essays collected in this volume, and on my own experience in working on the papers of Robert Boyle and others, my initial thought was how relatively recent it is that archives have come to the fore in the history of science. In 1960, in a survey of Newton scholarship at a History of Science Society meeting, I. Bernard Cohen remarked: "What is especially astonishing about the huge number of Newtonian articles and books is the general failure of the authors to consult the manuscript collections." This comment is quoted by Rupert and Marie Boas Hall in the Foreword to the 1978 reprint of their 1962 anthology, *Unpublished Scientific Papers of Isaac Newton*,[1] and I think the Halls were real pioneers in seeing the potential of manuscript material to transform our understanding of seventeenth-century science, first through that book and then through the edition of *The Correspondence of Henry Oldenburg* that they brought out between 1965 and 1986.[2] A comparable figure was Richard S. Westfall, who

This paper stems from the remarks that I made at the conclusion of the conference devoted to 'Archival Afterlives' held at the Royal Society on 2 June 2015, and I am grateful to the editors of the volume for inviting me undertake this role. The comments that I made on that occasion have been largely retained, but I have adapted them to take account of the final versions of the papers that have been submitted for publication in this volume and have added a few further reflections. I am grateful to the editors and to Richard Serjeantson for their comments on my text.

Notes & Records = Notes and Records of the Royal Society

1 A.R. and M.B. Hall, eds., *Unpublished Scientific Papers of Isaac Newton* (Paperback edition, Cambridge: Cambridge University Press, 1978), p. v.

2 A.R. and M.B. Hall, eds., *The Correspondence of Henry Oldenburg* (13 vols., Madison and Milwaukee: University of Wisconsin Press, 1965–73; London: Mansell and Taylor & Francis, 1975–86). I specifically remarked on the Halls' pioneering role in my obituary for Marie Boas Hall in *The Times*, 20 March 2009; see also Frank James' account of A.R. and M.B. Hall in *Biographical Memoirs of Fellows of the British Academy*, vol. 11 (published online, 2012), 353–408, 371, and passim.

 | DOI 10.1163/9789004324305_011

also published some key documents in the 1950s and 1960s, including the notorious list of his sins that Newton compiled in shorthand in 1662; later, Westfall was to make extensive use of manuscripts in preparing his magisterial life of Newton, *Never at Rest*, of 1980.[3] This was also the era when other manuscript-based projects connected with Newton emerged, at the hands slightly earlier of H.W. Turnbull and D.T. Whiteside, and rather later of Alan Shapiro.[4]

Of course, historians of science were not unique in this. As Richard Serjeantson reminds us in his chapter in this volume, it was in a comparable period that Peter Beal conceived and implemented his magisterial catalogue of English Literary Manuscripts, the first volumes of which came out in 1980, thus reflecting an interest in early manuscript testimonies to literary texts that was exemplified by work in the 1970s like that of Peter J. Croft on autograph poetry, or the collaborative edition of *Poems on Affairs of State* of the early Augustan period.[5] There are also fields in which historians made extensive use of manuscript sources at an even earlier date, for instance the history of education, as exemplified by W.T. Costello as early as 1958, while manuscripts are perhaps surprisingly prominent in the bibliography to Hugh Kearney's *Scholars and Gentlemen* of 1970.[6]

On the other hand, perhaps partly because of their pioneering status, scholars who made use of manuscript material in this way were often prone to think less deeply than they might have done about the nature of the archives that they used. In the case of the history of science, this is symbolised by Westfall's 1956 publication of a fragment of Robert Boyle's essay 'Of the Atomicall Philosophy', since he had not studied the surrounding archive fully enough to realise that far more leaves of the text survived than those that he published, and a similar

3 R.S. Westfall, "Short-writing and the State of Newton's Conscience," *Notes & Records* 18 (1963): 10–16, and *Never at Rest: A Biography of Isaac Newton* (Cambridge: University Press, 1980).

4 H.W. Turnbull et al.,eds., *The Correspondence of Isaac Newton* (7 vols., Cambridge: University Press, 1959–77); D.T. Whiteside, ed., *The Mathematical Papers of Isaac Newton* (8 vols., Cambridge: University Press, 1967–81); Alan E. Shapiro, ed., *The Optical Papers of Isaac Newton*, vol. 1 (Cambridge: University Press, 1984). See also J.E. McGuire and Martin Tamny, eds., *Certain Philosophical Questions: Newton's Trinity Notebook* (Cambridge: University Press, 1983).

5 Peter Beal, *Index of English Literary Manuscripts, Volume 1, 1450–1625* (2 vols., London: Mansell, 1980); P.J. Croft, *Autograph Poetry in the English Language* (2 vols., London: Cassell, 1973); G. de F. Lord et al., eds., *Poems on Affairs of State* (7 vols., New Haven: Yale University Press, 1963–75).

6 W.T. Costello, *The Scholastic Curriculum at Early Seventeenth-century Cambridge* (Cambridge, Mass.: Harvard University Press, 1958); H.F. Kearney, *Scholars and Gentlemen: Universities and Society in Pre-industrial Britain 1500–1700* (London: Faber & Faber, 1970).

mishap befell Marie Boas Hall in her *Robert Boyle on Natural Philosophy* of 1965.[7] Even with the magisterial *Correspondence of Henry Oldenburg* and other publications with which Rupert Hall was associated, such as *The Correspondence of Isaac Newton*, it might be felt that the archives from which the content of such editions were abstracted were rather taken for granted and not subjected to as searching scrutiny as they deserved in their own right.[8]

The 1980s saw new and important developments, perhaps most notably the Hartlib Papers Project, launched at huge expense and with much publicity in 1987, a venture that did for the first time take an archive as its unit of study. It was also novel for its use of digital technology as its mode of publication, something that has now become commonplace.[9] Yet, as I said more than once in print after the project was completed, I always felt that those involved in the Hartlib Papers Project showed a lack of curiosity about the nature of the archive to which all their effort was devoted—how it was compiled and organised, and how it had been handed down to posterity—issues that were likely to be more significant in terms of its content than the organisers of the project seem to have realised.[10] This has now finally been rectified by Leigh Penman in a paper that was presented at the conference in June 2015 and has since been published in *Book History*, which provides exactly the analysis that I was calling for.[11] At last we have a full understanding of what happened to the archive

7 R.S. Westfall, "Unpublished Boyle Papers Relating to Scientific Method," *Annals of Science* 12 (1956): 63–73, 103–17, on 111–13, and see Michael Hunter and Edward B. Davis, eds., *The Works of Robert Boyle* (14 vols., London: Pickering & Chatto, 1999–2000), vol. 13, XL–XLI and 225ff.; M.B. Hall, *Robert Boyle on Natural Philosophy* (Bloomington: Indiana University Press, 1965), 252–5, an edition of part only of Boyle's "Enquirys and Experiments about Electricall Bodys": see Michael Hunter, ed., *Robert Boyle's 'Heads' and 'Inquiries'* (London: Robert Boyle Project Occasional Papers No. 1, 2005), XI and 12–15.

8 I was therefore not entirely surprised that, when he reviewed *Archives of the Scientific Revolution* in *Isis*, 90 (1999): 804, Alan Shapiro could not see much point in the book. For comparable remarks to those that I have made here from a different perspective, see Harold Love, *The Culture and Commerce of Texts: Scribal Publication in Seventeenth-century England* (Amherst: University of Massachusetts Press, 1998; originally published at Oxford in 1993), esp. chapter 1.

9 See the online edition, Mark Greengrass, Michael Leslie and Michael Hannon, eds., *The Hartlib Papers* (Sheffield: HRI Online Publications, 2013), https://www.hrionline.ac.uk/hartlib.

10 Michael Hunter, "Whither Editing?" *Studies in History and Philosophy of Science* 34 (2003): 805–20, on 812–18; *Editing Early Modern Texts: An Introduction to Principles and Practice* (Basingstoke: Palgrave Macmillan, 2007), 52–4.

11 Leigh Penman, "*Omnium Exposita Rapinæ*: The Afterlives of the Papers of Samuel Hartlib," *Book History* 19 (2016): 1–65.

in Hartlib's final years, in the aftermath of his death and thereafter, and our understanding of it will be transformed accordingly. In archival terms it is a strange quirk of fortune that the crucial Bundle 72, abstracted from the rest of the deposit by George Henry Turnbull, which forms the basis of many of Leigh's conclusions, was never made available electronically with the rest of the archive. Now, there is work to be done on the formation of the archive and on the role of the scribes and others who contributed to it in Hartlib's lifetime.

It is also worth mentioning another project of about the same period, the edition of *The Correspondence of John Flamsteed*, inaugurated by Eric Forbes and brought to fruition after his death by Lesley Murdin and Frances Willmoth; it comprises three hefty volumes, the first published in 1995 and the last in 2002.[12] In many ways, this is an excellent work, a model example of a collected *Correspondence*, yet it seemed to me that it shared some of the limitations of the Oldenburg and Newton correspondence editions in lacking curiosity about its archival basis. By definition a *Correspondence* is an artificial construct, cutting across archival boundaries to collect together all the letters to or from a specific individual. It seems to me that this makes it all the more imperative to explain as clearly as possible just what this archival basis is, and the implications of this for its content, something that the Flamsteed edition often failed to do, as I pointed out in a review essay.[13] I there noted the haphazardness of survival of some of the key documents on which the edition was based, meaning that our knowledge of certain epistolary exchanges is very one-sided, for instance those between Flamsteed and Richard Towneley or William Molyneux.

This brings me to my own work on the archive of Robert Boyle at the Royal Society, which began in 1986 and perhaps reached its climax with the publication of the revised version of my catalogue of the archive and commentary on it in *The Boyle Papers* of 2007.[14] Of course, I had had an apprenticeship prior to this, working in the 1970s on the papers of John Aubrey, John Evelyn and the Royal Society, though I have to admit that at that point I shared the relative lack of interest in the nature of these archives in their own right that I have observed

12 E.G. Forbes, Lesley Murdin and Frances Willmoth, eds., *The Correspondence of John Flamsteed, First Astronomer Royal* (3 vols., Bristol: Institute of Physics Publishing, 1995–2002).

13 Hunter, "Whither Editing", 806–10.

14 Michael Hunter, with contributions by Edward B. Davis, Harriet Knight, Charles Littleton and Lawrence M. Principe, *The Boyle Papers: Understanding the Manuscripts of Robert Boyle* (Aldershot: Ashgate, 2007); the catalogue had originally been published as *Letters and Papers of Robert Boyle: A Guide to the Manuscripts and Microfilm* (Bethesda, Md.: University Publications of America, 1992).

in the earlier scholars on whom I have already commented.[15] The first archive that I had to catalogue for myself was that of the astrologer, Samuel Jeake, and his father, also Samuel Jeake, of Rye, and I think that this made me think more deeply about the nature of such deposits than hitherto, thus providing a useful training for my work on Boyle which started soon afterwards.[16] The Boyle archive is huge and unwieldy, once described as "uncatalogued and uncataloguable,"[17] but I felt that, as I and my research assistants worked through it, I gradually came to understand it in various respects. One crucial break-through, as is expounded in *The Boyle Papers*, was the realisation that Boyle's habitual use of amanuenses made it possible to identify chronological strata within the archive; another was the extent to which Boyle's own inventories of his papers illustrated how he organised and used them during his lifetime.[18]

There was also evidence concerning the fate of the papers after his death, including the very perceptive archival work of William Wotton, who was due to write Boyle's biography but never did; and then the more damaging work of Wotton's successors, Henry Miles and Thomas Birch, whose depredations (often well-intended, as we will see) I was able to chronicle in detail, helped particularly by Lawrence M. Principe. We discovered that, large as Boyle's archive remains today, it was once much larger, and by tabulating Miles' listings of Boyle material in the light of our knowledge of Boyle's corpus, we were able to document what is now missing.[19] Equally significant were the implications

15 For the principal products of the work in question see my *John Aubrey and the Realm of Learning* (London: Duckworth, 1975); *Science and Society in Restoration England* (Cambridge: University Press, 1981); *Establishing the New Science: the Experience of the Early Royal Society* (Woodbridge: Boydell Press, 1989); and *Science and the Shape of Orthodoxy: Intellectual Change in Late Seventeenth-century Britain* (Woodbridge: Boydell Press, 1995).

16 Michael Hunter and Annabel Gregory, eds., *An Astrological Diary of the Seventeenth Century: Samuel Jeake of Rye 1652–99* (Oxford: Clarendon Press, 1988) and Michael Hunter, Giles Mandelbrote, Richard Ovenden and Nigel Smith, eds., *A Radical's Books: The Library Catalogue of Samuel Jeake of Rye 1623–90* (Woodbridge: D.S. Brewer, 1999), esp. 333–44.

17 M.A. Stewart, "The Authenticity of Robert Boyle's Anonymous Writings on Reason," *Bodleian Library Quarterly* 10 (1978–82): 280–9, on 283.

18 See Hunter, *Boyle Papers*, esp. 44–57; Michael Hunter, "Mapping the Mind of Robert Boyle: The Evidence of the Boyle Papers," in Michael Hunter, ed., *Archives of the Scientific Revolution: The Formation and Exchange of Ideas in Seventeenth-century Europe* (Woodbridge: Boydell Press, 1998), 121–36.

19 Michael Hunter and Lawrence M. Principe, "The Lost Papers of Robert Boyle," *Annals of Science* 60 (2003), 269–311, reprinted in slightly extended form in *The Boyle Papers*, 73–135; on Wotton, see ibid., esp. 23–5, 40–1.

for Boyle's correspondence, since *The Correspondence of Robert Boyle* includes references to letters known from lists by Wotton, Miles and Birch once to have existed, but which are now lost, on a scale which as far as I am aware is virtually unparalleled: indeed, I now slightly wish that I had published a paper setting out my findings and their significance for the understanding of archives more generally, rather than simply deploying this material to enhance the edition itself.[20] These are matters to which I will return later in this paper.

My work on Boyle received significant funding from the Leverhulme Trust, among other bodies, and one component of their grant paid for an attempt to broaden the scope of the project by considering how a similar treatment might be given to other scientists of Boyle's period. The result was a conference held at the Royal Society in 1996 entitled "Archives of the Scientific Revolution," the content of which was published in book form two years later.[21] At the conference, various scholars applied a comparable approach to the papers of Newton, Galileo, Christiaan Huygens, Sir William Petty, Marcello Malpighi, Gottfried Wilhelm Leibniz and Ismael Bouillau; there were also chapters on the Hartlib Papers and on the archives of the Royal Society and the Académie des Sciences. I would like to think that this helped to establish the field to which the current volume is devoted, though of course during the two decades that have since elapsed such studies have become more and more prevalent, encouraged not least by the "material turn" that has taken such a hold in literary studies, exemplified by the work of scholars like James Daybell, and perhaps particularly his book, *The Material Letter in Early Modern England* (2012).[22] Indeed, this is a trend on the implications of which it is worth reflecting: it does seem to be the case that, as digital technologies mean that more and more of our contact with texts is on-line and virtual, a stress on materiality forms a natural reaction to this.[23]

20 See Michael Hunter, Antonio Clericuzio and Lawrence M. Principe, eds., *The Correspondence of Robert Boyle* (6 vols., London: Pickering & Chatto, 2001), vol. 1, xxv–xxxi, vol. 6, 397–414, and passim.

21 Hunter, *Archives of the Scientific Revolution*.

22 James Daybell, *The Material Letter in Early Modern England: Manuscript Letters and the Culture and Practices of Letter-Writing, 1512–1635* (Basingstoke: Palgrave Macmillan UK, 2012).

23 See Daybell, *The Material Letter in Early Modern England*. I am grateful to Elizabeth Yale for her comments on this matter. See the comparable comments of David Pearson in his "Introduction" to Giles Mandelbrote and Barry Taylor, eds., *Libraries within the Library: The Origins of the British Library's Printed Collections* (London: British Library, 2009), 1–7, on 3.

Turning to "Archival Afterlives," I think there can be no doubt that, although from a much earlier date manuscripts had been preserved more or less effectively in what might be called traditional types of repository—family archives and the archives of great houses, or the holdings of colleges, schools and even governmental and commercial institutions—it is arguably the case that the seventeenth century was a formative period in terms of attitudes to the papers of intellectuals and their preservation. I made this point in my introduction to *Archives of the Scientific Revolution*, and it has since been argued more fully by Elizabeth Yale in an important study published in *Book History* in 2009, which she has extended in her book, *Sociable Knowledge*, of 2016.[24] Elizabeth gave her article the subtitle "How Early Modern Naturalists Invented the Archive," and there is a great deal of truth in this, though she has herself written elsewhere about parallel developments in other realms, including those of politics and administration.[25]

Richard Serjeantson in his chapter in this book documents the role in the transmission of manuscripts of editors and of "intelligencers," who actively pursued items that seemed to them significant and were often instrumental in preserving them. One key figure here was Samuel Hartlib, as is illustrated by Richard's use of references to relevant material in Hartlib's *Ephemerides*. Equally revealing is Carol Pal's essay on Hartlib's significance as a source of information about female intellectuals of the period, who would otherwise have been rendered almost invisible by their tangential role in relation to the male-dominated institutions of the day. As she points out, Hartlib's archive was very much a working one, and through it we gain important insights into the activity of various female figures. Among these were Dorothy Moore, and crucial light is shed on the educational, political and religious issues that concerned her. Perhaps most significant was Katherine, Lady Ranelagh, who Carol sees almost as a partner to Hartlib in her activity as an intelligencer on medical, religious and political topics, including such matters as Robert Wood's pioneering advocacy of decimalisation.

Another crucial figure was the virtuoso and antiquary, John Aubrey, who was more overtly concerned with preservation than Hartlib had been, and whose

24 Michael Hunter, "Introduction," in *Archives of the Scientific Revolution*, 1–20, on 5–6; Elizabeth Yale, "With Slips and Scraps: How Early Modern Naturalists Invented the Archive," *Book History* 12 (2009): 1–36; Elizabeth Yale, *Sociable Knowledge: Natural History and the Nation in Early Modern Britain* (Philadelphia: University of Pennsylvania Press, 2016), ch. 6.

25 Elizabeth Yale, "The History of Archives: The State of the Discipline," *Book History* 18 (2015): 332–59.

repeated plaintive calls for the conservation of such material Elizabeth Yale quotes at length in the paper already cited. So, too, does Sara Dry in her perceptive recent account of *The Newton Papers* and their tribulations.[26] It is also revealing how central to Aubrey's classic work, *Brief Lives*, was an interest in the papers of the figures who form its subject: he frequently included as part of his biographies a heading "Scripsit," which listed not only printed but also manuscript works by the author in question.[27] Recently, Kate Bennett has reminded us of Aubrey's personal preservationist role in her magisterial edition of *Brief Lives*, arguing for the significance of the manuscripts of *Brief Lives* themselves as a "paper museum," in which Aubrey incorporated autograph letters and documents in the handwriting of his subjects or their close associates. On Aubrey's description of the work as "a rarity worth it's weight in Gold" she comments how this "claims for the *Brief Lives* manuscripts a primary value as unique and curious material objects."[28]

Aubrey is equally notable for his insistence on the role of institutions as repositories in which important written materials could be preserved, reflecting a yearning for permanence in the aftermath of the disruptions caused by the Civil War and its aftermath which, as I have argued elsewhere, was one of the prime stimuli to the foundation of the Royal Society.[29] Aubrey even suggested that the writings of his friend, James Bovey, might appropriately be placed there, although Bovey had never had any association with the society.[30] It is revealing that the society commissioned a complete copy of Aubrey's *Natural History of Wiltshire*, and Aubrey also lay behind certain attempts made by the society to investigate and preserve the papers of deceased scientists, including Samuel Foster and Thomas Merry; in addition, he was associated with the project which actually came to fruition of editing and publishing the papers of Jeremiah Horrocks.[31] However, it was not the Royal Society that

26 See note 24; Sarah Dry, *The Newton Papers: The Strange and True Odyssey of Isaac Newton's Manuscripts* (Oxford: University Press, 2014), pp. 14–15, 186.

27 John Aubrey, *Brief Lives*, ed. Kate Bennett (2 vols., Oxford: University Press, 2015), vol. 1, 49–50, 169–70 and passim.

28 Ibid., vol. 1, LXXVI–LXXVII.

29 Hunter, *Establishing the New Science*, 1–15.

30 Aubrey, *Brief Lives*, vol. 1, 467–8.

31 For *The Natural History of Wiltshire*, see Hunter, *John Aubrey*, 88–9; Yale, *Sociable Knowledge*, 136–8. For Foster see Thomas Birch, *The History of the Royal Society* (4 vols., London, 1756–7), vol. 2, 410, vol. 3, 316–17 (and Aubrey, *Brief Lives*, vol. 2, 1468–9); for Merry, Birch, *History*, vol. 4, 223–4, 296, 326–7 (and Aubrey, *Brief Lives*, 1713–14). On the Horrocks papers and the edition of them see Birch, *History*, esp. vol. 1, 386, 395, 412–14, 422, 456–7, 470–1, 473; vol. 2, 7, 48, 359, 480; vol. 3, 58, and *Jeremiae Horroccii … Opera Posthuma* (London, 1672).

became the recipient of Aubrey's own manuscript remains but a different institution, the museum founded at Oxford by Elias Ashmole, which opened in 1683. This reflected a similar wish to provide a permanent centre for science, formulated in comparable circumstances: it is perhaps revealing that in Ashmole's negotiations with the University of Oxford about setting up the museum he made reference to what might happen to its content in the event of "civil wars or like calamities."[32] Aubrey saw the Ashmolean as providing the ideal repository for his archive, and, though this image was slightly tarnished by an early burglary and by the museum's willingness to lend manuscripts to figures like his brother, William, who borrowed and never returned Liber B of Aubrey's *Wiltshire Antiquities* in 1703, it is true that Aubrey's manuscripts have survived surprisingly intact as a result of his prescience.[33]

A contrasting case is provided by Martin Lister, the subject of Anna Marie Roos' chapter in this volume. Lister donated a large number of printed books and some important antiquities and other objects to the Ashmolean during his lifetime, as was recorded alongside Aubrey's donation in a lavish inscription still preserved there; in addition, he bequeathed the copper plates used in his classic work on conchology to the museum at his death in 1712.[34] In his lifetime bequests, Lister had evident ulterior motives in terms of ingratiating himself with Ashmole and hence with the university in pursuit of an honorary degree, but what is significant here is the clear sense that his gifts reveal as to what kind of material it was appropriate to preserve and what it was not. What he did *not* consider appropriate for such treatment was his personal archive, which was only rescued from the "pastrycooks oven" by William Huddesford in the late eighteenth century, in circumstances which Anna Marie recounts in her essay. Lister's attitude is revealing, and it is salutary to remember that attitudes like his have been in evidence ever since: not all have agreed with Aubrey's relentless presumption of the desirability of preservation at all costs. In Lister's case, it is slightly ironic that a good deal of Anna Marie Roos' evidence about the actual process of image-creation by Lister and his daughters, of which she is able to give an almost tactile account, depends on the use of material among the papers that were so fortuitously preserved as a result of

32 Hunter, *Establishing the New Science*, 9–10.

33 For the burglary see Anthony Powell, *John Aubrey and his Friends* (revised edn., London: Mercury Books, 1963), 217–18; for William Aubrey and Liber B see Hunter, *John Aubrey*, 91, and for his gift to the Ashmolean in general, ibid., 87, 90–1.

34 See Anna Marie Roos, *Web of Nature: Martin Lister (1639–1712, the First Arachnologist* (Leiden: Brill, 2011), 9–10, 238, 240, 247, 260–2, 265, 266–9, 273n., 390–1, 418. For a listing of the Lister MSS now in the Bodleian see ibid., 427–8.

Huddesford's prescience—an episode that occurred, incidentally, at much the same time as Robert Boyle's huge archive reached the Royal Society, where it has remained since its donation in 1769.

In the early eighteenth century, the lead in preservationist activity was taken less by institutions like the Royal Society than by one man, Sir Hans Sloane, who vacuumed up a vast quantity of archival material in addition to the massive range of natural and artificial objects that he accumulated during his career and that is now forming the subject of the "Reconstructing Sloane" Project. Here, it is important to remember the extent to which Sloane's massive collections took shape through his cannibalization of earlier holdings. An example of this is provided by his relations with the earlier collector, William Courten, which are documented in Anna Marie Roos' chapter, in this case including actual specimens. Further evidence is provided in Arnold Hunt's chapter, which uses the case of James Petiver to illustrate the extent to which the Sloane collection exemplifies the theme of "archives within the archive." Hunt reflects tellingly on the aspersions that have often been cast on Petiver, both socially and intellectually; he illustrates how closely Petiver and Sloane worked together and the extent to which Sloane adopted and deployed Petiver's system for tabulating his specimens, often by reference to his published *Gazophylacium Naturae*, as is revealed by collating texts and specimens in the different institutions (the British Library and the Natural History Museum) in which these are now to be found. Sloane's accumulation of earlier assemblages is further illustrated by his activity as a book-collector, as has been most fully studied by the Sloane Printed Books Project, the work of which is usefully summarized by Alison Walker in her chapter.

Hence there is a fruitful history to be written about how the various assemblages on which we are retrospectively dependent were actually accumulated, and the papers presented at the conference and printed in this volume make an important contribution to this. Yet equally significant is another theme that arises in many of the papers, namely the *uses* to which it was thought appropriate for the material that was preserved in this way to be put. In 'Scarlet Letters', Vera Keller takes us back to the seventeenth-century Royal Society and to an interesting instance of archival preservation that was originally utilitarian in motivation—in terms of the society's interest in the History of Trades and, within this, in the production of superior, highly coloured dyes. As she shows, the society was able to access papers on the subject compiled by Sir Theodore de Mayerne in the early Stuart period, which were made available to it by his godson, Sir Theodore de Vaux. In addition to dyeing, de Vaux produced material on various other topics, and both were the subject of sustained study by Robert Hooke and others. In all, it is an extraordinary and revealing episode, all the more so due to the rivalry between de Vaux and another of Mayerne's descendents, Sir John Colladon, whose share of the manuscripts was to end up in the collection of Sloane.

It was not only in this instance that the Royal Society was significant for the records it preserved of contemporary technology. The same is also true of another part of its technological programme, namely the records of agrarian practice in different parts of England kept by the so-called Georgical Committee; these still survive among the society's Classified Papers and they formed the subject of an important study by R.V. Lennard in the 1930s.[35] Such records might be seen as examples of collaborative research bearing a slight resemblance to the Human Genome Project, the preservation of records of which by the Wellcome Trust and others forms the subject of Victoria Sloyan's chapter, and in this connection it is appropriate to mention another early initiative -- namely the documentation of smallpox inoculation that the society collected in the 1720s under the auspices of James Jurin, on which Andrea Rusnock has been able to draw in various publications, including her useful, if partial, edition of Jurin's letters.[36] Such examples remind us of the truly significant role that the Royal Society played in the preservation of data of this kind in its early years.

Similar conclusions can be drawn concerning Sloane, who not only accumulated manuscripts and books but also made use of them. Alison Walker's paper illustrates such utilitarianism on his part by focusing particularly on the annotated volumes that he acquired from earlier holdings, using available clues to suggest *why* he collected some of the things he did. In particular, he was in search of material from printed pharmacopeia and the like that would be of use in his medical practice (in the case of Dr Luke Rugeley, he evidently hoped to find an eye remedy for which Rugeley had been famous, though in the end he had to purchase this—ironically finding it unremarkable). Elizabeth Yale complements this by illustrating how Sloane indulged in what she calls "the politics of the archive" through his use of manuscripts, including items from his own collection. At meetings of the Royal Society he made a point of

35 R.V. Lennard, "English Agriculture under Charles II: The Evidence of the Royal Society's 'Enquiries,'" *Economic History Review* 4 (1932–4): 23–45.

36 Andrea Rusnock (ed.), *The Correspondence of James Jurin (1684–1750)* (Amsterdam: Rodopi, 1996) and "The Weight of Evidence and the Burden of Authority: Case Histories, Medical Statistics and Smallpox Inoculation," in Roy Porter (ed.), *Medicine in the Enlightenment* (Amsterdam: Rodopi, 1995), pp. 289–315. See also her *Vital Accounts: Quantifying Health and Population in Eighteenth-century England and France* (Cambridge: University Press, 2002).

presenting, and often publishing in the *Philosophical Transactions*, significant letters by naturalists and others, including Edward Lhuyd and earlier figures like Sir Thomas Browne, some of which he owned while others came from the manuscript holdings of others. He also preserved in his own collection the draft minutes of the Royal Society covering the period of his notorious feud with John Woodward, a medical Fellow like himself who virulently disagreed with Sloane and others on various topics: the rough version significantly complements the society's official minutes by showing some of the decisions that went into the exact wording selected for the final version.

The ultimate goal for Sloane's collection was to form the basis of the greatest public repository of all in the form of the British Museum, and this was indeed realised through the negotiations that led to the Act of Parliament that established the museum in 1753. On the other hand, partly because in retrospect this outcome seems so obvious, it struck me when reflecting on the papers that appear in this book that no-one has ever managed to answer the question of just how early in his collecting career Sloane decided what the fate of his collection should be—in other words, how long before he compiled his will in 1739 he gave thought to the issue of how others might benefit in the longer term from his huge assemblage of material.[37] In this connection, it is interesting that Arnold Hunt illustrates in relation to Petiver how there were suspicions at the time that Sloane's professed openness was restricted by his own authorial ambitions, in this case concerning his *Voyage to … Jamaica*, of which, although the first volume was published in 1707, the second did not follow for nearly twenty years.

I come last to the relationship between archives and publication, which arises in various chapters in this book. This is central to Richard Serjeantson's paper, since it was largely editors with publication in mind who played a crucial role in the descent of Bacon's manuscripts which he there meticulously reconstructs, doing justice for the first time to the Meautys portion as well as those associated with Sir William Boswell and William Rawley. Yet, as he points out, these efforts to retrieve Bacon's scattered manuscripts were motivated almost exclusively by a wish to get them into print, and once the goal of publication had been achieved, material in manuscript form was no longer perceived to be worth retaining. As he puts it, even in the hands of keen collectors, "the fact of a work's being printed still constituted a death-sentence to original

37 For some comments on this topic see Marjorie Caygill, "Sloane's Will and the Establishment of the British Museum," in Arthur MacGregor (ed.), *Sir Hans Sloane: Collector, Scientist, Antiquary, Founding Father of the British Museum* (London: British Museum Press, 1994), 45–68, on 46–7.

documents." This state of affairs was replicated time and again in the seventeenth- and early eighteenth-century context, since all too often the original manuscripts of correspondence and other items included in editions of the period no longer survive. A case in point is that of Isaac Barrow, of whom we have hardly any manuscripts because he was so efficient in getting his works into print. This was something that one commentator, Sir Robert Southwell, saw as a virtue, citing Barrow as a monitory example to Sir William Petty, of whose chaotic manuscripts he despaired: Barrow, he noted, left only a handful of manuscripts to his executor, "soe carefull had he beene to finish all he ever tooke in hand, either printing what he wrote, or leaving his thoughts and collections all ready for the Presse."[38] Only with the onset of collectors like Robert and Edward Harley, who actually valued manuscripts like Bacon's as relics of the great man in their own right, did this start to change. Indeed, as Sir Noel Malcolm presciently observed during the discussion that formed part of the June 2015 conference, this is a point of transition in the history of archives that we need better to understand -- from a phase when material was valued because it was perceived as useful, suitable as a contribution to current knowledge that might be added to and enhanced, to a rather different phase when the papers of a great man were preserved largely as an act of commemoration.

In this volume, a good example of the presumption that manuscripts should be put to current use when they were published is provided by Anna Marie Roos' account of the way in which William Huddesford dealt with the legacy of Martin Lister and Edward Lhuyd in the 1760s. His professed aim was to re-index their papers in order to make them useable in terms of "the most approved modern system," as he put it in his prospectus for his new edition of Lister's work;[39] he also edited the plates that he inherited, removing images from these in order to bring them into line with current knowledge on the subject. In other cases, documents were strategically edited when they were prepared for print, and Elizabeth Yale illustrates, in relation to the letters of Edward Lhuyd and others that Sloane published in *Philosophical Transactions*, how personal comments and the like were routinely omitted. Elsewhere, she has illustrated how this was also true of William Derham's 1718 edition of the *Philosophical Letters* of the naturalist, John Ray: Derham made assiduous

38 Quoted in Frances Harris, "Ireland as a Laboratory: The Archive of Sir William Petty," in Hunter, *Archives of the Scientific Revolution*, 73–90, on 89.

39 Anna Marie Roos has kindly confirmed that this is to be found in Box 3 of the Lister Ephemera at the Bodleian; there is also a copy in the Newberry Library, Chicago: see *Newberry Library Online Catalog*, accessed April 21, 2017, https://vufind.carli.illinois.edu/, searching on "William Huddesford."

use of a red pencil to indicate passages in Ray's letters to be omitted from the printed edition, evidently because he thought them irrelevant or demeaning – though mercifully the letters in question still survive, and subsequent editors have been able to do much to rectify this.[40]

An interesting analogy to Derham's concern about Ray is provided by the case of Robert Boyle, whose eighteenth-century editor, Thomas Birch, showed a comparable anxiety to defend Boyle's reputation, particularly from accusations of credulity and enthusiasm, the bugbears of fashionable eighteenth-century intellectuals. With the help of Henry Miles, Birch purged Boyle's correspondence of items that he deemed damaging from this point of view, notably letters from alchemists and religious radicals, which were marked in their inventories as "unintelligible" or "Enthusiastic" and which to a disproportionate extent no longer survive.[41] Miles and Birch also jettisoned most of the highly revealing casuistical manuscripts that formed part of Boyle's archive on the grounds that these were "not suited to the genius of the present age," though luckily in this case certain key treatises survive in a "shadow" archive, in this case that of Boyle's casuistical advisor, Thomas Barlow, now at Queen's College, Oxford.[42] The result of this tidying up of the archive by its eighteenth-century custodians was to have an unfortunate historiographical legacy, providing the basis for a misleading view of Boyle as unequivocally opposed to "enthusiasts" by J.R. Jacob and others which I have spent years trying to undo.[43] In addition, as with Derham's use of red pencil though with a more damaging effect since the items in question no longer survive at all, Miles and Birch also jettisoned letters that they evidently saw as trivial, for instance concerning Boyle's relations with his family, his landed estates or his dealings with printers. This too, I think, has had an unfortunate historiographical legacy, encouraging a view

40 See her paper in Carla Bittel, Elaine Leong, and Christine von Oertzen, eds., *Working with Paper: Gendered Practices in the History of Knowledge*, forthcoming from the Max Planck Institute, Berlin. For subsequent editions, see R.W.T. Gunther (ed.), *Further Correspondence of John Ray* (London: Ray Society, 1928) and Anna Marie Roos (ed.) *The Correspondence of Dr Martin Lister (1639–1712)*, vol. 1: *1662–1677* (Leiden: Brill, 2015).

41 Michael Hunter, *Robert Boyle (1627–1691): Scrupulosity and Science* (Woodbridge: Boydell Press, 2000), pp. 265–6.

42 Michael Hunter, "The Disquieted Mind in Casuistry and Natural Philosophy: Robert Boyle and Thomas Barlow," in Harald Braun and Edward Vallance, eds., *Contexts of Conscience in Early Modern Europe 1500–1700* (Basingstoke: Palgrave Macmillan, 2004), 82–99, reprinted in Michael Hunter, *Boyle Studies* (Farnham: Ashgate, 2015), 107–29.

43 Hunter, *Robert Boyle*, esp. 51–7, 63–4, and see also my *Boyle: Between God and Science* (New Haven and London: Yale University Press, 2009), 58, 261–2.

of the great man as more Olympian and detached from his context than was really the case.

The cases of Ray and Boyle illustrate a further aspect of the role of posthumous publication. With Ray, the posthumous publication of his *History of Insects* in 1710 and Derham's issue of a third edition of his *Three Physico-Theological Discourses* in 1713 can be seen as a utilitarian contribution to current knowledge like Huddesford's editions of Lister and Lhuyd. On the other hand, it might be argued that Derham's subsequent publications, continued after his death by his wife's nephew, George Scott, were rather different, being of material that was aimed rather at enhancing Ray's reputation, in the form of his letters and his records of his itineraries. But there were limits to what seemed appropriate for publication, as is revealed by Scott's letter book for 1757–8, which has recently come to light and is now in the Bodleian Library at Oxford.[44] For this shows his indecision about just how much it *was* appropriate to publish in this way, with material that seemed repetitive or trivial ultimately being rejected on the grounds that it failed to enhance Ray's reputation: thus, in addition to the manuscripts by Ray that were included in the 1760 volume for which he was responsible, Scott seems to have started to prepare for publication items by Ray that he ultimately decided were not worthy of print. Instead, the manuscripts in question were allowed to lapse quietly into family descent (together, luckily, with those that *had* been used), going from Scott to Robert Bodle and his daughters, and thus ending up at Netherton in Devon and Woolston Hall in Essex (in the latter case, incidentally, also including Hooke material that had been in Derham's hands).[45]

Here, there is an analogy with the case of Boyle which I had not fully understood prior to the June 2015 conference. One of the puzzling things about the material that Miles and Birch listed in their inventories but is now lost is that, in addition to losses that can accounted for by their desire to protect Boyle's reputation, there are others that cannot, since the material in their lists that no longer survives includes some interesting-sounding and quite complete sets of papers, for instance relating to Boyle's *The Usefulness of Natural Philosophy* or his largely lost "Dialogues concerning Heat and Flame." These were not published and they no longer survive, and my hypothesis is that, having abstracted them

---

44 Bodleian Library, Oxford, MS Eng. e. 3655, and see the discussion in Michael Hunter, "John Ray in Italy: Lost Manuscripts Rediscovered," *Notes & Records* 68 (2014): 93–109, on 95. For the editions, see William Derham, ed., *Philosophical Letters between the Late Learned Mr Ray and Several of his Ingenious Correspondents* (London, 1718) and William Derham and George Scott, eds., *Select Remains of the Learned John Ray* (London, 1760).

45 See Hunter, "John Ray in Italy," 94–5.

from the archive with a view to publishing them, Miles and Birch decided that they were too fragmentary to merit such treatment, since they would do nothing to enhance Boyle's reputation. As it was, only hitherto unpublished sections of Boyle's *The Christian Virtuoso* were printed for the first time in the final volume of Birch's 1744 edition of his *Works* (and needless to say the manuscripts of these, as well as the works that were discarded, have failed to survive).[46]

My final point stems from Victoria Sloyan's fascinating paper about the Human Genome Project archive and the extent to which archives can actually be created. I have already noted the analogues to this in some of the holdings that the Royal Society rather less self-consciously brought into being in the late seventeenth and early eighteenth centuries. But it now occurs to me that, in a digital age, we could not so much create as recreate archives that we know once to have existed but which no longer survive. I have already alluded to *The Correspondence of Robert Boyle* and the extent to which, within that, we were able to offer evidence of exchanges of letters that are now known only through listings of material later destroyed. In this and other, similar cases, such lost letters have in fact already started to appear in databases of correspondence of the period like *Early Modern Letters Online* or *Electronic Enlightenment*, meaning that they can be included in the reconstruction of epistolary networks that such resources facilitate.[47] A further potential source comprises the lists of the writings of eminent contemporaries compiled by intelligencers like Hartlib and Aubrey, to which Richard Serjeantson draws our attention in his chapter, while others have used the catalogues of nineteenth-century autograph dealers to identify letters or groups of letters that have long been irrevocably scattered or lost.[48] Using sources like these, there may be potential to go further than hitherto in reconstructing in digital form archives that have long since ceased to exist as physical entities. It is a tantalizing thought.

46 See Hunter and Principe, "Lost Papers of Boyle;" Hunter and Davis, *Works of Boyle*, vol. 12, XLIX–LXV, esp. LI and 367–530.

47 See *Early Modern Letters Online,* accessed April 21, 2017, http://emlo.bodleian.ox.ac.uk/blog/?page_id=907; and *Electronic Enlightenment,* accessed April 21, 2017, http://www.e-enlightenment.com/. The latter includes the lost letters from *The Correspondence of Robert Boyle,* while the former, for instance, includes the otherwise unknown letters listed in Martin Lister's diary of his journey to Montpellier in 1663–5 and itemised in the first 38 pp. of Roos, *Correspondence of Lister,* vol. 1.

48 See, for instance, Elizabethanne Boran, ed., *The Correspondence of James Ussher, 1600–1656* (3 vols., Dublin: Irish Manuscripts Commission, 2015), vol. 2, 649–50, 780–5; Noel Malcolm and Mikko Tolonen, "The Correspondence of Thomas Hobbes: Some New Items," *Historical Journal* 51 (2008): 481–495; see esp. 485. For much helpful background information, see A.N.L. Munby, *The Cult of the Autograph Letter in England* (London: Athlone Press, 1962).

Index